Einstein's Generation

Einstein's Generation

The Origins of the Relativity Revolution

RICHARD STALEY

THE UNIVERSITY OF CHICAGO PRESS CHICAGO AND LONDON

RICHARD STALEY is associate professor in the Department of the History of Science, University of Wisconsin–Madison.

The University of Chicago Press, Chicago 60637
The University of Chicago Press, Ltd., London

Printed in the United States of America
17 16 15 14 13 12 11 10 09 08 1 2 3 4 5

ISBN-13: 978-0-226-77056-7 (cloth)
ISBN-13: 978-0-226-77057-4 (paper)
ISBN-10: 0-226-77056-7 (cloth)
ISBN-10: 0-226-77057-5 (paper)

Library of Congress Cataloging-in-Publication Data
Staley, Richard.
Einstein's generation : the origins of the relativity revolution / Richard Staley.
p. cm.
Includes bibliographical references and index.
ISBN-13: 978-0-226-77056-7 (cloth: alk. paper)
ISBN-10: 0-226-77056-7 (cloth: alk. paper)
ISBN-13: 978-0-226-77057-4 (pbk.: alk. paper)
ISBN-10: 0-226-77057-5 (pbk.: alk. paper)
1. Physics—Social aspects—History—20th century. 2. Physics—Methodology.
3. Materials science. 4. Relativity (Physics)—History. I. Title.
QC7.S777 2008
530.09′04—dc 2008014155

⊗ The paper used in this publication meets the minimum requirements of the American National Standard for Information Sciences—Permanence of Paper for Printed Library Materials, ANSI Z39.48-1992.

Contents

Preface and Acknowledgments

Albert Einstein is instantly recognizable and for centuries to come will surely remain an important touchstone for both popular and scholarly understandings of physics and of the time in which he lived and worked. As some measure of that popular interest and perhaps also an early indication of my own later intellectual concerns, as a schoolboy I once decided that I wanted to read Einstein's original work on relativity. I visited the library of Monash University only to realize with some shock that it was written in German and that I would need still more than language skills to understand the pages I was holding in my hands. (If my memory is correct, I was leafing through the journal *Annalen der Physik*.) While in general we know Einstein extremely well, our understanding of the goals and endeavors of many of his contemporaries is far less comprehensive. Further, my earliest scholarly work on his colleague and friend Max Born showed me that to a large extent scholars had interpreted Born's 1909 work on a relativistic theory of the electron through the lens of a specific approach to relativity, identified with Einstein. As natural as that appeared and as valuable as those earlier historical studies were, I felt a subtly different approach was needed to understand how Born could have been so ambitious and yet shot down so quickly—for what his contemporaries saw as a critical if flawed contribution, and what historians described as exhibiting elementary failures to understand Einstein's theory. The questions at issue concerned both understanding why Born had taken up Einstein's earlier work in the particular way he did, and recognizing the assumptions underlying historians' perspectives nearly a century later. I came to feel that it is still somewhat difficult for historians to work with the full implications of what is a widely recognized fact: that knowledge is created and purveyed within a community. Both Born's and Einstein's work as individuals had to be recognized and utilized by their colleagues to count as knowledge, and that process within the community of their peers often involved subtle transformations in the meaning of the work concerned. If historians of the sciences face difficult challenges working with the communal fabric of scientific knowledge, popular images of science are still more strongly in the thrall of stories of individual heroes. This book addresses that general problematic. Rather than exploring the physics community through one of its most famous representatives, it reexamines the work in special relativity through which Einstein first became well known in Germany—by beginning with the broader community within which that work was founded. Examining the period from Albert Michelson's famous 1881 ether-drift experiment to the year that Einstein's work was incorporated in the first textbook devoted to relativity in 1911, we will find that relativity was far more than a theory alone and learn to question the historical framework within which it is usually presented as involving the overthrow of classical by modern physics.

The first grounds for the approach I have taken were developed in my work at the University of Cambridge on Max Born and the German physics community and a collaborative postdoctoral fellowship at the Whipple Museum for the History of Science oriented in large part around curating two innovative exhibitions: *Empires of Physics* and *1900: The New Age*. Throughout my studies and work in Cambridge, faculty and students at the History and Philosophy of Science Department provided an extraordinarily rich and stimulating environment. Among the many wonderful historians I had the chance to learn from there I would like to mention in particular the fellowship of Jenn Tucker, Jeff Hughes, Arne Hessenbruch, and Andy Warwick, as well as the uncommon inspiration that flowed from Simon Schaffer's generous scholarship (in all its remarkable facets). Working at the Whipple together with Simon, Jim Bennett, Otto Sibum, and Bob Brain helped me begin the task of integrating an understanding of the cultural and material dimensions of physics with the intellectual and disciplinary orientation of my earliest studies, and was important in shaping the narrative and interpretive framework I gave this book when it crystallized as a proposal in 1998. For their early support and encouragement of the project in that form, I would like to thank in addition Norton Wise, Sam Elworthy, Peter Galison, and Susan Abrams at the University of Chicago Press.

Over the following years, material support for different facets of my research was provided by the Max Planck Institute for the History of Science, the Pollock Award for Research in the History of Astronomy, and the Graduate School of the University of Wisconsin–Madison. I am also grateful to a large number of archives and libraries for their assistance, and for permission to use and cite their collections; the letter codes by which I refer to specific collections (and a list of collections consulted) is provided at the beginning of the bibliography. I owe a great debt to the earlier work of many scholars, and especially to the beauty of Martin Klein's writing, the challenge of Thomas Kuhn's reconstruction of the history of quantum theory, the comprehensive nitty-gritty of Arthur I. Miller's account of Einstein's special theory, and the historical tact exemplified throughout Russell McCormmach's body of work. My two fellowships in Otto Sibum's research group on the Experimental History of Science at the Max Planck Institute were particularly valuable. There I benefited especially from continued conversations with Otto, Charlotte Bigg, Suman Seth, and David Bloor. A large number of people have read some part or the whole of this manuscript. For the care they took (and also for the challenge of their criticism), I would like to thank my colleagues at the University of Wisconsin–Madison, and especially Gregg Mitman, Lynn Nyhart, Tom Broman, Mike Shank, and Ron Numbers; Xavier Roque and members of the Barcelona-based Inter-University Doctoral Program in the History of Science; Suman Seth and students at Cornell University; and Andrew Warwick, David Cassidy, Harry Collins, Olivier Darrigol, Michel Janssen, Charlotte Bigg, Simon Schaffer, Charles Crossett, Alexander Geppert, and Peter Susalla. For their editorial shepherding of a sometimes-unwieldy manuscript I am especially grateful to Catherine Rice, and to Jennifer Howard who has accompanied its latter progress with much understanding. I thank Dawn Hall for her careful copyediting.

Chapters 2 and 3 of *Einstein's Generation* draw upon earlier versions of my work on Michelson published in Staley, "Travelling Light" (2002) and Staley, "The Interferometer and the Spectroscope" (2003). Chapter 8 presents a revised version of Staley, "On the Histories of Relativity" (1998), and chapters 9 and 10 draw upon Staley, "On the Co-Creation of Classical and Modern Physics" (2005).

I am by training a historian of science, and that training has in large measure come through immersion in the materials of the period I have studied rather than through a formal education in physics or in history. The papers that we revisit here from Einstein, Born, Michelson, Kaufmann, and others have been my principal guide, and these scientists' lives a substantial inspiration. I hope this book will help bridge the humanities and the sciences, and although my concern is with understanding the relations between research physics and the society within which it is cultivated I have not assumed a mathematical or scientific background in the reader. Rather I have endeavored to present the central experimental work and conceptual features of the work discussed in a form accessible to any interested reader.

Finally, for their patience and love sharing my own life in the period this book has been such a major preoccupation, I thank Laura Radefeld, and most especially my wife Elisabeth Emter and our children Francesca and Griffin.

CHAPTER ONE

Introduction: Einstein's Generation

In late October 1911, a small group of eighteen physicists from six different nations and representing several different fields of research gathered around a single table at the Hôtel Métropole in Brussels. Young experts on radioactivity like Ernest Rutherford and Marie Curie sat side by side with long-established masters of mathematical theory or low temperature research like Henri Poincaré and Kamerlingh Onnes. A "confidential" invitation signed by the chemical industrialist Ernest Solvay had brought them together for what was described as a new kind of conference, a kind of "Scientific Council." Solvay's wealth may have been responsible for the occasion, but the guiding spirit was clearly the Berlin physical chemist Walther Nernst (identified in the return address). Solvay's invitation had convened this committee of eminent experts on the grounds that the principles of molecular theory and kinetic theory of matter were badly in need of revision—although many present knew very little about the precise subject matter to be discussed. Holed up for five days, Hendrik Antoon Lorentz first instructed his colleagues in the failure of classical theory to provide a satisfactory treatment of the radiation emitted by the "blackbody," an experimental system that was particularly interesting because it enabled extremely general conclusions to be drawn about the interchange of energy between matter and radiation. Then Max Planck described the successful theory he had found in 1900, identifying its radical implication of discrete energy quanta. Finally, Albert Einstein and others outlined more recent extensions of the new theory, especially to the specific heats of solids. But the many hours given to discussion would define the conference even more than the reports delivered. Three scientific secretaries made sure no passing comment was neglected, pasting handwritten notes from the participants into a manuscript of record that formed the basis for the published proceedings. Befitting their very different levels of engagement with the field before arriving in Brussels, participants expressed a range of assessments leaving it. Einstein said he'd learned nothing new. In contrast, the French mathematician Henri Poincaré positively reeled with shock. Already a strong proponent of a radical new mechanics based on electrodynamic theory and the principle of relativity, Poincaré's subsequent papers graphically registered the surprise of recognizing a still more remarkable challenge to mechanics in the new doctrine of energy.[1]

The first Solvay Council has long stood as a symbol of the emergence of modern physics, and its structure exemplifies a view that we now take for granted: that the verities of classical physics broke apart around 1900, being succeeded by the radical new quantum theory and relativity. Both this language and the conceptual divisions it represents are fundamental to our understanding of the achievements of turn-of-the-century physics. Indeed, physics texts sometimes split their subject into classical and modern chapters, and philosophers have elaborated the distinction between the classical, Newtonian verities of absolute space and time and deterministic explanations (appropriate to ordinary experience), and the modern relativistic space-time and fundamentally probabilistic theories required to understand the very fast and very small. This demarcation on the grounds of fundamental techniques, regimes of experience, and conceptual implications is commonly taken to reflect a historical periodization. For example, in the "Autobiographical Notes" he published in 1949, Albert Einstein gave a fascinating but in some respects rather crude sketch of the physics of his student days in the late 1890s. In spite of the fruitfulness of particulars, "dogmatic rigidity prevailed in matters of principles: In the beginning (if there was such a thing) God created Newton's laws of motion together with the necessary masses and forces. This is all; everything beyond this follows from the development of appropriate mathematical

1. The proceedings were originally published in Langevin and de Broglie, eds., *Théorie du rayonnement et les quanta* (1912). For studies see Mehra, *Solvay Conferences* (1975); Barkan, "Witches Sabbath" (1993); Marage and Wallenborn, eds., *Solvay Councils* (1999).

methods by means of deduction." Einstein was impressed by the accomplishments of mechanics—many in apparently distant fields—noting in particular that the "statistical theory of classical mechanics" had been able to deduce the basic laws of thermodynamics. A little later he writes also of "classical thermodynamics" as the only physical theory of universal content that he is convinced will never be overthrown.[2]

Einstein's views can be taken as characteristic of our present understanding of *classical physics* (indeed, many analysts have accepted this and similar references from other physicists without question). The term offers an appropriately general and inclusive description of a discipline bound by dogmatic certainties that extended from Newton to about 1900. It would not take many of us long to identify the chief architects of these extraordinary disciplinary transformations either, pointing to Einstein and Planck, both at the table in Brussels, and then looking toward others present, perhaps especially Marie Curie and Ernest Rutherford, for their involvement in the similarly exciting development of radioactivity.

This book opens the period around 1900 to new analysis. Focusing in the first instance on the development of special relativity, it aims to deepen our understanding of the relations between the material and conceptual dimensions of relativity physics and reexamine the relations between Einstein and his community. I trace scientific research in its transformations across experiments, instruments, and theory, and move between individual endeavors and science in international congresses to explore the way arcane results might carry the cultural freight of worldviews. My study will establish a new understanding of the origins, rise, and emergence of what we now describe as modern physics—one that departs significantly from the kind of narrative we have inherited from Einstein and others, while at the same time explaining its origins. Leaving the linear certainties of the customary account behind will require a complex investigative approach. In this Introduction I first describe in some detail the three interrelated thematic concerns that underlie this book before outlining its narrative structure.

Three investigative themes structure my engagement with physics circa 1900, the early development of relativity, and its histories. The first is to

2. Einstein, "Autobiographical Notes" (1949), 19 ("in the beginning"), 21 (statistical theory), and 33 (classical thermodynamics). The dogmatic rigidity Einstein describes was not representative of many of the physicists whose work he studied most closely, such as Mach, Poincaré, and Boltzmann.

offer a fresh approach to the contributions of two major figures. Setting the careers of Albert Michelson and Albert Einstein in new contexts by focusing on several revealing blind spots in current treatments of their work will deliver valuable insight into both these individuals and the communities they helped to configure. A second goal will be to generate new perspectives on disciplinary change by seeking critical breadth, both in our understanding of the nature and activities of the physics community and in our treatment of one of its main products, relativity. Historical accounts of the period have often privileged discovery narratives and theoretical perspectives (even when discussing key experiments). To disclose the more complex grounds originally at issue, I approach the material culture of relativity physics as an experimental and instrumental achievement and explore a rich variety of means through which different segments of the physics community sought to promote unity—intellectually and socially, across both national and disciplinary boundaries. The third structuring concern involves investigating the relations between new knowledge, what I shall call *disciplinary memory*, and disciplinary identity, especially as these are revealed in the fates of two important but very different international congresses. These themes are strongly interrelated, but teasing them apart will provide some orientation to the narrative to follow.

Careers in Context: Michelson and Einstein

First, consider individuals and the work and reputations of the two figures who provided the most important experimental and theoretical grounds for relativity. One helped shape the experimental agenda of physics in the United States in the period in which that discipline stepped out of the shadow of astronomy to become a significant academic presence. The second became the public face of physics after World War I, an image of intelligence and, as far as *Time* magazine is concerned, *the* person of the twentieth century. Comparatively speaking, both Albert Michelson and Albert Einstein have been subject to a great deal of attention. We know far more about each of them than about even some of their close colleagues, like Edward Morley and William Stratton who both wrote papers with Michelson, or Max Laue and Paul Ehrenfest who worked with Einstein. Nevertheless, there is a great asymmetry in the amount and character of the attention that Michelson and Einstein have received. As the cover that *Time* devoted to the century's greatest minds suggests—depicting a somewhat bored Einstein on Sigmund Freud's couch—Einstein is the one we really want to get to know.[3]

Focusing on a few major heroes is not wrong, and with a life as extraordinary as Einstein's there are great benefits to be obtained from examining his engagement with physics and the political and cultural environment in which science is enmeshed. Nevertheless, at least for the period from the 1890s through to the 1920s, a real cost has flowed from our abiding interest in Einstein and a few other central figures or events. Put briefly and only a little too simplistically, to a large extent Einstein's work has defined our understanding of the nature of the problems facing physics, and comparisons with Einstein's thought have provided a major means of analyzing his community. Treatments of Michelson and his ether-drift experiment provide an example.

In 1881 Michelson developed a novel experimental arrangement with light and mirrors that offered the possibility of testing a major hypothesis. Light was widely thought to propagate through some form of medium, described as the ether. Michelson's experiment was designed to test whether the effect of any ether wind created by the earth's motion through that medium could be discerned in the interference patterns created by two beams of light. Working with Edward Morley in 1887, the null result obtained from an improved version of this experiment was widely regarded as offering a significant challenge to optical and electromagnetic theory. Although there has been some interest in examining Michelson's earlier measurements of the velocity of light and the origins of his ether-drift design, most academic scholarship on Michelson's research has been dominated by the importance of the ether-drift experiment for the subsequent development of relativity. Analysts have been primarily concerned with investigating how different theorists responded to the potential challenge the null result provided to the existence of the ether, and especially with the question of the extent to which Einstein might have drawn on Michelson's experiment nearly twenty years later. The exception to this rule is one fine and detailed study of the ether-drift experiment that generates considerable interest by showing that it had a far longer and more complex history than is usually granted, resulting in a series of repetitions that

3. *Time*, 29 March (1999). The lack of interest is appropriate: the relationship between Einstein and Freud was marked by wariness on both sides. *Time* anointed Einstein in Golden, "Einstein" (1999).

spanned right into the 1920s. But as Loyd Swenson recognized, focusing on a single experiment could only ever offer a partial study of Michelson's interests.[4] This book explores one of the reasons why the 1887 ether-drift experiment could still be questioned more than thirty years after it was first performed: it was never in fact completed as planned.

Before and after they undertook their improved version of the ether-drift experiment in 1887, Michelson and Morley devoted much more time and resources to a related setup. It used the same basic technique Michelson had pioneered in 1881, but this time with the aim of providing a new standard of length based on the wavelength of light. Thus, when Michelson and Morley abandoned the ether-drift experiment after four days of observations and only part way through the yearlong protocols they outlined (in what was meant to be a preliminary publication), they did so in order to find a way of counting the number of wavelengths in a meter. That problem stood at the intersection of scientific interest and industrial concern; and their focus on it shows that Michelson was already convinced of the ether-drift result, but even more importantly, wanted to turn his experiment into a new instrument. Pursuing that endeavor in the period between 1886 and 1902, Michelson utilized similar apparatus in a wide range of different contexts. The most notable included providing a way to measure the dimensions of astronomical objects, developing a new form of spectroscopy, and determining the length of the international meter in terms of the wavelengths of cadmium light in 1893. The latter task was extremely demanding technically, and its physical significance is indicated by the fact that a single repetition of his measurement by two French physicists fourteen years later provides the major reason Michelson won the Nobel Prize in 1907. Accepting his award, Michelson celebrated the ether-drift experiment *despite* its null result and because it had led him to a new instrument, the interferometer. By recovering the logic of his experimental work and appreciating the uses Michelson found

4. As we shall see in chapters 2 and 3, comparatively little attention has been devoted to integrating Michelson's famous ether-drift experiment into either his earlier work on velocity of light or his activities in the period from 1886 through to the early 1900s. The most important commentaries bearing on Michelson include Gerald Holton's influential argument that Einstein did not rely closely on Michelson's experiment in his pathway to relativity, Swenson's study of ether-drift experiments, and the valuable social biography provided by Michelson's daughter. See Holton, "Einstein, Michelson, and the 'Crucial' Experiment" (1988); Holton, "More on Einstein, Michelson, and the 'Crucial' Experiment" (1993); Swenson, *Ethereal Aether* (1972); Swenson, "Michelson-Morley, Einstein, and Interferometry" (1988); Livingston, *Master of Light* (1973).

for his experiment, this book will restore his research to its most immediate scientific context and link Michelson to his community and contemporary industrial and cultural concerns in new ways. We will see later that in a number of unexpected respects this endeavor also leads us back to the history of relativity physics.

In the case of Einstein, with a plethora of biographies and fascinating studies of his developmental path and the nuances of his thought already in hand, a different approach is appropriate.[5] Rather than recovering something like his scientific self-portrait, understanding better how *others* engaged with Einstein's work has the most potential to deepen our understanding of his community. Accordingly, in this book we meet Einstein as most others did, gradually and selectively, focusing on what his work on relativity might mean for an earlier great reductionist and unifying program, electron theory (especially as that had been developed by H. A. Lorentz). It is well known that the first commentary on Einstein's 1905 paper on relativity purportedly disproved it with precise measures of the mass of fast-moving electrons. Appreciating more fully than we have hitherto the promise, coherence, and technical standards of Walter Kaufmann's research project is critical to understanding how Einstein's work was judged: Max Planck described Kaufmann's experiments as spelling life or death for different theories of the electron, and worked hard to ensure that relativity was not immediately dismissed on their account. But neither Planck nor Einstein addressed directly some of the most important critiques of relativity that came from adherents of a rival program to provide a new foundation for physics based on electromagnetic theory. Here we explore the demands that theorists like Paul Ehrenfest and Max Born made on relativity, as they respectively critiqued the theory and sought to extend it after 1905—by focusing on the electron. My studies will give a strong sense of what contemporaries regarded as the primary limitations to relativity theory.

Approaching Einstein as much through others as through his own concerns makes it possible to reexamine his emergence as a notable figure in the German physics community. Recently, several historians have highlighted the teleological nature and interpretative limitations of accounts

5. The standard biographies are Pais, "*Subtle Is the Lord*" (1982); Fölsing, *Einstein* (1997). Holton's work has been particularly influential, while Stachel has long provided the backbone to Einstein studies. See especially Holton, *Thematic Origins* (1988); Stachel, *Einstein from "B" to "Z"* (2002); Howard and Stachel, eds., *Einstein: The Formative Years* (2000). For an overview of recent studies, see Staley, "Interdisciplinary Atomism" (2003).

shaped primarily around Einstein's contributions. The historiographically critical perspective they display has very often been based on recovering the research traditions and conceptual insights informing the work of specific individuals, or members of distinctly different interpretive communities—examining in particular Henri Poincaré, H. A. Lorentz, Hermann Minkowski and German mathematicians, or the British physics community.[6] Here I shall extend the insights of such studies by following a number of actors from different segments of the German physics community long enough to discover the integrity of their concerns in building up, working with, and diverging from aspects of relativity. My cast of characters will be familiar to historians of relativity, but many features of their contributions have been neglected or treated solely from a perspective close to Einstein's. Suspending evaluative contrasts for a moment will allow a more subtle treatment of the dynamics of engagement than has been possible in the important works in which the rise of relativity in Germany has been described.[7] My approach will bring new facets to our understanding of the endeavors of Poincaré, Kaufmann, Alfred Bucherer and Adolf Bestelmeyer, Ehrenfest, Born, Planck, Lorentz, Minkowski, and Laue, as well as Einstein.

Chronologically these men are all members of Einstein's generation or that of his teachers, but note that they were by no means part of Einstein's cohort (at least until after he had become well known in Germany). We already have excellent biographical treatments of Einstein, detailed portraits of his personal relationships with friends, teachers, and first wife Mileva, and extremely subtle studies of his relations to intellectual forebears such as Mach, Lorentz, and Poincaré. None of the people I have chosen to examine here were Einstein's immediate colleagues. Nor, with the exception of Poincaré, Planck, and Lorentz, did they play particularly important roles in his development. However, all made significant contributions to the development of relativity, and for that reason played a

6. See Darrigol, "Poincaré's Criticism" (1995), 2; Darrigol, "Electrodynamic Origins" (1996); Darrigol, "Einstein-Poincaré" (2004); Katzir, "Poincaré's Relativistic Physics" (2005); Janssen, "Reconsidering a Scientific Revolution" (2002); Corry, "Minkowski and the Postulate of Relativity" (1997); Walter, "Minkowski, Mathematicians, and the Mathematical Theory of Relativity" (1999); Warwick, *Masters of Theory* (2003).

7. See especially Miller, *Einstein's Special Theory* (1981); Goldberg, *Understanding Relativity* (1984); Pyenson, *The Young Einstein* (1985); Pyenson, "The Relativity Revolution" (1987).

major role in shaping how his community understood Einstein. In addition, to an unusual extent these men understood themselves as part of a common era, a generational group witnessing radical change. Further, their initial distance from Einstein will prove an advantage in approaching some of the most difficult issues in the reception of innovative research: in several cases, whether or not these authors agreed with Einstein, changing attitudes to their work helped generate a distinctively new perspective on the nature and validity of Einstein's studies. Thus, rather than examining Einstein's formation I will attend here to the way a common understanding of the stature of his contributions emerged within his community. My study, then, continually moves between the individual and his/her community; but does so in contrasting ways. Examining Michelson, I first recover the investigator's perspective before exploring what his work meant to others, thereby shedding light equally on physics in America and the relations between theory and experiment in the international community. Einstein, however, I shall approach first through the often-critical perspectives of his colleagues.

Critical Breadth: Physics circa 1900 and Relativity Physics

Focusing on the work and reputation of Michelson and Einstein, great as they were, cannot deliver a full understanding of a community in a period of extraordinary change. The second major concern shaping this book involves two specific means of generating new perspectives on disciplinary change, in both cases by widening the unit of analysis. The first involves setting out to study physics circa 1900. And I mean the whole of physics. The desire for comprehensiveness is of course impossible to achieve, but stating it is important when as historians we usually select our subjects radically. Furthermore, approaching an entire discipline meaningfully is also feasible when physicists themselves set that aim. Around 1900 a few eminent physicists offered interpretive overviews of their field that noted the coincidence of the turn of the century with a period of rapid intellectual change. Even more importantly the French Physical Society convened an extraordinary conference with the goal of international participation and disciplinary completeness. I will turn to the inaugural International Congress of Physics soon. Appreciating its breadth—from metrology and mechanics through to biological and cosmic physics—is important to gaining perspective on a discipline.[8] Exploring the 1900 congress will provide a very fruitful means of setting the more specific fields of work that remain my primary concern—interferometry, the electron, and relativity—in the kind of broader contexts in which contemporaries saw them. But I seek also a comprehensive appreciation of the depth of labor in physics. That means examining the changing role of instrument makers in Michelson's research program in early chapters and exploring also the work required to put physics on display in the World's Fair of 1900. Fifty million people traveled to Paris to see the best the world had to offer in that year, and the Congress of Physics took place under its ambit. Among the travelers and exhibitors, the German government laboratories and individual German instrument makers combined to present a unified, representative display. The quality of their collective exhibitions stunned and worried observers from other countries. The Americans also put on collective displays, but in this case they did so to hide the weak spots left by only partial participation from their leading firms. The director of the Department of Machinery and Electricity also made an unusual request concerning a small retrospective "museum of science" in which original mechanical and electrical apparatus were to be exhibited. Francis Drake wanted the display to exemplify a *classical* architectural style. His report noted that it was extremely difficult "to convey to the decorators the idea that the exhibits which were to be housed would admit of classic treatment, although the pure classic would be too severe to produce a proper effect."[9] The episode indicates that concepts of classical and modern were being reworked on extremely wide cultural grounds. Terms generally used to point to ancient philosophies, languages, and culture, and widely invoked in "classical" and "modern" streams of secondary schooling, were now being applied to the arts, technology, and science. But it also shows that the attempt to ally extremely recent technologies with the canons of traditional forms sometimes met considerable cognitive resistance. It can remind us that the physics of the nineteenth century was decidedly modern before it became classical; indeed, within a brief, ten-year period, fields of research that had stood firmly at the forefront of progress would suddenly be recast as fruits of the past. Working

8. For institutional and interpretative surveys of the physics discipline in 1900 see Forman, Heilbron, and Weart, "Physics *circa* 1900" (1975); Hiebert, "*Fin-de-Siècle* Physics" (1979); Heilbron, "*Fin-de-Siècle* Physics" (1982); Porter, "Death of the Object" (1994).

9. Drake, "Report" (1901), 124–26 and 136.

to obtain critical breadth and depth—by examining material and cultural as well as intellectual dimensions to the work of physics, moving from individuals through to national and international endeavors—will offer an unusually rich framework from which to appreciate the various stages on which physicists worked for unity and sought to understand both intellectual and disciplinary change.

My second major strategy for offering new perspectives on disciplinary change has been implicit in much of the foregoing. Rather than exploring the history of relativity *theory*, a formulation that is an outcome of Einstein's involvement, my pursuit is a wider undertaking that can be described as relativity *physics*. As we have seen, as it is currently told the history of relativity centers on Einstein, tracing the genesis of a radical new theory in his long and private study of the electrodynamics of moving bodies. We have been taught to see the Michelson-Morley experiment as marginal to this study, to distinguish strongly Einstein's insight from the partial anticipations of forebears such as Lorentz and Poincaré, and to celebrate relativity as the overthrow of classical physics.[10] But in the first brief history of the subject (incorporated in an important research review of 1907), Einstein made the Michelson-Morley experiment a central event and wrote of the "union" of Lorentz's theory with the principle of relativity. A year later he told Arnold Sommerfeld, "If the Michelson-Morley experiment had not brought us into the greatest predicament, no one would have perceived the theory of relativity as a (half) salvation."[11] These comments raise important questions about Einstein's relations to the physics community in particular and concerning the relations between the empirical and theoretical dimensions of relativity in general.

Consider first the relations between an individual and his community. If we know Einstein's path to relativity did not rely on the Michelson-

10. In addition to the references focused on Einstein cited above, the works of Miller, Goldberg, and Pyenson have provided the most important broader histories of relativity in the German physics community (see notes 5 and 7 above). Setting the subject in the framework of intellectual and cultural history, Christopher Herbert has discerned a controversial "relativity movement" in Victorian literature that he argues provided a cultural context for the subsequent articulation of Einstein's theory (and for the repression of a disposition toward forms of philosophical relativism). The understanding of relativity physics I pursue here is consistent with facets of his argument, and should contribute to the still more comprehensive and subtle treatment that is required to build a convincing account of the cultural history of the relative and absolute. See Herbert, *Victorian Relativity* (2001).

11. The review was Einstein, "Relativitätsprinzip" (1907). Einstein wrote to Sommerfeld on 14 January 1908, in Einstein, *Collected Papers*, vol. 5 (1993), 86–89.

Morley experiment, why did he put so much emphasis on it for his colleagues? If it is the distinctions between Einstein's and Lorentz's work that are most important, what kind of union was this? In what sense did Einstein write of "us"? And why was the theory of relativity a half salvation only? Addressing questions of this nature brings *Einstein's Generation* into a complementary and critical relationship with an earlier scholarship that has been primarily concerned with the genesis of relativity and with demonstrating the uniqueness of Einstein's insight. Indeed, stepping aside from these traditional historiographical concerns will enable this study to show the antecedents to these lines of argument—within the physics community. This is because Einstein and his contemporaries were deeply concerned with understanding and shaping the relations between his work and the field it entered. In their work—after 1905—we can see how the themes of distinction and discovery were first generated and played out among physicists.

Turn now to the more general issue of the relations between theory and experiment. Involving as it did increasingly separate communities of theorists and experimentalists, this needs to be considered both epistemologically and in disciplinary terms. Understanding relativity as a primarily theoretical achievement and focusing fairly narrowly on new conceptual insights valorizes only some facets of the work involved in the complex development of understandings of the ether and the electron. When we explore relativity as an experimental and instrumental achievement we can see a whole network of engagements that are more or less strongly interrelated with the story of relativity that historians presently tell.[12] Consider the experimentalists who worked on ether and the electron, for example. Their work provided grounds for the development of important concepts, and also offered the primary empirical tests of relativity; and historians have most often been interested in establishing the probity of their results and analyzing theorists' responses to them. Yet in both fields, in addition to the inquiry about the relative motion of the earth through the ether that animated much discussion, experimentalists

12. The importance of a more thorough engagement with experiment in general was first emphasized by Shapin and Schaffer, *Leviathan and the Air-Pump* (1985); Franklin, *The Neglect of Experiment* (1986). Galison urges us to see metaphysics in machines and machines in metaphysics in his study of Einstein's awareness of clock synchronization technologies (Galison, *Einstein's Clocks* [2003]). The most important studies of the experimental testing of relativity and electron theory are Miller, *Einstein's Special Theory* (1981); Hon, "Experimental Error" (1995).

articulated a distinction between relative and absolute measures and experiments. That distinction involved different kinds of precision—relative precision within a given framework and absolute precision that involved independent measures of all the parameters concerned. Indeed, as far as Walter Kaufmann was concerned, the experiments on the electron that tested views of relative and absolute space and time, and hence might establish the validity of the principle of relativity, themselves had to rest on absolute measurements.

Experimental practice also engaged with central elements of Einstein's relativity theory in other ways that have not yet drawn attention. We saw that Michelson used the ether-drift apparatus as an instrument to provide an absolute determination of the length of the rigid rod that constituted the international meter in Sèvres. Confirming his measurement fourteen years later was widely taken to offer the first actual proof that the dimensions of that rod had remained unchanged. In Einstein's hands the interferometer might establish that all objects contract with motion; but in Michelson's hands it also demonstrated the invariance of the dimensions of the international meter (this is because Michelson's measurements take place within a frame of reference at rest with respect to the meter rod, while measurements taken from a moving frame of reference would show it to be contracted). In most cases, picking out elements like this does not show that our current understanding of the development of *Einstein's* relativity *theory* is fundamentally in error. But it does show that there was far more in play in relativity *physics* than conceptual advances alone, and that the terrain of the absolute and relative, the rigid and contractible holds hitherto unappreciated empirical, conceptual, and disciplinary dimensions. Thus, by emphasizing experimentalists' work and material culture this book contributes to the material history of the central concepts of Einstein's theory, light, rigid rods, and mirrors. It also demonstrates the historiographic costs of confining attention too narrowly to the concerns of theoreticians—we simply sacrifice our understanding of major features of the work of physics and the nature of the physics community.

New Knowledge, Disciplinary Memory, and Disciplinary Identity

I will illustrate something of how and why that particular sacrifice has been made by turning to the final investigative theme that structures *Einstein's Generation*: the relations between new knowledge, disciplinary memory, and disciplinary identity. Examining physicists' uses of historical resources in two related directions will enable this book to provide an unusually broad and general study of what I call *disciplinary memory*. The first direction concerns physicists' interpretive work to give meaning to innovative research, as they grapple with the relations between old and new techniques and interpret each in the light of the other. An examination of more or less elaborate "research histories" written by practitioners of relativity between 1905 and 1911 will focus my study of this facet of physicists' use of history. The second direction involves examining the ways that a somewhat looser sense of memory, embodied in canonical narratives of disciplinary development, helps inform physicists' sense of the nature, past, and future of their discipline, and thereby contributes to their identity as a community. Considering two highly unusual but very different international congresses will provide a particularly sharp understanding of the productive role of these two interrelated senses of disciplinary memory, as research histories and as disciplinary narratives.

First, Paris in early August 1900. Alfred Cornu would have felt extraordinarily gratified as he set his papers on the lectern and waited for the audience to settle. Eighteen months earlier as president of the French Physical Society he had sent letters to other national associations and journals floating the possibility of holding a conference in association with the upcoming Exposition Universelle in Paris. Receiving around one thousand expressions of interest had convinced Cornu and the French Society they had a historic opportunity to strive for international scope and disciplinary completeness. They commissioned eighty papers across the entire physics discipline, asking recognized experts from different nations to present surveys of recent progress in their specific fields. Looking at the audience in front of him, Cornu could see how fully his ambition had paid off. The podium was stuffed with dignitaries and officials—ministers of education and war among them—and the final tally of participants reached 836, including most eminent physicists of the day. Note that at this time, historians and sociologists have estimated the total number of working physicists as between 1,000 and 1,500.[13]

Cornu thought the sessions over the next week would form a true scientific monument, one that would ensure the congress "a durable memory,

13. This is the finding of Forman, Heilbron, and Weart, "Physics *circa* 1900" (1975). The congress reports were published as Poincaré and Guillaume, eds., *Rapports* and *Travaux*, 4 vols. (1900–1901).

an important role in the history of the progress of natural philosophy at the end of the nineteenth century." Its editors regarded the published proceedings as presenting the most complete image of any science ever given at one point of time; and its four volumes were sent to every participant for the cost of postage—I have found them in every big library I've visited.[14] Yet the 1900 congress has been almost completely forgotten by both physicists and historians of science. It no longer figures in physicists' image of their past, and has never before attracted sustained historical attention. Both facts cry out for explanation, and the contrast with the first Solvay conference, held in Brussels in 1911 and still celebrated today, could not be greater.

Einstein's Generation will conclude with a study of the Solvay Council that is devoted to demonstrating two important arguments. The first is that the conference was particularly significant not because of any new contributions to the development of quantum theory, but because it became the forum for the presentation of a novel understanding of the *past* of the physics discipline, one that turned on a fulcrum around 1900 and contrasted a classical era with the strange fruits of modern theory. Ironically, the novelty of this depiction—which we could say made a monument of previous theory, by describing it as classical—has been invisible to analysts, even as it has framed most subsequent accounts of the period. Second, I will argue that this understanding and its subsequent rapid propagation through the physics community were highly contingent on the particular time and place in which it was first offered. My study will show that even though it referred to the legacy of a long past that they all knew from their own experience, physicists did not describe, and could not have described, classical physics in this way just eleven years earlier in 1900. Further, the fact that this vision soon became widely shared has everything to do with the particular form that Walther Nernst gave his conference in a Brussels hotel.

Thus, by displacing relativity from the particular historical narrative in which it has traditionally been set, linked so closely to Einstein and read as a chapter in the overthrow of classical physics, this study of *Einstein's Generation* will give a broader and critically nuanced picture of physics in the period—one that allows a fuller explanation of the formation of that disciplinary narrative. At the same time as defining the content and scope

14. Cornu's comment is recorded in "Procès-verbaux" (1901), 6–7. One of the editors described the proceedings in Guillaume, "International Physical Congress" (1900).

of a new physics, we will see that it helped show the power of the new specialty, theoretical physics. Historians of science have shown that following the widespread emergence of academic physics laboratories in the mid- to late nineteenth century, the physics community had reached a point of increasing intellectual and disciplinary specialization by 1900—which this study of the 1900 congress will amply confirm. Most importantly, the period is regarded as one in which theoretical physics in Germany flowered intellectually and attained institutional maturity with the creation of new chairs for the subdiscipline, with Planck and Einstein as its most important representatives. Accordingly, a major study of the rise of theoretical physics in Germany by Christa Jungnickel and Russell McCormmach concludes with an account of relativity, quantum theory, and atomic theory largely focused on Einstein. But relying on traditional, theory-oriented historical approaches to these fields of work, in many respects the terms of their narrative already assume the condition they set out to analyze and document.[15] This book will only be able to address some facets of the emergence of theoretical physics; but it is important to note that only after having gained the critical breadth the present study achieves by examining material culture, experiment, and theory on an even footing, can we appreciate why we currently describe the period in these particular terms, and thereby gain a deeper understanding of just what it means to identify relativity physics as Einstein's relativity.

Narrative Structure

The present study spans the period from the 1870s to 1911 and moves between Michelson in the United States and the German physics community and Einstein via the International Congress of Physics in Paris, 1900. Its structure follows a largely chronological framework that allows a continual development of my overarching concerns with material culture and the relations between experiment and theory, and with disciplinary identity and the links between individual researchers and broader communities. Both pedagogical accounts of relativity and histories of its development often begin with the null result of Albert Michelson's experiment to detect the movement of the earth through the optical ether. Part I of *Einstein's*

15. See Jungnickel and McCormmach, *Intellectual Mastery*, vol. 1 (1986), preface; Jungnickel and McCormmach, *Intellectual Mastery*, vol. 2 (1986).

Generation follows Albert Michelson and the physics of light in two chapters that survey Michelson's early work on velocity of light and ether drift, his development of the interferometer, and his engagement in standards projects. Dealing with the period from the late 1870s up to 1900 these chapters explore the complex and productive relationships among theory, experiments, and instruments, and offer a more comprehensive, continuous, and coherent vision of Michelson's efforts than has been delivered by earlier accounts of ether drift as a crucial experiment. Here, through Michelson, we meet the American physics community and its instrument makers, both fruitfully poised between the more powerful disciplines of industrial mechanics and academic astronomy.

Part II switches focus radically to explore science on display in the 1900 World's Fair, and the physics discipline at work in the inaugural International Congress of Physics, in chapters 4 and 5 respectively. Taking the form of a tour through the exhibition halls before joining participants in the lecture theaters of the congress, my study of the state of physics in the year 1900 explores the interrelations between national, international, and disciplinary identities in two contexts: first as they are revealed in instrument displays and exhibition proceedings, and then in the reports physicists gave on interferometry, the velocity of light, and the newly discovered electron. Now we meet Michelson as French and other physicists perceived his contributions and explore the ways that a new particle troubled categorization and suggested new lines of research.

The three chapters of part III take up first the electron and then relativity, encountering Albert Einstein in the company of German physicists who sought to explore the possibility of founding the whole of physics on electromagnetic theory, principally by understanding the behavior of charged particles moving near the speed of light. Chapters 6 and 7 show that experimentalists like Walter Kaufmann and theorists like Paul Ehrenfest and Max Born noted Einstein's advocacy of the principle of relativity and his predictions for the relation between the mass and velocity of the electron. The critical dialogue that ensued eventually transformed their understanding of the power and limitations of their empirical techniques, conceptual and mathematical tools, and overarching goals. As chapter 8 demonstrates, however, relativity was initially seen as a collaborative endeavor. Focusing on the productive role that physicists' "research histories" played in clarifying new contributions will enable us to chart the gradual process through which Einstein came to be regarded as the primary author of a revolutionary theory. Here we meet the Michelson-Morley experiment in

Einstein's hands, as he stressed its role in the experimental justification and pedagogical derivations of relativity—at a time when experiments on the electron stood strongly against relativity.

Part IV widens perspective still further to examine the genesis of our most basic interpretive framework for understanding the disciplinary changes that occurred circa 1900. Historians have an extremely detailed understanding of the development of relativity and quantum theory, but we have very little sense of when and why contemporaries began to think in terms of a contrast between classical and modern theory. Chapter 9 takes up that question and demonstrates that far from representing consensus views of the past, the earliest invocations of concepts of *classical* theory came only in the late 1890s, responded to critical changes already long underway, and sought to advance controversial programs for the future. Our present concept of classical physics was largely shaped by the way proponents of relativity and quantum theory developed contrasts with prior physics, but the dynamics of this relationship differed significantly in the two cases. Chapter 10 shows how the very different fates of a contrast with classical theory in relativity and quantum theory were finally resolved in favor of a newly inclusive understanding of classical physics. This concluding chapter establishes the importance of the 1911 Solvay Council in shaping the disciplinary memory of physics, and documents the elisions that simultaneously obscured the complex historical grounds on which that memory was based. Both in propagating a distinctively new understanding of the classical past and in shaping an image of science in the early twentieth century, the proceedings of the council demonstrate the extent to which visions of the past and future of physics were co-created. Perhaps to our cost, those singular visions have deeply informed our understandings of the nature of physics circa 1900 as well as its history. They have celebrated relativity theory and quantum theory at the expense of the richer and more complex fields of relativity physics and radiation physics from which they stem.

The chapters of *Einstein's Generation* take several quite different approaches toward opening the physics community and its history to new analysis, designed to establish a multifaceted understanding of a variegated discipline. I study two major individuals (but focus in different ways on the question of how others' used their work); examine the social and disciplinary dimensions of international congresses; trace the trajectory of the electron as it met the principle of relativity (and follow it through the work of both experimentalists and theoreticians); explore scientists' participant histories as a neglected facet of research practices; and in order to understand the birth of "classical physics," study episodes in the forgotten history of a word. The chapters also explore the different voices of an unusually large range of figures intimately concerned with shaping the development of physics: an ambitious young experimentalist developing a new instrument; instrument makers marketing products across national borders; German experimental physicists evaluating worldviews in the traces that subatomic particles leave on photographic plates; theoreticians constructing new interpretations of electron theory and relativity physics. In each case we will see that considerable labor was undertaken to achieve different forms of unity: unity of instrumentation, disciplinary unity, national displays, unified conceptual foundations, common interpretations. But we will also see the work invested in creating distinctions, and the concomitant difficulties and advantages of crossing disciplinary and epistemological boundaries. *Einstein's Generation* will show that people from different segments of the physics community pursued a rich range of strategies of consolidation and diversification, seeking to shape both the work of physics and our understanding of its nature and history. It is not enough to recognize the particular coherence that Einstein's vision brought to physics. To understand the power and weaknesses of his discipline, we need to understand his generation.

PART I
From Ether Drift to Interferometry

Remembering Michelson

Most scientific research goes unnoticed, and the work that survives the generations usually does so thoroughly transformed. A few rare papers are read and reread decades or centuries after their original production, but in most cases physics students learn the fundaments of their subject in treatments tailored to present pedagogical purposes, and it is only as they move toward active research that they encounter original papers in anything like their own terms. Experiments too are carefully reworked so their principles can be readily demonstrated in front of the class, or important techniques can be learned through practical exercises at the bench. Popular accounts of a given field often emerge first from those physicists who wish to frame its broader implications for others, and reveal sometimes uneasy, sometimes brilliantly successful alliances of pedagogical, historical, and contemporary interests. Journalists dig and quarry, historians mine and delve. Sometimes their endeavor will be to understand an individual scientist's intentions within the whole cloth of their career and work. Often their interest will be more glancing than that. And this is true also of research scientists, who engage their colleagues' papers with the primary aim of going beyond them.

What, then, remains of the work of someone like Albert A. Michelson, America's first Nobel Laureate in physics and the inaugural professor of physics at the University of Chicago? While experts in optics could draw on different resources, most people's knowledge of Michelson undoubtedly stems from introductory physics texts, popularizations of modern science, or biographies. They are likely to have come across one or two diagrams of an experiment Michelson carried out with Edward W. Morley in 1887, with mirrors mounted on a stone block floating in a mercury bath (for an example, see fig. 2.7). They will surely have met a schematic image of the pathway of two pencils of light traveling in different directions in relation to motion through ether; and they will have read a brief description of the challenge that the null result of the experiment posed for conceptions of the ether. Some may have seen interactive representations on the World Wide Web, or worked with a model of the apparatus as well. It is most often part of a historical and pedagogical pathway to the development of Einstein's relativity, and if the student has learned any more of Michelson, it may be that he was a brilliant experimentalist but was somewhat conservative, never himself accepting relativity—and even that in the 1890s he thought that physics was done, bar the last few decimal places. If our reader or student knows any more of Einstein, thanks to the influential work of the historian Gerald Holton, it might be that despite their common linkage, Michelson's experiment was actually not that important for Einstein, who found other ways to the conviction that the electrodynamics of his time was fundamentally flawed.[1]

If I return to the work of Michelson in great detail in following chapters and plumb that of Einstein and others too, it is not because I want to fight this process of selective winnowing, which is after all integral to the progress of physics. But I do want to document and understand it. We

1. For a recent pedagogical treatment of Michelson's work, see Serway and Jewett, *Physics for Scientists and Engineers, with Modern Physics* (2004), 1248–50. The Michelson-Morley experiment is analyzed after a discussion of the Galilean principle of relativity and sets the stage for Einstein's principle of relativity. Michelson's interferometer is actually first introduced in a previous chapter on interference of light waves, where it is described together with the two applications of Fourier Transform Infrared Spectroscopy and the Laser Interferometer Gravitational Wave Observatory (LIGO), on 1194–96. Those specializing in optics would acquire considerably more familiarity with Michelson's work than others. For example Born and Wolf, *Principles of Optics* (1999) briefly mentions ether drift and Michelson's determination of the meter in wavelengths but discusses his method of analyzing spectral lines in some detail, see 334–36 and 352–59. Michelson's famous assertion that the more important fundamental laws had already been discovered was in Michelson, *Light Waves* (1902), 23–24. On Einstein's independence from Michelson's experiment, see especially Holton, "Einstein, Michelson, and the 'Crucial' Experiment" (1988); Holton, "More on Einstein, Michelson, and the 'Crucial' Experiment" (1993), but also his foreword to Swenson, *Ethereal Aether* (1972).

need to recognize clearly the active interplay of current interests, memory, and neglect in the development of innovative research and pedagogy, with their continual reworking of the past, and to understand also how deeply this process in the scientific community shapes the image of science. I will show here how central it has been to our understanding of Einstein's early achievements and to our present perspective on the history of his discipline. That history matters to physicists. When the sociologists Harry Collins and Trevor Pinch drew on some of our best historical accounts in a chapter titled "Two Experiments That 'Proved' the Theory of Relativity," for a general audience, their treatment received sharp responses from Cornell University and Ohio State University physicists who questioned the assessment of key experiments in some detail. David Mermin, Kurt Gottfried, and Kenneth Wilson argued that a broader complex of issues was relevant to the acceptance of special relativity (pointing, for example, to experiments on the electron in addition to the Michelson-Morley experiment). Among the many fascinating features of the exchange that ensued, I want to note two in particular. The first is that long engagement in their discipline gave these working physicists a strong enough sense of physics circa 1900 to at least begin a critique of the relevant chapter as history; and second, that this was important to all involved because a proper understanding of the process of science was at issue.[2] That too is why these events matter to me as a historian; but this book will show that a still more comprehensive awareness of the subtle and varied relations between theory and experiment is required to understand two apparently contradictory features of the rise of relativity.

I will argue here that experiment and experimentalists were *more* significant to the development of relativity physics than is usually appreciated, but also that Einstein's special theory came to be widely accepted by 1911 without *any* experiment being regarded as offering uncontroversial and definitive proof of his approach. Making this argument will rest

2. The book was Collins and Pinch, *The Golem*, 1st ed. (1993). Responses were Mermin, "Sustaining Myth" (1996); Mermin, "Golemization" (1996); Gottfried and Wilson, "Science as a Cultural Construct" (1997). Collins and Pinch had focused on Michelson-Morley and eclipse results in relation to special and general relativity respectively. They discussed the exchange and analyzed textbook treatments of the Michelson-Morley experiments in Collins and Pinch, *The Golem*, 2nd ed. (1998), 151–80. The episode was part of the "science wars"; see especially Labinger and Collins, eds., *The One Culture?* (2001). Matthew Stanley and Alistair Sponsel have recently offered a far stronger understanding of Eddington's work. Stanley, "An Expedition to Heal the Wounds of War" (2003); Sponsel, "Constructing a 'Revolution in Science'" (2002).

in part on broadening our appreciation of the nature of experimentalists' contributions, by paying as much attention to both the empirical methods and theoretical discourse of physicists like Michelson and Walter Kaufmann as to the results they delivered on the ether and the electron respectively. Reconstructing something close to Michelson's perspective on his research in the first part of my study will require moving between the diverse intellectual imperatives of different research goals; the social relations that shaped his engagement with mentors and instrument makers; and the material stuff of instruments, apparatus, and measuring techniques. While many features of this exploration shed significant light on what it meant to build a career in physics in nineteenth-century America and are valuable for this reason alone, later chapters will show that several of the concerns Michelson pursued can also be identified in Kaufmann's research in the early twentieth century, and thereby became part of the testing of relativity—even in experiments on the electron. This is particularly true of their common interest in moving from relative to absolute measures and in optical techniques of measurement. I begin here a study of long-standing themes that cross experiment and theory in unexpected ways.

As well as recovering the breadth of work at play in the development of relativity, making my argument for the varied significance of experiment will depend on closely tracking the transmission of scientific research through time. This means examining both the origins of innovative experiments, instruments, and theory (the discovery stories that are the bread and butter of many accounts of the sciences), and the way later physicists take up and depart from the diverse resources they find in particular research. The main burden of chapter 2 will be a study of the origins of Michelson's ether-drift experiment, while chapter 3 will explore what Michelson and his contemporaries made of it in the years up to 1900. This more neglected inquiry into when and why specific results, methods, and interpretations survive over time affords considerable insight into subtle transformations in their meaning, as well as into their propagation in broader circles.

This study will underline the fact that the obverse of disciplinary memory is the *neglect* of substantial features of the ongoing work of physics, facets that although integral to the creative process of scientific research are nevertheless overshadowed by images and narratives that cannot help but celebrate their subject selectively. Clearly, just which techniques *and* which stories survive is critical to both the practice and image of physics, creatively intermeshed as we will find these facets to be. Knowing more about what was done circa 1900 will certainly expand our appreciation of the past; but only knowing why we remember what we now know—and have forgotten so much else—will we be able to explain why, with physicists, most people in the Western world think that our physical worldview wrestled free of a classical past just then, reaching a new modernity in the early twentieth century.

CHAPTER TWO

Albert Michelson, the Velocity of Light, and Ether Drift

An "Absorbing Interest": Measuring the Velocity of Light

The terse but revealing biographical sketch that Albert Abraham Michelson wrote for publication with the Nobel Prize papers in 1907 will help orient us to his early career. It is often noted (usually with surprise) that the prize was awarded for Michelson's invention of the interferometer and the spectroscopic and metrological work carried out with it rather than for the now better-known ether-drift experiment. In its entirety Michelson's sketch gives a valuable picture of the organic growth of his scientific concerns, one that helps explain why Michelson regarded the ether-drift experiment as a transitional achievement and springboard for other inquiries. His youngest daughter, Dorothy Michelson Livingston, has given a much fuller account of Michelson's life, describing the social background of an immigrant picking a distinguished path through the opportunities of nineteenth-century America. And Loyd Swenson provides a fascinating study of his work on ether drift. Detailing the contested history of the experiment over a span of thirty years—far longer than we shall follow it—Swenson explicitly sought to avoid the simplifications (and occasional outright falsifications) of approaches concentrating on the advent of Einstein and relativity. Yet Swenson later allowed that his study remained limited by its focus on an *experiment*.[1] So now let Michelson convey his own sense of his achievements:

> Albert A. Michelson
>
> Born at Strelno, Prussia, December 19, 1852. Educated in the public schools of San Francisco. In 1869, after matriculating from the San Francisco High School, obtained an appointment from President U.S. GRANT to the U.S. Naval Academy. Graduated as Ensign U.S. Navy in 1873, and made a two years' cruise in the West Indies, after which was ordered to the Naval Academy as Instructor in Physics and Chemistry, under the direction of Admiral (then Commander) W. T. SAMPSON.

Michelson continues in this terse style to speak of his absorbing interest in the measurement of the velocity of light and repetition of the "celebrated experiment of FOUCAULT," obtaining results as good as any hitherto with cobbled together apparatus and meager funds. Soon he had gone still further, delivering a value with an estimated uncertainty of only 60 km/s. Michelson describes traveling to Europe to study in the laboratories of "HELMHOLTZ and QUINCKE in Germany, and of CORNU, MASCART, and LIPPMAN [*sic*] in France." Then he turns to his ether-drift experiment: "During the winter of 1880 a method was devised for testing the Relative Motion of the Earth and the Ether—which resulted in the invention of the Interferometer. The method was tried first in the Physikalische Institut at Berlin and afterwards at the Physikalische Observatorium at Potsdam, but gave a negative result."[2]

The sketch is revealing. Its rare descriptive touches highlight Michelson's uncommon taste for physics and pride at winning good results from humble circumstances, while the litany of names shows his awareness of the powerful figures that nurtured and rewarded his achievements. This chapter takes up Ensign Michelson's story from his first engagement in velocity of light experiments in order to explore how he moved from measuring that velocity to devising an ether-drift experiment—and inventing

1. See Livingston, *Master of Light* (1973); Swenson, *Ethereal Aether* (1972), xii. Swenson indicated that he was aware that a focus on interferometry and instrumentation would have further enriched his study in Swenson, "Michelson-Morley, Einstein, and Interferometry" (1988). The third major scholarly contribution to our understanding of Michelson is provided by the essays of Goldberg and Stuewer, eds., *The Michelson Era* (1988).

2. Michelson, "Biographical Sketch" (1909). Later versions have modulated the telegraphic brevity of the original publication.

FIGURE 2.1 Michelson offered a sophisticated self-portrait as a cadet at the Naval Academy, depicting himself with a canvas and mirror, engaged in the act of painting. Courtesy AIP Emilio Segrè Visual Archives.

an *instrument.* It is striking that Michelson describes his new method as resulting in the invention of the interferometer, before he mentions its formal result. Taking seriously Michelson's focus on methods and apparatus will help me argue that among the give and take of his research on velocity of light, Michelson found material resources that became central to the design of the ether-drift experiment and intellectual and practical concerns that marked what he made of his instrument over the next fifteen years. To understand both the genesis of the ether-drift apparatus and the uses to which Michelson later put it, we need to look beyond the experiment itself.

Michelson studied and then taught at the Department of Physics and Chemistry at the U.S. Naval Academy in Annapolis in a period of pedagogical reform that exposed students to more intense study as one means of applying science to military aims. The problem of measuring the velocity of light was a standard topic. Indeed, as a first classman Michelson's final examination in physics asked for a description of the apparatus with which the French scientist Léon Foucault had determined the velocity of light (in 1862), a statement of the results obtained for light propagating in a medium denser than air, and a demonstration that they complied with the undulatory (wave) theory of light using the physical explanation of single refraction.[3] Michelson's taste for such problems was widely recognized. At his commencement, Superintendent John L. Worden told him, "If you'd give less attention to scientific things and more to your naval gunnery, there might come a time when you would know enough to be of some use to your country."[4] Despite Worden's rebuke, the navy had a place for Michelson's preoccupations. When he returned from the two years' cruise completing his studies as a midshipman in 1875, Michelson was asked to join his former teachers as an instructor in the academy. Kathryn Olesko has contended that Annapolis pedagogy was the most important factor shaping the evolution of Michelson's investigative style, particularly for the emphasis Commander Sampson placed on understanding experimental errors. But we will soon see that the role of Simon Newcomb as a mentor and critic may have been even more significant (and was certainly more challenging).[5] Michelson initially led students in recitation classes, but in 1877–78 was asked to give a lecture on optics—and decided to set up a version of Foucault's apparatus.

Demonstration experiments had long been a staple of physics teaching, only recently supplemented by offering students practical experience in the laboratory. Michelson began teaching in a period associated equally with rapid intellectual and institutional expansion, increasing disciplinary differentiation, and new methods of teaching experiment.[6] Reviews of one of the first laboratory manuals illustrate the situation. Published by the MIT professor Edward Pickering in 1873, the book offered recipes for a whole series of experiments to be performed by students. Its reviewers noted the novelty of this endeavor and compared chemistry and physics to explain why laboratories had come later to the latter. In particular, the more complex and delicate apparatus required meant the costs

3. See Livingston, *Master of Light* (1973), 46.

4. Quoted in the *New York Times*, 10 May 1931, 3, as cited in Kevles, *The Physicists* (1995), 28. Worden had commanded the *Monitor* on its famous engagement with the *Merrimac*.

5. Olesko, "Michelson" (1988).

6. On the rise of laboratory teaching, see Gooday, "Precision Measurement" (1990). On the physics discipline in the U.S., see Kevles, *The Physicists* (1995).

and risks of introducing large numbers of students to practical manipulation in physics seemed far greater.[7] The demand side of the equation was as important (even if reviewers left it unmentioned). It was only in the second half of the century that the growing industrial uses of electric telegraphy and the new dynamo had brought a new theater for the skills and knowledge that could be imparted in a laboratory. These were the central reasons for the influx of new students into the classroom, bringing unprecedented career goals and ultimately establishing physics as a new profession rather than the province of academic elites alone. Later, Michelson defined his interest in light against the more common attractions of electricity and magnetism; his earliest work was to distinguish him for his research rather than for his teaching.

Our most comprehensive study of the physics discipline in the United States has long been Daniel Kevles's *The Physicists*. Paying a great deal of attention to the social setting of the new laboratories and the various means by which physics was integrated into civic life in the United States, Kevles gives a rich portrait of what the discipline meant to its professors, to university presidents, and to congressmen. For Kevles, Michelson is an exceptional figure. Like the experimentalist H. A. Rowland and the mathematical physicist J. W. Gibbs, he rose above the humdrum reality of a discipline that lagged behind its European counterparts (especially in research productivity and significance) and lacked the institutional framework to reach toward higher standards.[8] Exploring the nature of Michelson's experiments will help me offer a different perspective, one that explains why Michelson could rapidly establish a viable research program

7. Pickering, *Elements of Physical Manipulation*, Parts I and II (1873 and 1876). Individual sections are divided into a treatment of Apparatus followed by a discussion of the Experiment to be performed. The comments on apparatus come in a review that noted also that in the beginning of the century physics had been an undifferentiated tract roamed across at will by chemists and mathematicians, whereas the subjects of light, sound, heat, and electricity now covered such extensive grounds that distinct chairs of chemistry and physics existed within universities and colleges. ECP-HUA, HUG 1690.23F E. C. Pickering Scrapbooks 1866–1877 & 1877–1887, Volume 1; Anonymous review, Jan. 1874 [source unknown], 182–84, quote on 182. A second reviewer wrote: "There can be little doubt that oral teaching is that which is best suited to students who are beginning experimental work of any sort, and that as much may often be learnt in five minutes by seeing another perform an experiment as would be acquired in as many hours with the aid of a book alone to explain the construction of the apparatus." A. W. R., "Pickering's 'Physical Manipulation'" (1874).

8. Kevles, *The Physicists* (1995), esp. chaps. 3 and 4, "The Flaws of American Physics" and "Pure Science and Practical Politics."

and stand out among his colleagues. Tracing networks of patronage and practical instrumentation I will maintain that Michelson's engagement with the existing strengths of the astronomical community and their instrument makers was critical to his early success. In chapter 3 the point will be broadened. There, examining work on light and mechanics across science and industry in the United States will demonstrate that both Rowland and Michelson won a distinctive place for physics between (and above) the more established scientific and industrial strengths of astronomy and precision mechanics largely by co-opting several lines of investigation initiated by astronomers. Following models set within the observatory, both physicists pursued research projects that addressed astronomical as well as physical concerns—and often did so by setting new standards, especially in optical research.

Rather than transforming Foucault's arrangement into a demonstration experiment or bringing it from the lectern to classroom benches for students to perform themselves, Michelson expanded scale and took the experiment outdoors. He tried out a new design that might allow him to rival previous determinations and offer a new contribution to research. Then he brought in precision instrumentation to make that possibility stick.

Michelson was aided by a recent study of velocity of light measurements by the French physicist Alfred Cornu. In 1867 Cornu had become professor of physics at his alma mater the École Polytechnique. There, joining what was already a well-established tradition in France, he specialized in experimental optics and showed a particular interest in addressing questions of concern to both physics and astronomy. Cornu reviewed contrasting velocity of light arrangements employed by his compatriots Foucault and Armand Hippolyte Louis Fizeau before beginning experiments along Fizeau's lines in 1871. Twenty-two years earlier, Fizeau had passed a beam of light through a revolving toothed wheel and sent it over a distance of 8.6 kilometers before a mirror returned the light along its original path. The method relies on the fact that when the toothed wheel revolves rapidly enough and the light travels far enough, the light traverses its double path in the time taken for the wheel's motion to bring another gap—rather than a tooth—into position. Then the observer sees a steady glow. As Cornu's research showed, a key disadvantage concerned the gradual and indistinct transition from darkness to light that showed when the wheel was turning rapidly enough. In contrast, Foucault had used a revolving mirror to send light toward a concave mirror and return

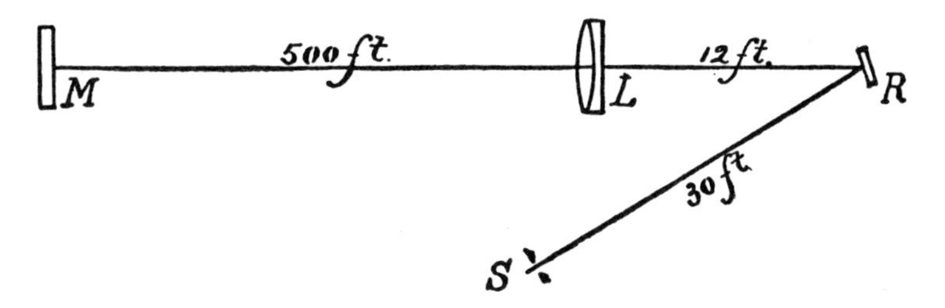

FIGURE 2.2 Michelson's 1878 velocity of light arrangement. The sun's rays pass through the slit *S* to the revolving mirror *R* and are reflected to a fixed plane mirror *M*. An image of the slit is formed on the surface of *M* by means of the lens *L*, and the light is reflected to retrace its path. When *R* is at rest the return image coincides with the slit itself; when *R* rotates rapidly the image is displaced through an angle twice the displacement of the mirror during the time required for the light to travel from *R* to *M* and back again. Michelson, "Velocity of Light" (1878), fig. 1.

it over a total path of only 20 meters. Michelson saw a way of improving Foucault's method to combine the advantages of both.[9]

His most important change involved replacing the concave mirror with a plane mirror and lens (fig. 2.2). This made it possible to use any distance between the two mirrors (rivaling the principal advantage of Fizeau's method) and to increase the distance between the revolving mirror and measuring scale. When spinning rapidly enough, the mirror turns while the light is in motion, hence deflecting the returning beam at a particular angle. Measuring the distance traveled, the frequency of the mirror's revolution and the angle of deflection enables a determination of the velocity of light. Early trials on Naval Academy grounds confirmed that the new arrangement improved the accuracy of the experiment considerably. With a distance of 500 feet between the mirrors, a radius of measurement of 15 to 30 feet, and the mirror making 130 turns per second, Michelson could increase the final displacement from the 0.7 mm Foucault obtained to about 5 millimeters.[10]

9. See Cornu, "Vitesse de la lumière" (1873); Cornu, "Vitesse de la lumière" (1874); Foucault, "Vitesse de la lumière" (1878 [1862]). Michelson's idea for an improvement came in November 1877, when his understanding of Foucault's experiment was still secondhand. In 1879, Newcomb sent Michelson a copy of the recent edition of Foucault's works, drawing Michelson to agree that the account of the experiments was very meager. See Michelson to Newcomb, 29 December 1878 and 16 January 1879, Reingold, ed., *Science* (1964), 283–85. On Fizeau's research, see Frercks, "Creativity and Technology" (2000).

10. One page only remains of an enthusiastic report on the progress of Michelson's plan, sent to Newcomb under the letterhead of the Department of Physics and Chemistry of the U.S. Naval Academy (perhaps written by Sampson) and dated 25 March 1878: "Thinking you

Nevertheless, a deflection that small raised a particular difficulty that should be noted carefully—it later proved central to Michelson's ability to develop a new instrument. In order to determine the position of a returning beam that was so close to the original without breaking the light path with his head, Foucault inserted a piece of glass and observed the faint *reflection* of the returning ray. Michelson briefly used an improved version, partially silvering the glass to strengthen the reflection (fig. 2.3).[11] The expedient proved temporary and has escaped the attention of previous historians. But it provides a significant material link between Michelson's earliest experimental work and his later ether-drift apparatus.

Michelson originally took up Foucault's experiment because of its pedagogical interest, but he was unusually fortunate. The topic brought him squarely into one of the most significant research projects underway in the U.S. astronomical community, pursued ever more vigorously by Simon Newcomb. While employed on a probationary basis as an astronomical computer in the Nautical Almanac Office in 1857, Newcomb had studied at the Lawrence Scientific School at Harvard University. In the early days of the Civil War he became professor of mathematics and astronomy at the U.S. Naval Observatory, and was appointed director of the Nautical Almanac Office in 1877. Having been a member of the National Academy of Sciences since 1869, he was by then one of the most respected and influential scientists in the United States. In the year Michelson met him Newcomb became still more widely known through the publication of his book *Popular Astronomy*. Commander Sampson put the two in contact, sending the astronomer an enthusiastic report of Michelson's plans in March. A month later Michelson wrote describing his arrangement and

would be interested to know how Michelson's plan for measuring the velocity of light is coming on, I can tell you that it promises entire success. The original plan has been considerably changed so that *any* distance can be used. The arrangement admits of such precise adjustment that I think that when we have arranged to count the revolutions of the mirror the results will be good. The large photo-heliostat silvered on its front face is used as the fixed mirror. The rotating mirror, also silvered on one face, is a little more than one inch in diameter. At a preliminary trial on Saturday with a distance of about 250 ft and about 125 revolutions we obtained a deviation of 1/25 inch. The fixed mirror is now placed at a mile distance and the mirror will be given a velocity of 200 turns." Simon Newcomb Papers, File 45 (SNP-LC 45), General Correspondence Jan. 23, 1856–Dec. 31, 1879. Also quoted in Livingston, *Master of Light* (1973), 54. The earliest published discussion was a letter Michelson wrote to the editors of the *American Journal of Science* in May, with a brief description and schematic drawing of the arrangement. Michelson, "Measuring the Velocity of Light" (1878).

11. Michelson, "Velocity of Light" (1878), 75–76.

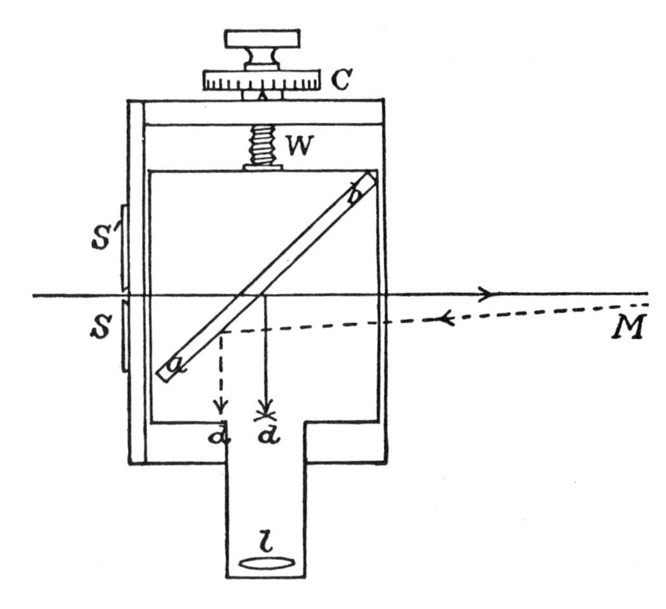

FIGURE 2.3 The slit and micrometer Michelson used in 1878. *S* is the slit; *ab* is a part-silvered piece of plane glass. The light proceeds in the direction *SM* and is returned in the contrary direction, part being reflected from *ab*, forming an image of the slit made to coincide with crosshairs at *d*. When the mirror revolves, the image is displaced to *d'*. The displacement could be measured by using the screw *w* to move the mirror *ab*, *d*, and *I* the lens for viewing the image in the direction *S'S*, until the displaced image again coincided with the crosshairs. The distance moved was read on the divided circle *c*. Michelson, "Velocity of Light" (1878), fig. 3.

asking for advice.[12] Michelson and Newcomb quickly formed an intense working relationship that strongly shaped Michelson's early development

12. On Newcomb, see Moyer, *A Scientist's Voice in American Culture* (1992). For Sampson's initiative, see note 10 above. Michelson wrote to Newcomb on 26 April 1878, describing, among other things, the recent breakage of a mirror (subsequently replaced by one constructed for $10). Their correspondence is held in a variety of archives: the Simon Newcomb Papers and the Naval Historical Foundation Collection both in the Library of Congress (SNP-LC and NHFC-LC), and the Records of the Department of the Navy, Records of the Naval Observatory, Record Group 78 in the National Archives and Records Administration (USNO-RG78-NARA). Much of their correspondence is published in Reingold, ed., *Science* (1964), and Livingston, *Master of Light* (1973) quotes some additional letters. Together these sources provide a very thorough but still incomplete coverage of the letters held. Copies of much of this material are also held in the Michelson Collection in the Nimitz Library of the U.S. Naval Academy (MC-USNA), which now holds material initially gathered for the Michelson Museum in China Lake (for an early guide to that collection, see Plum, "The Michelson Museum" [1954]). The Nimitz Library (and not the Library of Congress) now also holds the material and notes Livingston gathered in writing her biography.

as a physicist and the problems he took up in the next decades of his research. Newcomb's first letter revealed his appreciation for Michelson's design and shows a New World dimension that marked many responses to Michelson's achievements: "To have obtained so large a deviation from apparatus so extremely simple seems to me a triumph, upon which you ought to be most heartily congratulated. So far as I know it is the first actual experiment of this kind ever made on this side of the Atlantic."[13] Newcomb had begun planning such an experiment several years earlier, intending to use a very large distance between the two mirrors (of 1 to 3 miles). That would allow the revolving mirror to rotate more slowly, reducing the possibility of disturbing the image with the air currents it created. But he also expressed a willingness to step back: "If any one else will make the experiment with all necessary certainty and accuracy, I have no desire to take more than a subordinate part in it. All I want is a result." Newcomb asked Michelson to tell him when the experiment was again operating successfully, "that I may go down and see it with my own eyes."[14]

"All I Want Is a Result"

Newcomb sought an accurate value for the velocity of light in order to provide an authoritative determination of the "solar parallax," or equivalently the distance between the sun and the earth.[15] Since the latter is the basic unit of measurement for many astronomical calculations, his aim was as fundamental as rewriting the astronomical tables. Working in the office of the Nautical Almanac and at the Naval Observatory, Newcomb had analyzed the theoretical models and observational data on the relative positions of the earth, moon, and planets for over twenty years. Knowing that the accuracy of all astronomical data depends upon exact knowledge of the earth's position, he decided to reevaluate the reference system of the stars and the astronomical constants—and took advantage of the government and public attention that solar parallax received, with the transit of Venus across the face of the sun in 1874.[16] France, Russia, Britain, Germany, and the United States all mounted extensive

13. Newcomb to Michelson, 30 April 1878, as cited in Reingold, ed., *Science* (1964), 279.

14. Ibid.

15. The solar parallax is the angle subtended by the mean equatorial radius of the earth at the center of the sun, at the mean distance between the earth and the sun.

16. See Norberg, "Newcomb's Early Astronomical Career" (1978), 224.

expeditions to observe the event from spatially distant locations. Newcomb helped kick start the U.S. expeditions in 1870 and served as secretary to the commission that carried them out. But even while providing the most prominent public face of what became an extremely well-known and sometimes controversial campaign, behind the scenes Newcomb was increasingly committed to the alternative possibility of determining the solar parallax by combining terrestrial measures of the velocity of light with the constant of aberration.[17]

Newcomb had first advocated repeating Foucault's measurement on a larger scale in an 1867 paper on the distance between the earth and the sun. In 1876 he began drawing up concrete plans for a revolving mirror arrangement with the University of Pennsylvania physicist George Barker. At this stage they envisaged carrying out the experiment in Pittsburgh, or Menlo Park, New Jersey; Thomas Edison had offered the facilities of his shop if Newcomb agreed to make the measurement there.[18] Newcomb's applications for support and funding highlighted the poor state of astronomical tables presently at use around the world. He told the Secretary of the Navy the tables "are so far from perfect that scarcely any result in the national Ephemerides can be accepted as definitive when the greatest precision is required. Even in operations so simple as running a boundary line, it generally happens that the correction of astronomical data is a considerable part of the labor of calculation."[19] The transit of Venus expeditions had absorbed the enormous sum of $177,000 between 1871 and 1874, and would need still more funds to reduce the observations.[20]

17. Canales has described the very public difficulties posed by the "black drop" effect; for a more detailed account of the tensions underlying the U.S. campaign and the contrast and similarities between what have often been described as "astronomical" and "terrestrial" measures, see Canales, "Photogenic Venus" (2002); Staley, "Conspiracies of Proof" (2007).

18. Newcomb gave a history to velocity of light measurements and his own involvement in Newcomb, "Measures" (1882), 120. In his collaboration with Barker, Newcomb would provide the mirror while Barker looked after the engine to drive it. See Joseph Wharton to Simon Newcomb, 17, 24, and 26 July and 4 September 1876, and Barker to Newcomb, 26 February 1876, 30 April 1877, and 28 April 1878, all in SNP-LC 57, Subject File: "Eclipses, Electricity and Light," "Velocity of Light" folder; and Barker to Newcomb, 22 August 1879, NHFC-LC 22. A leading innovator in silver-on-glass technologies, Henry Draper silvered the mirror for Newcomb, Draper to Newcomb 17 and 20 May 1877, in SNP-LC 57, "Velocity of Light." This correspondence clearly indicates Michelson did not rekindle Newcomb's interest in the topic, as Jaffe believes. Jaffe, *Michelson and the Speed of Light* (1960), 54.

19. Newcomb to Com. R. W. Thompson, Secretary of the Navy, 12 December 1878, in SNP-LC 57.

20. In fact, reduction of the results was clearly hampered by Newcomb's lack of attention and the misplacement of funds before the task was handed over to William Harkness in 1882.

Newcomb contrasted time-consuming and costly transit observations with a relatively inexpensive experiment; he even promised that an expedition abroad to measure the next transit of Venus in 1882 would be unnecessary if his measurements could be made in time. There was a weak link in existing experiments: Foucault and Cornu's determinations differed by nearly 1 percent, and it was not clear which result should be trusted (Foucault's figure was 298,000 km/s while Cornu's result was 299,990 km/s). But there were broader objectives to the experiment also. Newcomb told the National Academy of Science it might deliver knowledge respecting "the qualities of the luminiferous ether and the relation between light, electricity and other visible forces."[21] Early velocity of light measures almost always mixed the two justifications of astronomical measures and ethereal inquiries.

By late 1878, Newcomb's collaboration with Barker had given way to more extensive plans. Now Newcomb wanted the experiment included in the Naval Observatory budget (estimating he needed $7,000 to do the work in Washington). By showing the possibility of reaching a definitive result, Michelson's trials helped Newcomb promote his own venture, despite an earlier admission to Rear Admiral Daniel Amman that Michelson's plans might be better than his own.[22] Clearly Newcomb was not yet ready to take a subordinate role and trust the certainty and accuracy of *whatever* result Michelson obtained. Later he wrote, "before the reliability of Mr. Michelson's work had been established, the preparations for the present determination had been so far advanced that it was not deemed advisable to make any change in them on account of what Mr. Michelson had done."[23]

21. This was in addition to what he described as the "more generally recognized" importance of the value for determining the distance of the sun. Newcomb, "On a Proposed Modification of Foucault's Method of Measuring the Velocity of Light," communicated to the National Academy of Sciences, April 1878. In SNP-LC 57.

22. Newcomb to Rear Admiral Daniel Amman, 5 June 1878, SNP-LC 57. The letter was an official report on Michelson's plan, in which Newcomb stated that it was well worth the encouragement of the Bureau of Navigation and the Congress. In a later letter Newcomb told the Secretary of the Navy that he had originally been loath to ask for government funding for a project he considered uncertain, but that Michelson's recent experiments placed the success of the proposed plan "beyond serious doubt." Newcomb to Thompson, 12 December 1878 (see note 19).

23. Newcomb, "Measures" (1882), 120. He added that Michelson's ability had led Newcomb to request that Michelson be detailed to assist him in his own experiments. Michelson acted in this capacity until September 1880.

For his part, Newcomb helped Michelson by securing further support within the navy and offering advice on experimental protocols. After the equipment failures of his first trials Michelson initially looked to Congress for funding, but in July a gift of $2,000 from his father-in-law allowed him to order new instruments and start preparing an improved measurement.[24] By the time the American Association for the Advancement of Science (AAAS) met late in 1878, Michelson could give a detailed report on his preliminary work. Echoing Newcomb's rationale for the determination, Michelson wrote that it seemed surprising only three scientists had sought to obtain the velocity of light experimentally, considering its importance "as one of the simplest and most accurate means of ascertaining the distance of the sun from the earth."[25] He detailed a preliminary series of ten observations from apparatus "adapted from the material found in the Laboratory of the Naval School." The results varied from 184,500 to 188,820 miles per second with a mean value of 186,508 miles per second. With a greater distance between the mirrors and better apparatus, he expected to obtain the correct result "within a few miles."[26]

One of the most important improvements Michelson made is richly symbolic of the wider implications of his project, substituting terrestrial measures for astronomical observations. The lens Michelson used in his first trials had been fabricated for transit of Venus observations (and, following Newcomb's directions was ground for photographic rather than visual observations). Now Michelson ordered a larger lens with a greater focal length from the noted telescope makers Alvan Clark and Sons in Cambridgeport, Boston. His calculations showed that with a lens of 7 inches in diameter and a focal length of 160 feet, he would obtain about thirty times as much light and still get a displacement four times as great.[27] Michelson made his first observations with the new lens on 30 January

24. See Michelson to Newcomb, 18 December 1878, in Reingold, ed., *Science* (1964), 280–81, and Michelson, "Velocity of Light" (1879), 116.

25. Michelson, "Velocity of Light" (1878), 71, also reprinted as Michelson, "Velocity of Light" (1928).

26. Michelson, "Velocity of Light" (1878), 77.

27. Michelson to Newcomb, 18 December 1878, in Reingold, ed., *Science* (1964), 280–81. In fact, the lens Clark and Sons provided was 8 inches in diameter with a focal length of 150 feet. Michelson, "Velocity of Light" (1879), 116. The transit lens was 5 inches in diameter with a focal length of 39 feet. Other concerns included replacing the bellows used to drive the mirror with a turbine wheel and keeping the rotation constant for enough time to take observations. On the bellows, see Michelson, "Velocity of Light" (1878), 74. On the constancy of rotation, see Michelson, "Velocity of Light" (1879), 123–24.

1879, and began his final series of observations on 5 June, taking readings in the mornings and evenings of eighteen days before bringing the trials to a conclusion on 2 July. He used a light path of approximately 2,000 feet, and now the returning beam was displaced some 5 cm. This made it possible to measure the position of the beam without the observer's head getting in the way. So Michelson dispensed with the half-silvered mirror he had employed previously, using instead a combined slit and scale.[28] Now he obtained a value of 299,944 ± 50 km/s for the velocity of light, quite close to Cornu's result.[29] As Olesko has shown, it is important to note that after Michelson's attention to the precision of his instruments came an equally critical concern with the accuracy of his data: now he made quantitative corrections for constant errors, used graphical analysis and interpolation to measure the increments of his micrometer, and applied the method of least squares to his data to establish the limits to the reliability.[30]

Having obtained a number kept Michelson one step ahead of the man who had so quickly become his principal mentor. Newcomb's first effort to have the navy pay was unsuccessful, so he turned again to the National Academy of Sciences. In January 1879, their select committee recommended that Newcomb apply to Congress for funding.[31] By this time Michelson was involved in Newcomb's project, in addition to completing his own. Visiting Cambridge in early January (probably in order to test and pick up his lens), Michelson discussed the relative merits of performing Newcomb's experiment in Washington or Cambridge with Edward Pickering, who had moved from teaching laboratory physics to become director of the Harvard College Observatory.[32] Finally, with the backing

28. See Michelson, "Velocity of Light" (1879), 116.

29. Ibid.

30. Olesko, "Michelson" (1988), 124. As Olesko shows, Michelson drew on the model provided by the 1873 English translation of Kohlrausch's *Introduction to Physical Measurement*, which became a required text in laboratory instruction for Naval cadets in 1878–79. His use of graphical analysis and interpolation also went further than Kohlrausch.

31. Report of the select committee referred to consider "The plan proposed by Prof. Newcomb for measuring the velocity of light, and the means for carrying it into effect," SNP-LC 57.

32. Obtaining a good line of sight over a long distance was a difficulty, and Cambridge offered several possibilities. However, later Newcomb wrote to Pickering to explain that the inconvenience involved in working elsewhere rendered Washington preferable. See Pickering to Newcomb, 6 January 1879, HCO-HUA UAV 630.14 Harvard College Observatory/E. C. Pickering, Director/Letterbooks (outgoing)/B1, February 1877–July 1880, and Newcomb to Pickering, 12 March 1880. HCO-HUA UA V 630.17.7.

of both the National Academy and the Secretary of the Navy, Newcomb succeeded in getting an appropriation of $5,000 from Congress on 3 March 1879. He began trials in 1880 and detailed Michelson to the Nautical Almanac to enlist his aid. In March 1880, Newcomb sent Michelson to Boston to test the apparatus being constructed by Alvan Clark and Sons at their workshop, and the young scientist then participated as an observer in the first four months of experimental work in Washington. But Michelson left for Europe on 13 September, before Newcomb's experiments were concluded.

For all the benefits his navy career had bestowed—an unusually deep education in science, followed by the opportunity to teach physics and considerable institutional and practical support for his research—Michelson evidently found his position as instructor at the Naval Academy restrictive. His involvement in Newcomb's work, for example, took place under orders, and he was still required to serve at sea periodically, with a major stint falling due. Looking for appropriate professional security, Michelson first sought an appointment as professor of mathematics and astronomy in the navy (the position Newcomb held). But it soon became clear this would involve a test of proficiency in both subjects of its title, so Michelson let it be known that he was interested in an academic professorship, declaring himself neither mathematician or astronomer, but a physicist looking for a fair position "with a respectable physical laboratory, and if possible also a respectable salary."[33] In just a few years Michelson had built up both a scientific reputation and a range of valuable personal contacts with navy and academic scientists. Many of these had been established at least in part through his extensive consultations to ensure the instruments he used met the highest possible standards. Albert Mayer of the Stevens Institute and William Rogers and Edward Pickering of the Harvard College Observatory had all provided advice or instrumental assistance, and Michelson could draw on still others through Newcomb's patronage. When an academic position was not forthcoming rapidly, he obtained leave of absence from the navy.[34] Travel was part of the ideal education of a developing scientist—with experience and perhaps a doctorate in the research laboratories of Europe conferring important

33. Michelson to A. M. Mayer, 26 June 1880, in Reingold, ed., *Science* (1964), 286–87.

34. For Michelson's search for a position, see Michelson to Mayer, ibid.; Wolcott Gibbs to Julius Hilgard, 19 July 1880, in Livingston, *Master of Light* (1973), 65–66; Barker to Newcomb, 3 December 1880, in SNP-LC 15; and Michelson to Pickering, 7 July 1880, ECP-HUA, HUG 1690.15, E. C. Pickering Private Letters 1850–83. While Michelson was in Europe, Barker was

advantages on budding academics. Now Michelson used the opportunity of travel to escape the duties pressed upon him in the navy and to further define himself as a research scientist in his own right. Leaving with Newcomb's project still in progress probably made something of a point about Michelson's overriding wish to serve his own priorities.

When Newcomb completed his observations and it became clear his results would differ from Michelson's—being as much as 200 km/s lower (and hence even further away from Cornu's figure)—the worried astronomer interrogated the conditions under which both experiments had been performed, to track down and try to eliminate the reason for discord. His new experience suggested potential weak points in Michelson's 1879 account of his experimental procedures, so Newcomb engaged in an extensive theoretical deconstruction of Michelson's work and as far as possible a material reconstruction of the arrangement. Newcomb's endeavor indicates revealing aspects of what scientists, sociologists, and historians have stressed is the complex task of assessing the value of another scientists research in the light of the fact that innovative experiments (and theory) are the product of a local culture, embedded in a complex range of practices, many of which will be tacit. One corollary is that replication or the use of similar theoretical approaches may not be possible on the basis of following printed recipes of the kind found in published papers, requiring rather that the culture of the original worksite be shared—an achievement sometimes only possible through direct personal contact. Studies of recent laser interferometry and precision optics in Fraunhofer's day, of the heat-measuring skills James Joule learned as a brewer and employed in his determinations of the mechanical equivalent of heat, and of the cultivation of theory in Cambridge have highlighted the integral role in science of skills and forms of knowledge that escape ready communication, whether through the scientist's or tradesman's interest in secrecy, because their salience or their nature has not been (re)cognized, or for other reasons.[35] Newcomb's assessment of Michelson required teasing apart the

---

instrumental in obtaining a position for him at the Case School of Applied Science in Cleveland. See Barker to John Stockwell, 22 March 1881, in the Case Western Reserve University Archives, quoted in Livingston, *Master of Light* (1973), 89–90.

35. Polanyi first drew attention to the issue that Collins highlighted in studies of replication, using the term "enculturational" learning to highlight the subtle information that requires personal contact for its transmission in contrast to algorithmic models of learning. Sibum's and Jackson's historical studies have helped deepen our understanding of the transmission and transformation of tacit skills between tradesmen, scientists, and (in Sibum's case)

complex of apparatus, protocols, and guiding assumptions the younger scientist had deployed.

Had it been possible for Newcomb to replicate Michelson's experiment he would certainly have done so. But Michelson's apparatus had been dispersed, and sometimes even disassembled in the process of checking it against standard measures. So Newcomb's letters followed Michelson first to Berlin, then to Heidelberg and Paris, with insistent questions about his instruments and procedures. On the basis of the replies he received Newcomb reassembled as much of Michelson's apparatus as he could, and then asked yet more questions about how they had been used.[36] Newcomb's particular critical concerns with distance measurement opened a leitmotif for Michelson's future research and provide an initial source for Michelson's later concern with standards of length. In addition, the correspondence between the two men makes visible a number of features of physics in the period: the working network that existed between particular American experimentalists, the instrument makers and standards they trusted, and a kind of economy of precision instruments that brought them together.

Michelson had used a scale provided by William Rogers, who as the foremost American authority on length standards was then engaged in comparisons of the meter and imperial yard.[37] The scale had been returned to Rogers for rechecking. Michelson advised Newcomb that the mirror had been made by the Fauth firm of Washington, but he would recommend using Alvan Clark instead.[38] The nagging problem of the whereabouts of Michelson's revolving mirror was finally solved when it

---

historians; Warwick has underlined the need to see theory in similar terms to experiment; Collins has recently provided a helpful categorization of different forms of tacit knowledge. See Polanyi, *Personal Knowledge* (1958); Collins, *Changing Order* (1985); Collins, "Tacit Knowledge, Trust, and the Q of Sapphire" (2001); Jackson, *Spectrum of Belief* (2000), esp. 10–14; Sibum, "Reworking the Mechanical Value of Heat" (1995); Warwick, *Masters of Theory* (2003).

36. Newcomb's concern contradicts a common misperception of the singular nature of Michelson's work that has been propagated by physicists close to Michelson. Referring to velocity of light measurements, for example, Harvey B. Lemon wrote, "neither the young man of 26 nor the old man of 79 has ever had a rival. It is remarkable testimony to the world-wide confidence in his ability, honesty, and judgment that during his lifetime no one ever attempted to repeat his experiments or check his results on this subject excepting himself." In the foreword to Michelson, *Studies in Optics* (1995 [1927]), vii–viii.

37. I will consider Rogers's work more fully in chapter 3. See also Evans and Warner, "Precision Engineering" (1988); Morley, "William Augustus Rogers" (1902).

38. Michelson to Newcomb, 15 July 1882, in Reingold, ed., *Science* (1964), 303–4.

was found much later in the home of his father-in-law.[39] The central issue of Newcomb's inquiries was one of distance measurements, both of small increments and great distances. Newcomb, for example, wondered from which point on the revolving mirror Michelson had measured, the center of the mirror as the publication assumed, or from the top of the mirror. Michelson conceded it was the latter and the appropriate correction hadn't been made. His number went down a few kilometers per second.[40] Then there was the question of the tape Michelson had used measuring the baseline. Had measurements been made from the metal ring at the end, or the first division, and by whom? Newcomb's results differed from Michelson's when making a rough comparison with the standard meter scale (Michelson had used Wurdemann's copy of the standard yard), and so he compared his tape against Michelson's, independent of any hypothesis concerning the scales of measure.[41] These exchanges show a common but careful recourse to standards, which, however, was not sufficient to answer the questions at issue. For Newcomb the instruments needed to be treated as particular: he needed to examine the actual micrometer Michelson had used; and they needed to be treated as personal: he needed to know how the person making the measurement used the particular instrument. But even after all the work of reconstruction, there still remained an "embarrassing" discrepancy, as Newcomb put it, a difference in results that threatened to be "simply unaccountable." Early in these exchanges Newcomb took the embarrassment on his own shoulders; but he proved himself a tenacious accountant, as we shall see.[42]

39. Ibid.

40. See Newcomb to Michelson, 15 August 1881 and Michelson to Newcomb, 29 August 1881, in bid., 296–97 and 298–300, on 297 and 298. Two corrections were made in 1882, bringing the previous result down by 34 kilometers to 299,910 ± 50 kilometers per second. Michelson, "Supplementary Measures" (1882), 243–44. Note that in 1879 Michelson had sent his tape measure to William Rogers for comparison. Michelson, "Velocity of Light" (1879), 140.

41. Newcomb to Michelson, 2 May 1881, in Reingold, ed., *Science* (1964), 290–91, on 291. On Wurdemann, see Turner, "William Würdemann" (1991).

42. Newcomb wrote, "I have been much embarrassed at the result I am going to get for the velocity of light" (Newcomb to Michelson, 2 May 1881, in Reingold, ed., *Science* [1964], 290–91, on 290). A month and a half later he wrote, "if it turns out that there is no mistake in the Coast Survey distance the difference of results will be simply unaccountable." Newcomb to Michelson, 23 June 1881, in Reingold, ed., *Science* (1964), 292–93, on 293. Reporting to the National Academy in 1881, he allowed that he hesitated to publish the results of his first three series of results, because of their divergence from Michelson's value, but highlighted Michelson's measurement of the distance the light traveled: "his description of this process

A Quantity "Quite Too Small to Be Observed": On the Origins of an Experiment

In the middle of these extended negotiations Michelson sent letters reporting his idea and then execution of an ether-drift experiment in Berlin and Potsdam in November of 1880 and May 1881 respectively.[43] Newcomb helped find funding for this experiment and outlined his views on first-order and second-order measurements, yet was evidently far more interested in how to reconcile velocity of light results than the ether-drift experiment. But Michelson's new plan addressed experimentally the theoretical assumption that there *was* a constant speed of light. In particular, given the assumption that the earth moved through a stationary ether, the experiment was designed to detect whether orientation or situation influenced the propagation of light. Different velocity of light results might have been due in part to the orientation of the light path in relation to the earth's motion through a stationary ether (reflecting either the earth's daily rotation or its passage around the sun), or to differences in the height of the experiment above sea level and hence to exposure to different degrees of ether drag (if the ether was found to be carried with the earth in its immediate vicinity, but stationary at greater distances from its surface).[44] The experiment thus underlines Michelson's thorough concern with the fundamental bases for the experimental study of light, and his concentration on something Newcomb referred to in funding applications but left subordinate in his actual work, "the qualities of the luminiferous ether."

Michelson's decision to focus on this particular problem was clearly stimulated by taking as a challenge comments James Clerk Maxwell made in a March 1879 letter to the new director of the Nautical Almanac, David Peck Todd (subsequently published in *Nature*). Maxwell stated that the velocity of the earth with respect to the ether alters the velocity of light by a quantity depending on the square of the ratio of the earth's velocity

---

is not so explicit that a scientific jury could pronounce the measure absolutely free from the possibility of error." Newcomb, "Report of Progress" (1881), 18.

43. See letters of Michelson to Newcomb, 22 November 1880, and Newcomb to Michelson 2 May 1881 and 2 June 1881, in Reingold, ed., *Science* (1964), 287–88, 290–92. The experiment was published in Michelson, "Relative Motion" (1881).

44. With the assumptions Michelson made in 1881, there would be a difference of approximately 5 meters per second between the velocity of light in a north-south and east-west orientation.

to that of light

$$\left(\frac{v_E}{c}\right)^2,$$

which he described as "quite too small to be observed."[45] Newcomb had a similar view. Responding to Michelson's plan, the astronomer wrote that given the minuteness of second-order effects he still hoped for the detection of a first-order effect

$$\left(\frac{v_E}{c}\right),$$

and thought measuring wavelengths through a transparent ruled plate offered the most likely way of disclosing one. This was the "crucial experiment" he would like to see tried. Thus the problem Michelson approached was well recognized in the office of the Nautical Almanac and could even draw the epithet "crucial"; but unlike others Michelson thought a second-order effect was "easily measurable."[46] Interference phenomena gave him this confidence.

A number of Foucault's colleagues and students in France had developed extremely precise measures of the speed of light through different media, pioneering the field of interferential refractometry in experiments that explored the implications of the wave theory of light and probed the relations between light, the medium through which it moved (the ether), and matter. Their instrumental techniques relied on measuring the displacement of the interference fringes of light and darkness that are revealed when light from a common source is split up, sent over pathways that differ in length or in the media that they traverse, and recombined. Edward Pickering drew Michelson's attention to experiments of this kind in August 1879, referring him to Mascart's studies of the refraction of gases.[47] Éleuthère Élie Nicolas Mascart had developed apparatus

45. Maxwell, "Letter to David Peck Todd, 19 March 1879" (1880).

46. Newcomb's view was conveyed in letters to Michelson of 2 May and 2 June 1881 in Reingold, ed., *Science* (1964), 290–91, 292, and 296–97. Newcomb thought Michelson's arrangement "very beautiful" but unlikely to furnish results due to the compensation effected by the light's return along its own path. Subsequently he agreed that a second order effect should be appreciable, "theoretically at least," but was skeptical about realizing it: "Still I cannot feel sure but what some little action may come in to nullify the effect of so minute a cause and it seems to me we ought to be able to devise some way of getting a result which would depend only on quantities to first order." Michelson's comment on ease of measurement is in Michelson, "Relative Motion" (1881), 121.

47. See Michelson to Newcomb, 29 December 1878 and 16 January 1879, in Reingold, ed., *Science* (1964), 283–85, and Pickering to Michelson, n.d., circa 21 August 1879, HCO-HUA,

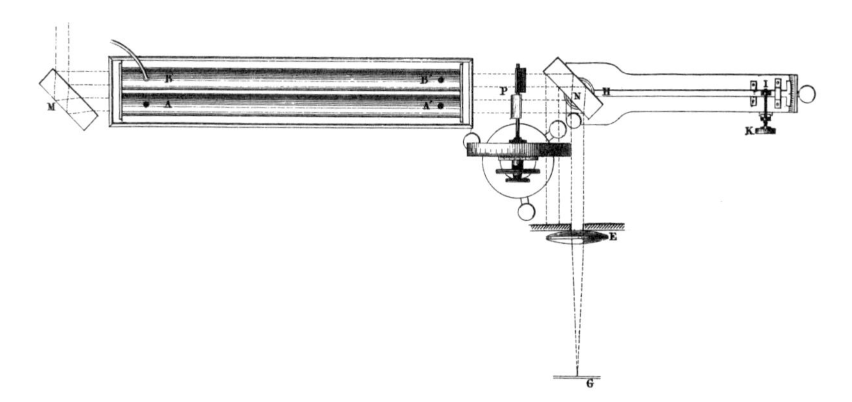

FIGURE 2.4 Jamin's refractometer, shown in plan. Two equally thick plane parallel glass plates *M* and *N* are silvered on their rear surfaces. Light from an extended source is incident on *M* at about 45° and gives rise to two beams, one reflected from the front surface of *M* and rear surface of *N*, the other reflected from the rear surface of *M* and the front surface of *N*; the two are recombined to give interference patterns in the focal plane of the telescope *E*. Chambers of gases or liquids are inserted in *AA′* and *BB′*, and *P* carries compensator plates. Jamin, "L'indice de réfraction de l'eau" (1858).

pioneered by Jules Jamin, which employed an arrangement of parallel mirrors to split a beam of light into two components that traveled different pathways parallel to each other and in close proximity. The light was then recombined, forming interference patterns that could be observed in a telescope (fig. 2.4). Placing different gases in each arm, relative differences in the velocity of the light could be measured that were of the order of fractions of the wavelength of light (thereby indicating the refractive index of the different gases). That kind of precision would suffice. So in French refractometry, Michelson knew a measuring technique accurate enough to address Maxwell's query; he came up with a geometrical arrangement that would reveal any differences in velocity due to orientation to the ether.

Maxwell had outlined the problem that all known terrestrial methods of measuring the velocity involved returning the light over its path, leading to a compensation that minimized the effect sought. So Michelson

---

UAV 630.14 Harvard College Observatory/E. C. Pickering, Director/Letterbooks (outgoing)/B1, Feb 1877–July 1880. Pickering wrote down references for experiments on the index of refraction of gases and the compressibility of water that he had discussed with Michelson earlier, and mailed Michelson some of his own articles. The references were to Mascart, "Réfraction des gaz" (1874); Mascart, "Dispersion des gaz" (1874); Mascart, "Réfraction de le'eau comprimée" (1874).

looked for a way to send two rays on different round journeys perpendicular to each other. An instrument could then be oriented so that one light beam traveled in the direction of the earth's motion in its orbit (east-west) while the other traveled perpendicular to this (north-south), thus maximizing the difference in velocity due to any ether wind created by the earth's motion. (Michelson initially thought incorrectly that light traveling north-south would not suffer any ether drag.) Changing the orientation of the instrument would then indicate any change in the velocity of the light in the different arms, through a shifting of the interference patterns created when the light was recombined.

Historians have speculated on the provenance of the half-silvered mirror that Michelson eventually deployed to exploit this geometry, but a little-noticed 1928 paper shows Michelson had the geometry before he knew how to realize it. At first he tried a technique that might have been suggested by Fizeau's well-known refractometer experiment on the dragging of light with a moving medium. Like Fizeau, Michelson used a masked lens to form two pencils of light, but then—in a step unprecedented in refractometer experiments—introduced a wedge-shaped mirror to send them at right angles. Mirrors were placed normal to return the two rays along their pathways, where they were brought together again on passing a second time through the lens. But this arrangement proved very difficult to work with, something Michelson later described as fortunate, because it led him to develop his own interferometer with the much simpler device of a part-silvered mirror splitting the beam (fig. 2.5).[48]

Without being aware of this sequence of events, historians have attributed the mirror to Michelson's awareness of the work of Fizeau, or his familiarity with the naval sextant.[49] Indeed, had Michelson gone very far with the arrangement illustrated in his "FIG. 2," following Fizeau he would have interposed a part-silvered mirror between the source of light and lens in order to observe the light fringes without disturbing light from the source. Following this kind of pathway may in fact have drawn his attention to the device. But as we have seen, Michelson's publications show

48. See Michelson, "The Michelson-Morley Experiment" (1928), 343. Previous scholars have not noticed Michelson's comments.

49. On Fizeau as a source, see Swenson, *Ethereal Aether* (1972), 61–62. On the sextant, see Stapleton, "Context of Science" (1988), 16. Others have described it as following from a study of Jamin's apparatus. Shankland, "Michelson-Morley Experiment" (1964), 18–19; Livingston, *Master of Light* (1973), 77.

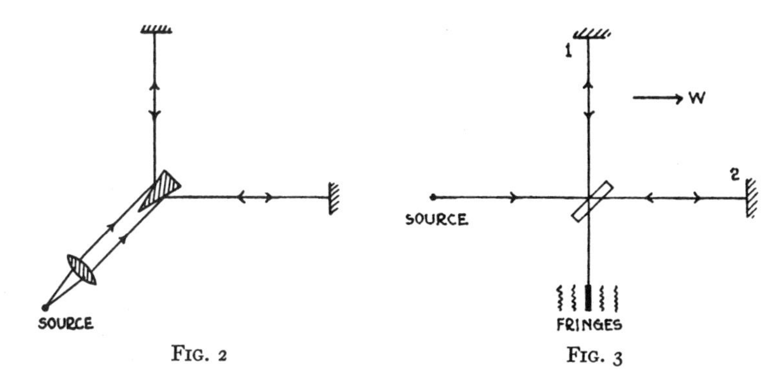

FIGURE 2.5 The two arrangements Michelson trialed in his first attempt to produce interference of light beams that travel parallel and transverse to the velocity *w* of the apparatus through the ether. His FIG. 2 depicts an unsuccessful method using a masked lens and wedge of glass. Michelson, "The Michelson-Morley Experiment" (1928), 343.

he already knew the part-silvered mirror rather well. It had formed a temporary expedient in his velocity of light experiments, abandoned when Michelson achieved a large enough displacement to measure the returning beam without disturbing the original. But it is likely that in 1878 he observed the cross of light the device produced, and decided in 1881 to use both arms in a new, tabletop velocity of light measurement. The continuity between the two arrangements is emphasized by the fact that in both cases the partially silvered mirror was carried on a micrometer screw. In 1878 this allowed measurement of the displacement; in 1881 Michelson used it to adjust the width, position, or direction of the fringes (fig. 2.6). Thus, paying close attention to instruments and apparatus shows practical connections invisible to historians focused on the ether-drift experiment alone: Michelson's instrumental breakthrough came from pushing to the center of the apparatus a device that had played a necessary but ancillary role in both his earlier experimental work and previous refractometer arrangements. The partially silvered mirror, oriented at 45° to the source of light, was an extremely simple and elegant solution to Michelson's need to separate light rays so widely.

Very soon after arriving in Berlin, having passed through London and Paris en route (without meeting any scientists in the latter city), Michelson described his plan to Hermann von Helmholtz, director of the Physical Laboratory in the University of Berlin. Initially Helmholtz believed

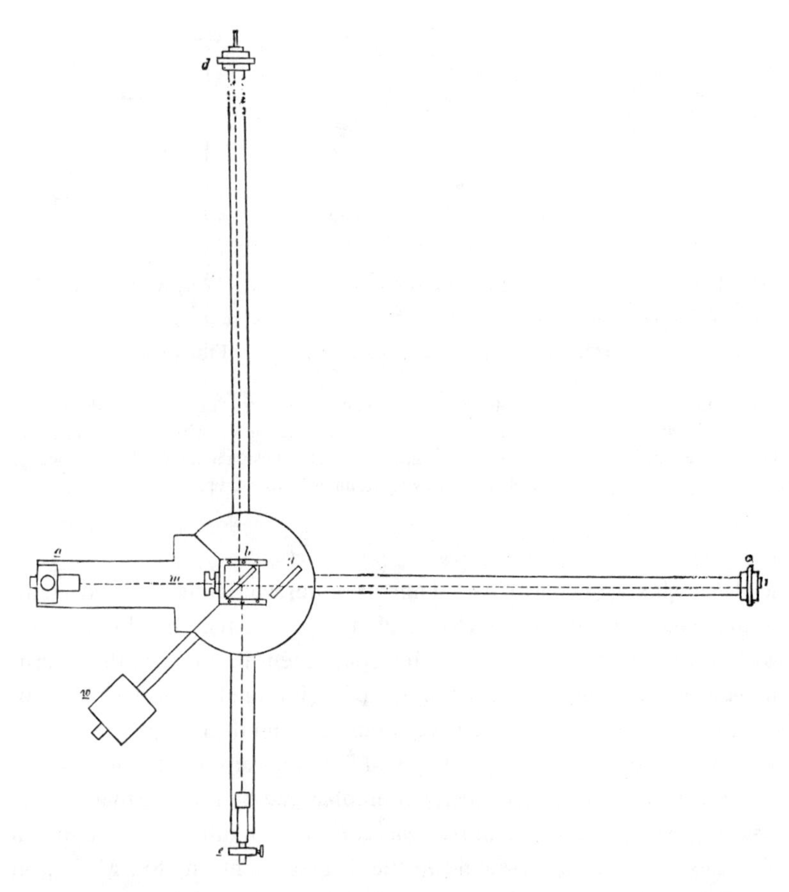

FIGURE 2.6 Plan of the 1881 ether-drift apparatus. The geometry is now well known: light from lamp *a* passes through the partially silvered mirror *b*, part being transmitted to the mirror *c* and part being reflected to the mirror *d*. From these the light is reflected to *b*, where transmission and reflection again occur, with the interference fringes being observed through the telescope at *e*. Note especially the similarity between the housing of the part-silvered mirror here and that in fig. 2.3. *g* is a compensating glass cut from the same piece as *b*; *m* is a micrometer screw that moves the plate *b* in the direction *bc*; *w* is a counterpoise. Michelson, "Relative Motion" (1881), 123.

temperature effects would swamp any observation of the expected displacement of the interference fringes.[50] Although he thought the laboratory lacked the necessary facilities (specifically a room that could be kept

50. Michelson to Newcomb, 22 November 1880, in Reingold, ed., *Science* (1964), 287–88. Haubold, Haubold, and Pyenson contradict Livingston in stating that Michelson studied in Paris before arriving in Berlin. This seems unlikely given that he left the United States in

at a constant temperature), Helmholtz offered to host Michelson's work, but since he could not provide any funding, thought Michelson should wait to carry out the experiment in the United States. But Michelson was moving too fast for that, and was also confident enough to back his own judgment about the conditions required (in the event a cardboard shielding was enough to deal with temperature effects). Both in regard to velocity of light and his present proposal, Michelson was quick to consult more experienced physicists but found his ability to conceptualize a suitable and relatively simple arrangement often outstripped that of his colleagues. (Evidently Michelson's tour of Europe was not to be that of a student, at least in regard to experimental work. He reported to Newcomb that he found the work in the laboratory class very elementary, but was using his time to attend Helmholtz's lectures on theoretical physics and study mathematics and mechanics at home.[51]) Michelson had also learned quickly that getting the experiment actually underway was the most practical way of making the advantages and limitations of his apparatus apparent. In this case Newcomb's contacts with Alexander Graham Bell helped Michelson find another patron. Bell gave Michelson $2,000 from the proceeds of his Volta award, which made it possible to order an instrument from Schmidt and Haensch so Michelson could attempt the experiment in Berlin. Michelson's letter to C. A. Bell (who was responsible for administering the foundation Bell had created to disburse the proceeds of the award) illustrates the young man's view about his reputation at the time:

> I took the liberty of questioning the prudence of the offer which Prof. Bell was kind enough to make, for as I then said, I was young and therefore liable to err, and my entire reputation as a "scientist" was based upon a single research, which fortunately was successful.
>
> Since, however, he was pleased to differ with me, I gratefully acknowledge the generous gift.
>
> I have not however undertaken this work without consulting several prominent scientific men. The answer they all give is that if successful the success would be grand; but that it is at present impossible to say whether the experiments are practicable or not. All, however, advise me to try. So that, with this

---

mid-September and arrived in Berlin later in that month. Haubold, Haubold, and Pyenson, "Michelson's First Ether-Drift Experiment" (1988), 43–44.

51. Michelson to Newcomb, 22 November 1880, in Reingold, ed., *Science* (1964), 287–88.

> assurance and that of Prof. Bell, that in either case he will consider the money well spent, I shall begin the work with greater confidence.[52]

Michelson was intensely aware of his reputation—and of the singularity of his experimental work on velocity of light.

"What May We Not Expect from One Made as Sensitive as Possible!"

Quite famously, the location in which the ether-drift experiment was performed proved critical, and it was not temperature but vibrations that provided the most serious obstacle to carrying it out. Setting up the apparatus on a stone pier in Helmholtz's laboratory in the middle of Berlin, Michelson found vibrations made it impossible to observe interference fringes, even in the middle of the night. This was a spectacular indication of the instrument's sensitivity and perhaps something of a disaster for the laboratory, which in the mid-1870s had spent the unusually large sum of 310,000 marks on its foundations, and had many isolated piers as a protection from both internal and external vibrations.[53] Seeking a quieter location, Michelson obtained permission to use the Astrophysicalisches Observatorium then being built in nearby Potsdam. He found the isolated piers there were also insufficient, but was able to obtain useful observations in the cellar whose circular walls formed the foundations for the great equatorial. When he came to write his report this series of relocations illustrated the sensitivity of the apparatus—vividly, if not precisely.[54]

Michelson worked through the winter. A major challenge came from the way the metal arms of the instrument bent during rotation; it had to be returned to the workshop. Once it came back, a large displacement of the fringes in one particular direction revealed an asymmetry in construction that led Michelson to discard the readings taken in that direction. Some slight cause like the springing of the tin lantern by heating often led to a sudden change in the position of the fringes, which necessitated the

52. Michelson to C. A. Bell, 22 December 1880, Alexander Graham Bell Papers, National Geographic Society Archives, Washington, DC, as quoted in Livingston, *Master of Light* (1973), 76.

53. See Forman, Heilbron, and Weart, "Physics *circa* 1900" (1975), 111.

54. Michelson, "Relative Motion" (1881), 124.

abandonment of that series of observations. Concluding the experiment in April 1881, Michelson published a table of results showing four different series of observations with the instrument turned 90° between each, and readings taken for five revolutions of the instrument in each position. Having expected to detect the earth's motion through the ether, he was evidently disappointed to report a null reading, deciding that the small displacements he did record were simply errors of experiment.[55] But this finding did not lead Michelson to doubt the existence of the ether. He announced rather that the hypothesis of the *stationary* ether was incorrect, and abandoned for the moment any ideas he may have had of tracking the ether currents. Nevertheless he did have two particular products from this experimental work: a result, and a novel form of refractor, a new piece of apparatus. In the 1881 paper he wrote: "It will be observed that this apparatus can very easily be made to serve as an 'interferential refractor,' and has the two important advantages of small cost and wide separation of the two pencils [of light]."[56] Something of Michelson's excited but unfocused sense of its future potential is suggested in his comment to Bell that, given the extreme sensitivity of an instrument "constructed expressly to avoid sensitiveness what may we not expect from one made as sensitive as possible! It seems to me that such a one may possibly excel the microphone."[57] Soon afterward Michelson undertook a careful theoretical study of the nature of the interference fringes produced, because previous studies did not match the form of the fringes he observed.[58]

At about the time Michelson completed the experiment he also heard he had obtained the first professorship of physics at the Case School of Applied Science in Cleveland (largely through the patronage of Newcomb's first collaborator on velocity of light measurements, George Barker). The trustees gave Michelson permission to finish his studies in

55. Ibid., 127–28.

56. Ibid., 123–24. Two decades later (on Michelson's advice) Gale used a modification of Michelson's arrangement for experiments on the refractive index of air at different pressures. His article also provides an overview of earlier research in the field. Gale, "Density and Index of Refraction of Air" (1902).

57. Michelson to Alexander Graham Bell, 17 April 1881, in Reingold, ed., *Science* (1964), 288–90. Michelson gave a demonstration of his apparatus to Bell on his return to the United States that was presumably more successful than the demonstration to Cornu discussed below. See Livingston, *Master of Light* (1973), 97, citing a letter from George Barker to A. G. Bell, undated, Alexander Graham Bell Papers.

58. Michelson, "Interference Phenomena in a New Form of Refractometer" (1882).

Paris before traveling to Cleveland, and provided $7,500 for instruments for the new laboratory.[59] This enabled Michelson to take advantage of the excellence of European and particularly French instrument makers to start his career directing a laboratory. But Michelson already had his most important instrument in his possession, taking the apparatus Schmidt and Haensch had constructed with him, as he went from Berlin to Heidelberg in the summer of 1881 and then to Paris in the fall. Yet his attempts to display the interferential refractometer there showed it didn't yet travel well.

Michelson was well prepared for his European sojourn, carrying letters of introduction to scientists in London, Berlin, and Paris from Newcomb, Pickering, and Rogers. Since all previous experimental measurements of the velocity of light had been made in France, and the principal predecessor to his new instrument also stemmed from France (with its rich tradition in optical physics), his visit to Paris was particularly important. There Michelson met Gabriel Lippmann at the Sorbonne, Alfred Cornu at the École Polytechnic, and two pioneers of interferential refractometry in Jules Jamin (also at the École Polytechnic) and Éleuthère Mascart, professor of physics at the Collège de France and the author of a first-order ether-drift experiment that had yielded the expected negative result.[60] Michelson's determination of the velocity of light had already won him a reputation in the city of light (and apparently led his hosts to expect a much older man), but Cornu was skeptical about the prospects of his new apparatus. He told Michelson, "that experiment was tried fifty years ago by the celebrated Fresnel and it does not work."[61] Interestingly, according to later accounts Michelson's disappointment at the null reading led him to refrain from mentioning the purpose for which the refractor had been designed. Rather than highlighting the result and an experiment, he described the apparatus more as a new instrument. Livingston

59. Livingston, *Master of Light* (1973), 89–91.

60. Michelson had written asking Pickering for letters for Fizeau, Cornu, and Jamin. Never having met the French physicists, since Paris was under siege at the time he had traveled Europe, Pickering passed this request on to Gibbs. Pickering to Gibbs, 16 September, HCO-HUA, UAV 630.14 Harvard College Observatory/E. C. Pickering, Director/Letterbooks (outgoing)/B2, July 1880–October 1883. See also Pickering to Michelson, 8 and 16 September 1880, Pickering to A. C. Ranyard, 8 September 1880, Pickering to A. Auwer, 8 September 1880.

61. Michelson, unpublished transcript of the opening remarks of his lecture, "Application of Interference Methods to Astronomy," at the Carnegie Institution of Washington, DC, 25 April 1923 (photocopy AAMC), reported in Livingston, *Master of Light* (1973), 87–88.

reports that he preferred to show the tricks it could perform rather than its failures.[62] But even this proved difficult. Confirming Cornu's skepticism, it took four days of constant work before Michelson managed to produce the interference fringes he had seen in Berlin and Potsdam. By then the temperamental nature of the refractor must have worried him. It is commonly thought that Michelson needed that long to realize that with a white light source (probably a candle), he was not using as homogeneous a source of light as he had in his earlier trials with sodium.[63] But this is unlikely to be the whole story, since Michelson used sodium to set up the apparatus in Potsdam but then conducted the experiment with the light of a lamp: he already had some experience with the character of different light sources.

The credibility of Michelson's result was also questioned subsequently, when the French scientist M. A. Potier and the Dutch mathematical physicist Hendrik Antoon Lorentz showed that Michelson's calculations had been in error and the apparatus needed a significantly longer path length to detect the second-order effect Michelson sought.[64] The mistake is an unusual one for a master in the U.S. Navy used to dealing with the motion of a ship at sea. Michelson had neglected to take into account the small effect of an ether wind on the motion of the light beam perpendicular to the earth's rotation (equivalent to traveling across a current [not without a current as he had assumed] rather than with and against a current). Clearly the value of both the ether-drift experiment and the interferential refractor apparatus eroded quite quickly on leaving Potsdam; it would take some work if Michelson was to enhance his reputation by adding the luster of a second experiment—or an instrument—to his determination of the velocity of light; which had itself been questioned by Newcomb.

Repetition and Different Cultures of Experimentation

When Michelson returned to the United States, he once again took up velocity of light measurements at Newcomb's urging and with funding

62. Ibid., 88. Note that on Michelson's behalf Cornu communicated a note (in French) giving a brief description of the refractor and the experiment to *Comptus rendus*. The paper acknowledged the error pointed out by Potier discussed below. Michelson, "Mouvement relatif" (1882).

63. Livingston, *Master of Light* (1973), 87–88.

64. Lorentz, "Mouvement de la terre" (1937 [1886]), 204–8.

Newcomb provided from the Bache fund of the National Academy of Sciences. But Michelson made it clear that despite his willingness to repeat his earlier work, he would only do so on his own terms in Cleveland rather than in Washington.[65] On this occasion Michelson also tested the velocity of red light and blue light (in addition to the sunlight tested previously), and the velocity of light traveling through carbon disulphide in addition to air. This was important because the Scottish researchers J. Young and G. Forbes had recently obtained results exceeding those of both Cornu and Michelson, and contended that the velocity of blue light was as much as 2 percent faster than that of red light.[66] With a baseline run along the rails of the New York, Chicago and St. Louis Railways, Michelson used the same equipment he had previously, with a few minor differences in method designed to improve performance. Eventually Newcomb and Michelson published values rather closer to one another than they had achieved previously (now falling inside their respective error estimates), in reports bound together in the same volume of the Naval Observatory Papers. Newcomb gave two figures, one using only results supposed to be nearly free from constant errors, of 299,860 (with a discordance of less than 10 km/s between the separate measures) and one including all determinations, of 299,810 ± 50 km/s. Michelson's figure from his Cleveland trials was 299,853 ± 60 km/s and he corrected his Annapolis result to read 299,910 km/s. It was crucial for Newcomb that it be made clear that Michelson's repetition (arranged solely because of the discord between their previous results) was an independent experiment, despite the considerable collaboration that had taken place earlier, Michelson's participation in Newcomb's work, and the fact that his funding came in part through Newcomb. In an introductory note to Michelson's paper, Newcomb wrote, "no instructions and suggestions had been sent to him except such as related to the investigation of possible sources of error in the application of his method."[67] Celebratory accounts have often presented

65. He would only be able to work for a month or so at a time in Washington, a period that might not even be sufficient to get the apparatus in working order. See Michelson to Newcomb, 8 June 1882, in Reingold, ed., *Science* (1964), 301–2.

66. See Young and Forbes, "Velocity of White and Coloured Light" (1881). Michelson and Rayleigh responded in Rayleigh, "Velocity of Light" (1881); Michelson, "Letter to the Editor" (1881).

67. Newcomb, "Introductory Note," to Michelson, "Supplementary Measures" (1882), 235; Newcomb, "Measures" (1882), 202. J. E. Hilgard wrote to Newcomb on 14 November 1884, enclosing a sketch of a title for the Michelson supplement that should be satisfactory to the Bache fund and to Michelson. MC-USNA, X-12.2 and X-12.3.

Michelson's results as definitive; but it is important to note that in Newcomb's final comparison, Newcomb offered his own best value as the "most probable result." As we shall see in chapter 5, many contemporaries undoubtedly regarded Newcomb's experiment and discussion as the most authoritative American treatment.

Pushed by Newcomb, Michelson's extended work on velocity of light provided a particularly deep initiation in the rigors of experimental practice and the vulnerability of results. The apparent simplicity—and singularity—of his first trial and experimental performance had been followed by work under Newcomb's leadership, and then by a critical interrogation that retrospectively exposed all aspects of his earlier experiment to scrutiny. Thus an unusually demanding accountability accompanied the benefits of Newcomb's patronage (and Michelson later admitted it would have been helpful if Newcomb had been able to study his first paper thoroughly before it reached the press).[68] Repeating the experiment with improved apparatus in Annapolis and Cleveland undoubtedly helped Michelson refine his empirical skills and approach to writing up experiments. Despite the value of this experience I will argue below that the way Michelson approached later research also bears the marks of a reaction against such repetitions. Note also that while repeating velocity of light experiments helped bring U.S. results within a narrow range, exposing the ether-drift apparatus to different conditions played a more complex role in the evolution of the experiment. While the first move from Berlin to Potsdam enabled the experiment to be performed at all, the second excursion from Potsdam to Paris made its shortcomings dramatically visible.

The presence of a highly active culture of experimentation in velocity of light measurements—with performances by different scientists in France, the United States, and Britain, employing a variety of arrangements with constant refinements of technique—highlights the absence of a similar culture around the ether-drift experiment in this period. Despite considerable theoretical interest in the properties of the ether, and the concentration on optical techniques in general and refractometry in particular in France, we have no records of people repeating Michelson's experiment, let alone engaging in the detailed process of challenge and reconstruction that took place between Newcomb and Michelson. In the first years of his tenure at Case, Michelson spent his time teaching

68. See Michelson to Newcomb, 29 August 1881, in Reingold, ed., *Science* (1964), 298–300 on 300.

and consolidating velocity of light measurements, but evidently wanted to make more of his Potsdam experiment. In 1884 he attended the annual meeting of the British Association for the Advancement of Science (BAAS) in Montreal. Both Lord Rayleigh and Sir William Thomson (later Lord Kelvin) made the trip from Britain, and Michelson also traveled to hear a series of lectures Thomson delivered in Baltimore in October.[69] At the BAAS meeting, Michelson presented a paper on the velocity of light in carbon disulphide, and Rayleigh spoke in Michelson's favor in his dispute with Young and Forbes over the velocity of blue and red light.[70] But Michelson also made sure Rayleigh and Thomson knew of his ether-drift work, again canvassing opinions on its significance. Soon afterward he wrote asking Gibbs for his opinion about theoretical aspects of the ether-drift experiment; his letter indicates that the two giants of British physics had told Michelson the best course of action would be to repeat Fizeau's 1851 experiment on the dragging of light in water, followed by his Potsdam experiment.[71]

The two experiments stood in potential opposition. The Fizeau experiment was designed to detect differences in the velocity of light passing through a material medium that was itself in motion. Fizeau had used a refractometer to shine two beams of light in opposite directions through water flowing through narrow pipes. The experiment therefore tested the extent to which the ether was dragged along by moving matter. Fizeau found only a fractional, partial drag, which depended on the index of refraction of water in a way widely taken to support Fresnel's hypothesis of a stationary or stagnant ether. The most natural interpretation of a negative result to the ether-drift experiment, however, was that the ether

69. Kelvin, *The Baltimore Lectures* (1904).

70. Michelson, "Velocity of Light in CS2" (1884). Like Michelson, Rayleigh had responded to Young and Forbes's claims previously. Rayleigh, "Velocity of Light" (1881). For a recent account, see Pippard, "Dark on a Light Subject" (1999). For a description of the relationship between Michelson and Rayleigh, first built around their interest in the velocity of light and Rayleigh's work on the relations between wave and group velocities, see Shankland, "Rayleigh and Michelson" (1967).

71. See Michelson's discussion of his plans in a letter to Josiah Willard Gibbs, Michelson to Gibbs, 15 December 1884, Gibbs Collection, Yale University Library, in Reingold, ed., *Science* (1964), 307–8. Also cited in Livingston, *Master of Light* (1973), 109–10. The papers were published as Michelson and Morley, "Motion of the Medium" (1886); Michelson and Morley, "Relative Motion" (1887). See also the correspondence between Michelson and Morley and William Thomson following the completion of the Fizeau experiment, Michelson and Morley to Thomson, 27 March 1886, and Thomson to Michelson and Morley, quoted in Thompson, *William Thomson* (1910), 857.

was not stationary but completely carried by the earth. If both were to be established, they created a conundrum. To carry out the experiments Michelson enlisted the aid of Edward Morley, professor of chemistry at the nearby Western Reserve University.[72] From the early 1870s, Morley's experimental research had been characterized by a similar commitment to precision work, especially determining the relative weights of oxygen and hydrogen, and he had extensive contacts with local industry, consulting for the iron industry, Standard Oil, the city gas works, a linseed oil company, and others. Like Michelson, Morley made the trip to Montreal and to the Baltimore lectures. Their work together began with Michelson setting up the repetition of Fizeau's ether-drag experiment in Morley's better-equipped laboratory in 1885. When Michelson suffered a nervous breakdown in mid-September, he turned over the apparatus and funds and asked Morley to finish the work. (Morley spoke of symptoms pointing to "softening of the brain" and was doubtful whether Michelson would ever be able to resume his career.) Preliminary trials showed Morley that the apparatus required significant modifications. He made new drawings and went to the instrument makers for new apparatus. Once Michelson had begun recovering in October, the physicist eagerly pressed Morley on his progress with the "fringes" and was soon giving advice. Returning from treatment in New York, he and Morley worked together to complete the experiment.[73] The outcome of sixty-six measurements with two different lengths of path and three different current flows was announced confidently: "The result of this work is therefore that the result announced by Fizeau is essentially correct; and that the luminiferous ether is entirely unaffected by the motion of the matter which it permeates."[74]

With this confirmation, Rayleigh once again wrote encouraging Michelson to repeat his Potsdam experiment. In a long reply Michelson described improvements to lengthen the light path and decrease the problem of vibrations, the two major deficiencies of the 1881 apparatus. Using

72. For a study of Fizeau's ether-drag experiment, see Frercks, "Fizeau's Research Program" (2005). On Morley, see Hamerla, *An American Scientist* (2006); Williams, *Edward Williams Morley* (1957).

73. See Edward W. Morley to Sardis Brewster Morley, 8 April 1885, 27 September 1885 ("softening"), 12 October and 19 November 1885, and 31 January 1886, and Michelson to Newcomb, 28 September 1885, Michelson to E. W. Morley, 12 ("fringes") and 23 October 1885, in Reingold, ed., *Science* (1964), 308–12. On Michelson's breakdown, which was a turning point in his increasingly poor relations with his wife and the faculty at Case, see Livingston, *Master of Light* (1973), 111–15.

74. Michelson and Morley, "Motion of the Medium" (1886), 386.

an array of mirrors to reflect the light repeatedly could solve the first problem, while Morley had suggested floating the apparatus in a trough of mercury, hence enabling it to revolve with minimal vibration.[75] But the cultural environment in which the experiment took place was as important to Michelson as these technical questions. He wrote:

> I have repeatedly tried to interest my scientific friends in this experiment without avail, and the reason for my never publishing the correction [due to Potier and Lorentz] was (I am ashamed to confess it) that I was discouraged by the slight attention the work received, and did not think it worth while.
>
> Your letter has however once more fired my enthusiasm and it has decided me to begin the work at once.[76]

Michelson and Morley's eventual publication included a diagram illustrating the apparatus and a graphical representation of the tiny fringe displacements they observed, following the apparatus as they slowly spun it on its axis on the morning and evenings of four different days in July 1887 (fig. 2.7). Their conclusion? "If there is any displacement due to the relative motion of the earth and the luminiferous ether, this cannot be much greater than 0.01 of the distance between the fringes." This was far less than the displacement expected from considering the earth's orbital motion; and they went on to demonstrate that the existing theories of Fresnel, Stokes, and Lorentz all failed to offer a consistent explanation of the relations between the moving earth and the luminiferous ether. The experiment thus ruptured the relations between phenomena and theory—but could still be regarded as unfinished work. Michelson and Morley had only taken observations over several days and compared these with the orbital motion of the earth. Combining orbital motion with the (unknown) motion of the solar system would modify the result, and it was just possible, though unlikely, that the resultant velocity at the time they observed was small. For this reason, they wrote, "The experiment will be repeated at intervals of three months, and thus all uncertainty will be avoided." A second limitation was raised in a supplement, admitting the possibility that in less protected environments than a basement

75. E. W. Morley to S. B. Morley, 17 April 1887, in Reingold, ed., *Science* (1964), 312–13.

76. Michelson to Rayleigh, 6 March 1887, RA-HAFB, also quoted in Livingston, *Master of Light* (1973), 123–26.

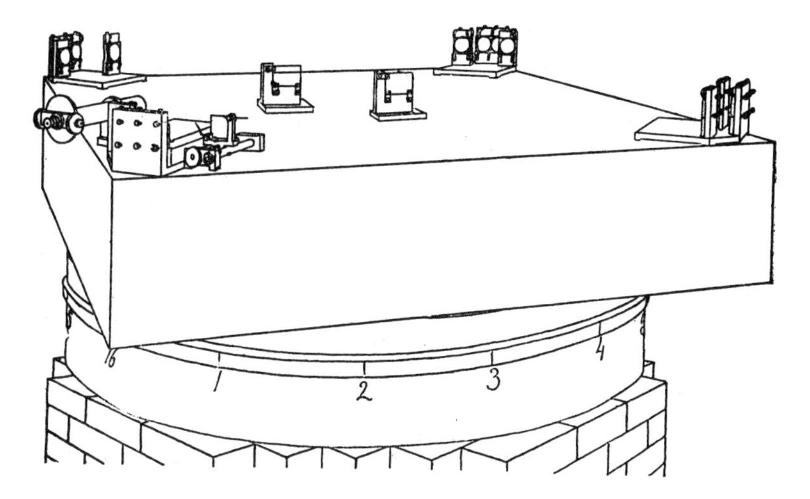

FIGURE 2.7 Perspective of the Michelson-Morley repetition of the Potsdam experiment. The half-silvered mirror is in the center, the light path ran between sets of four mirrors at each corner of the apparatus, and the telescope and a mirror on a micrometer screw are shown on the front left corner. The stone block rested on a wooden float contained in a cast-iron trough. Michelson and Morley, "Relative Motion" (1887).

laboratory—above sea level in locations like an exposed mountain peak—the relative motion of the earth through the ether might be perceptible.[77]

Significantly, both experiments involved replicating previous results. Michelson and Morley did so with a particular ideology, displayed in the careful terms in which they justified their endeavor. They wrote, for example, that the uncertainty of the results and interpretation of Fizeau's experiments "must be our excuse for its replication."[78] Implicit is the thought that ideally such an experiment should only need to be executed once. Comments Morley made explaining Michelson's nervous breakdown illustrate this even more clearly. Morley wrote to his father: "I can only guess at the stresses which brought about his illness. Overwork—and the ruthless discipline with which he drove himself to a task he felt

77. Michelson and Morley, "Relative Motion" (1887), 340–41. Michelson had described this difficulty in his letter to Rayleigh on 6 March 1887.

78. Michelson and Morley, "Motion of the Medium" (1886), 380. In the case of the second experiment, they wrote that they sought a theoretical effect that was much too large to be masked by experimental errors in the light of the oversights that had undermined the result of the earlier attempt. Michelson and Morley, "Relative Motion" (1887), 335.

must be done with such perfection that it could never again be called into question."[79] The remark suggests just how deeply the experience of Newcomb's insistent questioning haunted Michelson, with his desire to enter an active field with a definitive result. The extent to which Michelson sought—or was at least ready—to leave the ether-drift experiment as a singular performance is indicated by the fact that despite explicit recognition of important limitations, Michelson and Morley did not in fact complete further measurements over three month intervals, or carry the apparatus to a mountaintop.[80] This surely expresses confidence in the design and sensitivity of the experiment. Still, it remains a curious omission given the perceived importance of the result, the lengths to which Newcomb and Michelson were prepared to go consolidating measures of the velocity of light, and the tenacity with which Morley pursued his chemical meteorology by testing the oxygen content of air samples collected under different conditions. But as the following chapter will show, there were many reasons for moving on to work with other—related—problems.

Conclusion

Examining a critical period in Michelson's early career, this chapter has demonstrated that the ether-drift experiment emerged out of a complex of issues that Michelson first encountered in his work to provide a terrestrial determination of a constant central to astronomy; and we shall soon see that it quickly collapsed back into those concerns as Michelson worked to develop the apparatus at its heart into a new instrument. Histories of relativity focused primarily on theory, or even accounts of the ether-drift experiment as such, have focused too strongly on the result of the experiment to recognize continuities in method and apparatus that underlay its origins—and later uses. Exploring several productive tensions in Michelson's research has also conveyed a great deal about his development as a scientist.

79. E. W. Morley to S. B. Morley, 17 September. 1885, Edward W. Morley Papers, Library of Congress, Manuscript Division, cited in Livingston, *Master of Light* (1973), 112.

80. Michelson's next ether-drift paper was published ten years later, using a light path that extended up the side of the laboratory at the University of Chicago. In 1904 he pointed out difficulties plaguing a suggestion from Wilhelm Wien to measure ether drift with a one-way path. Michelson, "Relative Motion" (1897); Michelson, "Relative Motion" (1904).

To understand how Michelson developed telling experimental arrangements in both velocity and ether-drift measurements, note that his ability to imagine and trial new designs depended essentially on a readiness to adapt long-familiar apparatus to new purposes, whether a transit of Venus lens or a part-silvered mirror. Turning genial arrangements into decisive determinations then relied on marshaling a wide variety of resources—funding, instruments, and patronage—in the service of precision performances. Nevertheless, while often less readily visible, the relations between results—described in a Nobel address, celebrated in hagiographical histories, and still remembered now—and the methods that yielded them were always critical to Michelson. Further, we have seen that in important instances his contemporaries subjected those relations to severe scrutiny. Pushed by scientists determined to deliver an accurate value in the face of conflicting results, or to establish a theoretically significant conundrum like Newcomb or Rayleigh respectively, Michelson had to revisit and revise early results in the light of refined protocols. This was a trying process, and even as Michelson's reputation as a scientist reached beyond a single research, such repetitions opened up a significant tension between his determination to achieve singular results and his reliance on the stimulation of communal attention. Even as he entered fields of vibrant concern, Michelson was always and distinctively to strive for singularity; the following chapter will show that this marked his work with methods as much as it did his results.

To a large degree Michelson's disciplinary identity depended on the fact that astronomical interests dominated both the subjects of his inquiry and the community within which he moved. In many respects the character of his research in this early stage of his career reflects also the period in which Michelson worked. The significance of existing U.S. strengths in astronomy for the emergence of its physics discipline in the late nineteenth century has not yet been clearly recognized. Here we can see it in the associations Michelson formed, educated and teaching in a Naval institution in which astronomy and mathematics initially overshadowed the emerging identity of instruction in chemistry and physics. In the coming decades close links with astronomy were to strongly mark the research of both Michelson and Henry Rowland, but also to change significantly in character, shifting from early patronage relations of the kind this chapter has disclosed, to a later interdependence that recognized distinct disciplinary aims while sharing significant common ground. U.S. astronomers profited from the increasing disciplinary identity of physics by articulating a distinctive meld of *astro-physics* in research programs and new journals. For their part, we shall see that Michelson and Rowland—as physicists—both co-opted research programs in mechanics and optics that had initially been developed most strongly by astronomers, using them to prosecute distinctive new standards programs. In Michelson's case that depended on turning his ether-drift experiment to a new purpose.

CHAPTER THREE

Interferometers and Their Uses

The narrative arc of the previous chapter—moving from velocity of light to ether drift and repetitions—offers the kind of story that is generally told about Michelson, usually as a prelude to describing the role the experiment played in the refinement of H. A. Lorentz's theory of the electrodynamics of moving bodies, and later, the formation and justification of special relativity.[1] This chapter opens up a less familiar view of Michelson's achievements by focusing on the *instrumental* program Michelson initiated, rather than on the null result and the various implications drawn from it. Alongside the singularity we observed in the performance of the ether-drift experiment, other facets of Michelson's work attest to the elaboration of a new culture of instrumentation around the techniques of interferential refraction, and to a long-term continuity of research concerns. They crystallized around a third arrangement Michelson and Morley trialed in 1887, which was dedicated to turning light itself into a standard of length. That project was both technically demanding and led Michelson along new investigative pathways, in particular toward developing a new form of precision spectroscopy that promised to reveal hitherto unrealized spectral structure—but proved controversial in practice.

Drawing out the characteristics of the "culture of instrumentation" that Michelson pursued will deepen our understanding of both the scientist

1. See, for example, Darrigol, *Electrodynamics from Ampère to Einstein* (2000), chap. 8.

and his community. Noting that theory dominates many histories of science, historians have pointed out that tracing the fortunes of experiments or instruments offers new ways of periodizing scientific development and reveals significant features of the complex relations between the scientific community and broader cultural concerns.[2] Ever since telescopes, air pumps, and pendulums proved so richly fruitful objects of concern in the sixteenth and seventeenth centuries, at least some natural philosophers and scientists have centered significant research effort on instruments in particular. Perhaps they work to promote a distinctive new instrument (together with the observations it enables), to develop a set of investigations utilizing similar technical resources in different contexts, or to improve existing instrumentation. In a rich variety of ways, engagement with instruments helps configure both individual careers and the work of communities of researchers—and at any given time we might discern several diverse but interrelated cultures of practice crystallizing around and reshaping the use of specific investigative tools. After all, from beginning to end the design, construction, use, and application of these material objects all link scientists more or less closely with academic colleagues (sometimes in quite different fields), but just as importantly with instrument makers and commercial and industrial endeavors. Michelson's concerns developed in close relationship to those of his colleagues William Rogers and Henry Rowland. Highlighting the instruments these men developed and studying interrelations between their research, careers, and connections with industry, this chapter will help explain what made American physics distinctive in the late nineteenth century (and set the foundations for its development in the twentieth century also, when American physics has often been described as particularly "pragmatic").

Yet even in this group Michelson was unusual in the extent, coherence, and character of his interest in a particular group of instruments; examining his research through 1900 will help cultivate insight in several directions. First, exploring the diverse theaters in which Michelson used interferential techniques and the broad intellectual framework he gave his methods illuminates the varied strategies, practical and theoretical, he

2. Particularly important studies include Shapin and Schaffer, *Leviathan and the Air-Pump* (1985); Gooding, Pinch, and Schaffer, eds., *Uses of Experiment* (1989); Galison, *How Experiments End* (1987); Galison, *Image and Logic* (1997); Joerges and Shinn, eds., *Instrumentation* (2001).

employed to create and promote a new form of instrumentation. Thus Michelson's career offers a rare perspective on an experimentalist's work to achieve a form of instrumental unity—which in many respects bears comparison with more celebrated searches for theoretical unity. Second, Michelson pursued precision standards in a highly distinctive way, addressing the nexus between science and industry in an indirect but strategically significant manner. Analyzing his work on light complements earlier studies that have established the importance of metrology—the determination of appropriate units of measure—to scientists' involvement in fields such as electric telegraphy, the heat and power industries, and spectroscopic research. Understanding Michelson's research is particularly valuable because it demonstrates an avowedly extreme case: while standards of length were of great importance to industry in the United States, Michelson sought to establish natural and ultimate standards. His work to render light a standard of length shows a tension between practically useful contributions and highly refined (but potentially idiosyncratic) skills and measurements; and this tension ran through Michelson's relations with the physics community more generally.

There are still other fruits of our inquiry into the instrumental focus of Michelson's work. Michelson's research constitutes the state-of-the-art material realization of a number of elements that later became central to Einstein's conceptual approach to relativity—and will allow us to trace an intimate relation between the history of the screw, the experimental dimensions of relativity physics, and even the railroad. Historians have recently linked Einstein's innovations in kinematics to contemporary developments in clock standardization and the railroad. Briefly spelling out the respects in which clocks have been used to draw together relativity and railroads pedagogically and historically will help me outline my argument about the rich variety of ways in which the interferometer was significant to relativity physics, and Einstein's theory.

Introducing the measurement of time early in his 1905 paper, Einstein first notes that all propositions about time are always propositions about simultaneous events, and then offers an allusive elaboration to clarify the point. "If, for example," he writes, "I say that 'the train arrives here at 7 o'clock,' that means, more or less, 'the pointing of the small hand of my clock to 7 and the arrival of the train are simultaneous events.'" In his 1917 popular book on relativity, Einstein developed the pedagogical analogy still further by depicting railway travelers dropping stones, watching ravens, chasing light rays, and determining whether two lightning flashes occurred simultaneously.[3] Peter Galison has shown that the move is hardly accidental; the dual needs of coordinating railway times and determining longitude had made clock synchronization a pragmatic concern that crossed the desks of both the mathematician Henri Poincaré as head of the Bureau of Longitude in Paris in the late 1890s, and Einstein as a Patent Office clerk in Bern in the early 1900s. Thus we now know that the pedagogical and popular link between railroads and relativity that Einstein initiated has a historical complement significant to the genesis of the theory (even if Einstein did not reach his argument about relativity by contemplating railroad clocks alone).[4]

Einstein linked clocks and railroads in 1905 and 1917, but did little to elaborate on the "rigid bodies" or "measuring rods" that he described as fundamental to each and any theory of electrodynamics—apart from showing how they would contract in moving frames of reference.[5] Yet long before Einstein wrote, the interferometer had been related to measuring rods—and, indirectly, the railroad—by Michelson's use of the apparatus to determine the international meter. Michelson developed his approach at a time in which material standards of length were equally important to U.S. railroad manufacturers and spectroscopists. It is true that the specific way Michelson used light as a standard of length was then as far removed from the pragmatic needs of railway managers as Einstein's theory has always been—Michelson offered a level of accuracy the railway did not need. But this accuracy made his work central to relativity physics in several important respects, which later chapters will explore more fully. One concerns the importance of the instrument as ether-drift *experiment* (with a specific result) to other physicists in their work on the electrodynamics of moving bodies. We will begin our study of theorists' uses of Michelson's experiment here, and carry it further in subsequent chapters. A second concerns the extent to which interferometry

3. Einstein, "Elektrodynamik bewegter Körper" (1905), 893; Einstein, *Relativitätstheorie: Gemeinverständlich* (1917); and the English edition, Einstein, *Relativity* (1920).

4. Galison, *Einstein's Clocks* (2003), but see also Darrigol, "Poincaré's Criticism" (1995); Darrigol, "Einstein-Poincaré" (2004). For an outline of Einstein's path to relativity, see Stachel's editorial contribution "Einstein on the Theory of Relativity," in Einstein, *Collected Papers*, vol. 2 (1989), 253–74, also reprinted in; Stachel, *Einstein from 'B' to 'Z'* (2002). More recent investigative accounts include Rynasiewicz, "Construction of the Special Theory" (2000); Norton, "Einstein's Investigations" (2004); Rynasiewicz and Renn, "Einstein's *Annus mirabilis*" (2006).

5. Einstein, "Elektrodynamik bewegter Körper" (1905), 892.

in general embodied the level of accuracy on which the debate on electrodynamics depended. This will be emphasized in my examination of Walter Kaufmann's experiments on electron theory in chapter 6—which made important use of inteferometric measuring techniques. Third and fourth, we will find experimentalists, like theoreticians, elaborating important distinctions between relative and absolute measures and see that Michelson's work with light and mirrors provided the best test of the rigidity of the metal rod that embodied the international meter. And a fifth and final dimension concerns the pedagogical and justificatory role of the ether-drift experiment, which Einstein made significant use of from 1907.

The first part of this chapter establishes the framework that Michelson gave his research over nearly two decades. We explore in turn the basis for his instrumental program of research, the industrial and scientific context for work on standards of length, Michelson's relations with instrument makers, and the scientific ambition expressed in his dual campaign for a new class of instruments and accuracy in the extreme. Later sections consider how his community understood Michelson's work, examining the extent to which other physicists used his instruments and techniques from the 1880s to 1900. This study will display the gradual and fraught process by which the inteferometer secured its reputation for accuracy—setting the foundation for more specific studies of the significance of ether drift and interferometry for relativity in later chapters. The final section concludes our reconstruction of Michelson's perspective on his own work by exploring the way he associated the ether with scientific modernity circa 1900.

Instruments and Experiments

We need first to briefly revisit several of the experimental episodes discussed in chapter 2 in order to see how they could lead not just to new results or replications, or even to a new instrument, but to a new class of instrumentation. Recall that in 1881 Michelson assimilated his ether-drift apparatus with an existing instrument, writing that it could easily be made to serve as an "interferential refractor," while highlighting the low cost of the device and the great separation it enabled between the two beams of light employed. It is notable that Michelson's subsequent work to develop a new form of instrumentation relied at least as much on revisiting earlier refractometers and establishing clear relations with existing forms of instruments (especially telescopes and spectroscopes) as it depended on building novel devices. Initially his endeavor to fashion an instrument formed a subtext to Michelson's focus on specific experiments, but Michelson foregrounded the instrument as such ever more clearly over the next decade.

Although their results led in contrasting directions, and they are famous as *repetitions*, Michelson and Morley's experiments of 1886 and 1887 both provided an opportunity for significant technical development—along similar lines. Repeating Fizeau's 1851 ether-drag experiment, Michelson first considered Fizeau's original refractometer apparatus before settling on a new arrangement "after a number of trials." Just as he had in 1881, rather than using a lens that was masked except for two small apertures in order to send the light along two parallel pathways, Michelson and Morley again used a partially silvered mirror in combination with other, angled mirrors, to manipulate and separate more widely the two arms in which the light traveled (see fig. 3.1).[6] Repeating Michelson's ether-drift experiment in 1887, he and Morley used a different arrangement of mirrors to achieve the perpendicular form required, multiplied the number of mirrors to increase the path length, and transferred the micrometer movement from the partially silvered mirror in the center of the apparatus to a mirror at the end of one of the arms (fig. 3.2).

In addition to this kind of instrumental elaboration, the scientists began work with a simpler form that could be used as a comparator to establish light as a standard of length. In a paper published just after the ether-drift experiment, they described an arrangement more reminiscent of 1881. The instrument could be made far more stable since they had no need to spin it, and again they shifted the micrometer movement to the end of an arm. While formally more simple than the repetition experiments, its realization was significantly more challenging technically, and also exposed a cascade of new scientific problems. We will discuss this device more fully below (see fig. 3.4 for a diagram, and figs. 3.7–3.9 for illustrations of the final realization in 1892–93).[7]

Thus by 1887 Michelson had identified a continuity of apparatus across a number of experiments and had begun to work consciously with creating variations and modifications for different purposes. This was the technical heart of his advocacy of a kind of instrumental unity. For the next six years

6. Michelson and Morley, "Motion of the Medium" (1886), 380–81.

7. Michelson and Morley, "Wave Length of Sodium Light" (1887).

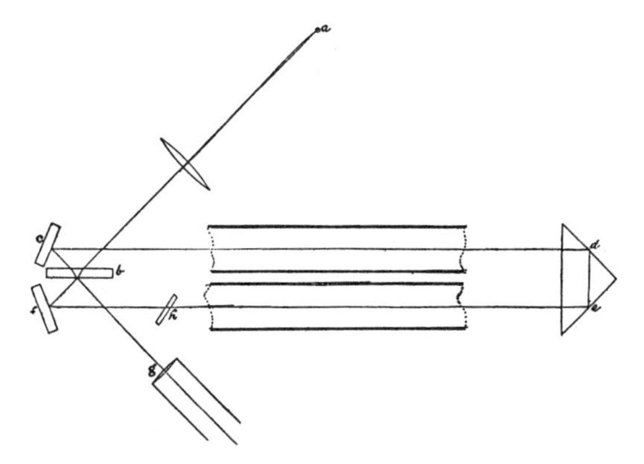

FIGURE 3.1 Plan of the refractometer used in Michelson and Morley's version of the Fizeau experiment. Light from the source *a* is divided on the half-silvered surface *b*, one part following the path *b c d e f b g* and the other the path *b f e d c b g*. Interference patterns are observed in the telescope *g*. Fizeau had used a masked lens similar to that shown in fig. 2.5 (as "FIG. 2"), where Michelson and Morley used *b*, *c*, *f*, and the compensating glass *h*. Michelson and Morley, "Motion of the Medium" (1886), 381.

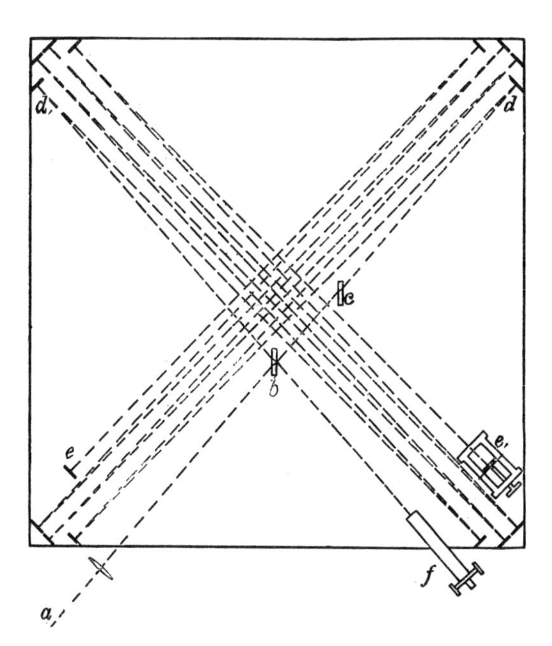

FIGURE 3.2 Plan of the refractometer used in Michelson and Morley's version of Michelson's ether-drift experiment. The telescope and micrometer movement are both at the lower right. Light from the source *a* is divided on the half-silvered surface *b*, one part following the path *b c d e* and the other the path *b d′ e′*, with multiple reflections in the mirrors *d*, *d′*, *e* and *e′* before being observed in the telescope *f*. Michelson and Morley, "Relative Motion" (1887), 338.

his work on standards of length absorbed most of his research efforts and led to many other applications of interferential methods that Michelson publicized as broadly as possible: if he could not interest others in the ether-drift experiment, he might interest them in an instrument. We will first explore the context that made Michelson's research on standards of length potentially valuable on several different fronts before examining the broader intellectual framework in which he set his work with light waves.

Standards of Length in Science and Industry

It is significant that immediately after the experiments of 1886 and 1887, at first in partnership with Morley and then alone, Michelson pursued the possibility of adapting the refractometer to enable the counting of light waves to serve as a standard of length. With this research Michelson brought a fundamental new tool to meet the practical concern with precise distance measurement so evident in his dialogue with Simon Newcomb. He also entered a field of critical interest to scientists, machine-tool firms, and industrial manufacturers—especially those involved with the railroad industry. Through the 1870s and 80s, standards of length were of broad and urgent concern in the United States.

Perhaps Simon Newcomb examined Michelson's distance measures so closely between 1879 and 1882 because his earlier experience with material standards of length had led him to expect trouble. Several years earlier, the Transit of Venus Commission had borrowed three steel rods from the Office of Standard Weights and Measures of the United States Coast and Geodetic Survey. In 1876 Newcomb found they did not correspond with the standard foot and meter rules of the respected machine tool manufacturers Darling, Brown & Sharpe, although these were consistent with each other.[8] He asked the Coast Survey to explain the discrepancy and

8. The bars were described as standard at 62 °F "very nearly," with a rate of expansion equal to 0.0000061 per degree F, and the two 10-foot rods were shorter than Darling, Brown & Sharpe measures by 0.10 inches in 35 feet (the third rod was 5 feet). See J. E. Hilgard to Rear Admiral C. H. Davis, Superintendent of the Naval Observatory, 10 March 1876; and Newcomb to Hilgard, 17 April 1876, NIST-RG167-NARA (Records of the National Institute of Standards and Technology, Records of the Office of Standard Weights and Measures, Correspondence of the Office, 1859–93) Volume 3 (1876–80) Entry 7 Box 3.

even hoped to carry out his velocity of light measurements with a precision sufficient to test the invariability of the standard of length.[9]

Letters between Darling, Brown & Sharpe and the Coast Survey show that the firm was also deeply concerned with the situation—despite the fact that they supplied the Coast Survey with materials for length standards, and had compared their own standards as recently as 1875.[10] Sending their standard meter to Julius E. Hilgard as superintendent of the Office for Weights and Measures, the firm emphasized the urgency of their concern: "You would very much oblige us if you could have this matter attended to at once as we cannot use our dividing machine till we receive the bar and your report in regard to the standards."[11] Knowing the relation between the U.S. standard and their own was crucial to setting up their dividing machines—which ruled the scales on all measuring instruments the firm produced. Within two years a further comparison was necessary. This time they described how difficult it was to determine what the correct standard was, stated they were providing standard measures for the cities and counties of Massachusetts, and took particular care to outline the protocols they followed to ensure the bars to be compared were at the

9. He initially hoped to reach a value with a probable error of between just 5 and 10 kilometers to enable such a test of invariability by later repetition. His conclusion offered suggestions to guide future physicists making a more precise determination, for just this reason: Newcomb, "Measures" (1882), 109 and 204–6.

10. See Darling, Brown & Sharpe to Hilgard, 29 June, 30 June, and 4 August 1875; Hilgard to Darling, Brown & Sharpe, 4 August 1875. Henry Dexter Sharpe comments that the sewing machine manufacture the firm had conducted since the 1860s relied upon Whitworth gauges. Discriminating use had revealed certain deficiencies, so the firm decided to create original standards: "a standard yard and meter, with a system of measuring products based on these new bases, which came to be known as *the B. & S. standards*. Growing out of Mr. Brown's studies, this whole system of original standards took years to perfect and was not completed until long after Mr. Brown's death in 1876." Sharpe, "Joseph R. Brown" (1949), 23. Hounshell discusses Brown & Sharpe in particular and the significance of sewing machine manufacture in general for the wider propagation of "armory practices" of interchangeable parts based on a rational jig and fixture system, and a rigid system of gauging. Hounshell, "The *System:* Theory and Practice" (1981). For an overview of machine tool firms in the period see Strassmann, *Risk and Technological Innovation* (1959), chap. 4.

11. Darling, Brown & Sharpe to Hilgard, 24 September 1877. NIST-RG167-NARA Entry 7 Box 3. In a letter to Hilgard on 16 October 1877 they wrote that "taking the new (to us) number 39.370 in. as the standard [i.e., of inches to the meter] there is only .00023 in. in a meter difference in our comparisons which perhaps is as close as can be expected. We shall now consider your comparison of our steel bar with the standard at Washington as correct and in our comparisons with it shall be able to detect errors as small as .000025 in."

same temperature.[12] In 1880 the Coast Survey took its turn as critic, requesting details about the construction of seventy-five steel bars the firm had supplied eleven years earlier for steel end measures. The answer cannot have pleased them. Darling, Brown & Sharpe replied:

> We find on referring to our books that the bars were hammered to straighten them, which with our present experience with such work, we should consider quite unfavorable in the permanency of their length. The steel should be made large enough to grind true without hammering. The expense would be more but there would not be so much strain, or if hammered it should be heated red hot afterwards before grinding.[13]

Clearly both the Coast Survey and Darling, Brown & Sharpe had begun to learn the hard way that the production and maintenance of reliable standards of length required unusual care.

But the chain of concern did not stop there. Indeed, by the late 1870s problems with standards of length were so endemic as to threaten what was increasingly regarded as a defining feature of the "American system of manufactures"—interchangeable parts. In 1876 the New York, Lake Erie and Western Railroad company went to the machine tool makers Pratt & Whitney after finding that nuts cut at one of their in-house shops would not fit bolts made at another. Purchasing the necessary taps and dies from other manufacturers led them to discover that nuts cut with the taps of one firm could not be screwed on a tap made by another—manufacturers had been working to different standards. The Master Car Builders' Association appointed a committee to select a firm of prominent standing to furnish standard U.S. thread screw gauges, and Pratt & Whitney were chosen for the task. The committee report describes the situation in colorful terms:

> Like Diogenes with his lamp, in search of an honest man, this company went to and fro in the land in search of a true inch, a true foot, or a true yard. They

12. Darling, Brown & Sharpe to Hilgard, 20 and 24 March 1879. See also their letters to Hilgard of 7, 12, 17, and 24 April and 13 May 1879. NIST-RG167-NARA Entry 7 Box 3. The firm is discussed in Evans, *Precision Engineering* (1989), 77–79.

13. See Assistant in Charge of the U.S. Coast and Geodetic Survey Office to Darling, Brown & Sharpe, 12 February 1880; Hilgard to Darling, Brown & Sharpe, 13 November 1869; and for the quote, Darling, Brown & Sharpe to Hilgard 14 February 1880. NIST-RG167-NARA Entry 7 Box 3.

> procured from different sources what they supposed were the most reliable standards of measurement, and found that none agreed. They had the same standards measured by what were considered the most reliable measuring machines and instruments in the country, and found that no two of these would measure the standard alike.[14]

In 1880 Francis Pratt wrote to the Coast Survey outlining their intent to manufacture a new standard and "get this matter settled once for all time, or at least so that everybody whose opinion is worth anything will admit that the bar is well authenticated and can be taken as a standard.... We are building the tools, and comparators, and hope before this year is out to show a set of size gauges that will bear criticism. Our customers are waiting for them."[15]

From 1860 on, Pratt & Whitney had run the gamut of guns, sewing machine, and then bicycle and automobile manufacturing. They regarded the provision of accurate gauges as the indispensable key to the growing industrial manufacture of interchangeable parts. The firm helped transfer practices pioneered within armories to an increasing range of industries, and based its reputation on the practical achievement of accuracy in industrial manufacture. The solution to problems like those in the nuts and taps of screw manufacture was high-precision limit gauges—"go" and "no-go" gauges that embodied the limits of tolerance acceptable, so that a machinist could test directly whether a particular part had the right dimensions. Providing such gauges depended in turn on establishing protocols to transfer measures from a reliable master standard.[16] In 1879, the same year that Albert Michelson went to William Rogers to recalibrate

14. Bond, ed., *Standards of Length* (1887), 67. The volume provides an account of the events in conjunction with other papers and is discussed below. Steven Usselman has studied the Master Car Builders' Association (MCBA) and emphasizes that railroad firms took a cooperative approach to many technical issues. His study of braking standards shows that in the 1880s the MCBA was able to reinforce its ability to enforce national standards by adopting structural changes that reflected the makeup of the industry. Usselman, "Air Brakes for Freight Trains" (1984); Usselman, "Patents Purloined" (1991).

15. F. A. Pratt to Prof. J. E. Denton, 17 August 1880, NIST-RG167-NARA Entry 7 Box 3. Note that in Britain in 1882 the Board of Trade requested the Royal Society to establish a committee to advise on improving procedures for comparing standards of length. As a result the Cambridge Scientific Instrument Company undertook a study of temperature regulation and comparer designs. Cattermole and Wolfe, *Horace Darwin's Shop* (1987), 195.

16. For a celebratory account of the company history, see Pratt & Whitney, *Accuracy for Seventy Years* (1930).

measures used in his velocity of light experiments, Pratt & Whitney also approached the Harvard astronomer.

William Rogers had first held a position as professor of mathematics and astronomy at Alfred University in western New York from 1859 to 1870, but also taught industrial mechanics there. From 1870 he worked as an assistant at the Harvard Observatory, where he was in charge of the newly erected 8-inch meridional circle. That was shortly before Edward Pickering became director of the Harvard Observatory and began programs of research that set that institution on the path of the *new astronomy*. By surveying the brightness of the stars photographically, Pickering helped establish a form of physical astronomy, or astrophysics, that was of growing importance in the United States. Spectroscopic study of the constitution of the stars soon followed photographic methods as a second major field of research in the new astronomy. Rogers's observational work led him to take up the problems of microscopy and ruling scales, and this in turn prompted a fundamental concern with precise distance measurements, and later, grating spectroscopy. As a result of finding (like Newcomb) that there were no well-defined relations between the yard and meter standards then in use in the United States, Rogers traveled across the Atlantic to undertake intercomparisons of the standard length measures of the United States, Britain, and the International Bureau of Weights and Measures in Sèvres. By the late 1870s he had become the foremost American authority on length standards.[17]

Pratt & Whitney sought to ensure success in their standards project by combining mechanical expertise with the credibility of both academic and government scientists. They asked Rogers to collaborate with their mechanical engineer George M. Bond (a graduate of Stevens Institute of Technology), and to work with Coast Survey scientists to conduct a series of comparisons. They wanted to establish the length of a particular standard but also to ascertain the most favorable procedures to make and repeat a variety of end and line measure comparisons, within limits of 1/50,000 of an inch. Rogers and Bond designed a universal "comparator" for this task, and two exemplars were made. One Rogers-Bond comparator was installed in the standards room of the Harvard Observatory, while

17. On Rogers, see Evans and Warner, "Precision Engineering" (1988); Morley, "William Augustus Rogers" (1902); Warner, "Rowland's Gratings" (1986); Brooks, "Towards the Perfect Screw Thread" (1993).

its pair stood in the Gauge Division of Pratt & Whitney (fig. 3.3).[18] But more importantly, Rogers pressed the Coast Survey to allow the critical comparison between the two Pratt & Whitney bars and the U.S. standard ("Bronze No. 11") to be made using the new comparator as well as the Coast Survey apparatus; with both John Clark of the Coast Survey and Rogers himself taking readings. This kind of control meant an unusual transparency, an assurance of continuity of protocols across the standards institution, the scientist, and the manufacturer. It resulted in master bars accurate to millionths of an inch.

Darling, Brown & Sharpe followed suit. The firm gave up their nickel-plated steel master standard as unreliable (having found its coefficient of expansion was unpredictable), and Mr. Darling spent many months considering the problems of length comparisons, took out patents on new reading scales, and like Rogers asked to visit the Coast Survey office in order to inspect their arrangements.[19] But while Darling sought transparency from the Coast Survey, he jealously guarded his ruling machine, even going so far as to protect it from the scrutiny of his partners within an off-limits room.[20] The Coast Survey placated their suppliers with an assurance of the value of their work and an order for materials for a set of standards to be sent out to agricultural colleges.

Thus, concern with standards of length moved from debates in the committee meetings of industrial associations, through private correspondence between different firms and government and academic scientists, and into the temperature-controlled rooms in which comparisons were made; and it soon led to the development of new practices and products

18. See Bond, ed., *Standards of Length* (1887), 67–68; Evans and Warner, "Precision Engineering" (1988), 6–8. Alvan Clark and Sons made an optically plane surface that Rogers used in testing the bedplate of the comparator. "Report of Professor Wm. A. Rogers" (1887), 7.

19. William Rogers gave up a nickel-plated steel master in 1882 for the same reason. See Darling, Brown & Sharpe to Hilgard, 3 and 27 April, and Darling to Hilgard 17 July and 8 and 15 August 1883, NIST-RG167-NARA NC-76, Entry 7 Box 4. On 3 April, Darling, Brown & Sharpe wrote: "there are *some things* to be done that require the greatest skill, greatest experience, and the very best means, and the comparison of a measure of length is one of those things."

20. See Evans, *Precision Engineering* (1989), 79; Rogers, "Practical Solution to the Perfect Screw Problem" (1884), 218. Oscar J. Beale eventually took over Darling's tasks, designing a measuring machine that was never offered for sale but "remained as a tool with which B. & S. gages were made and sold with a guaranty of accuracy to within *one ten-thousandth of an inch.*" Sharpe, "Joseph R. Brown" (1949), 23.

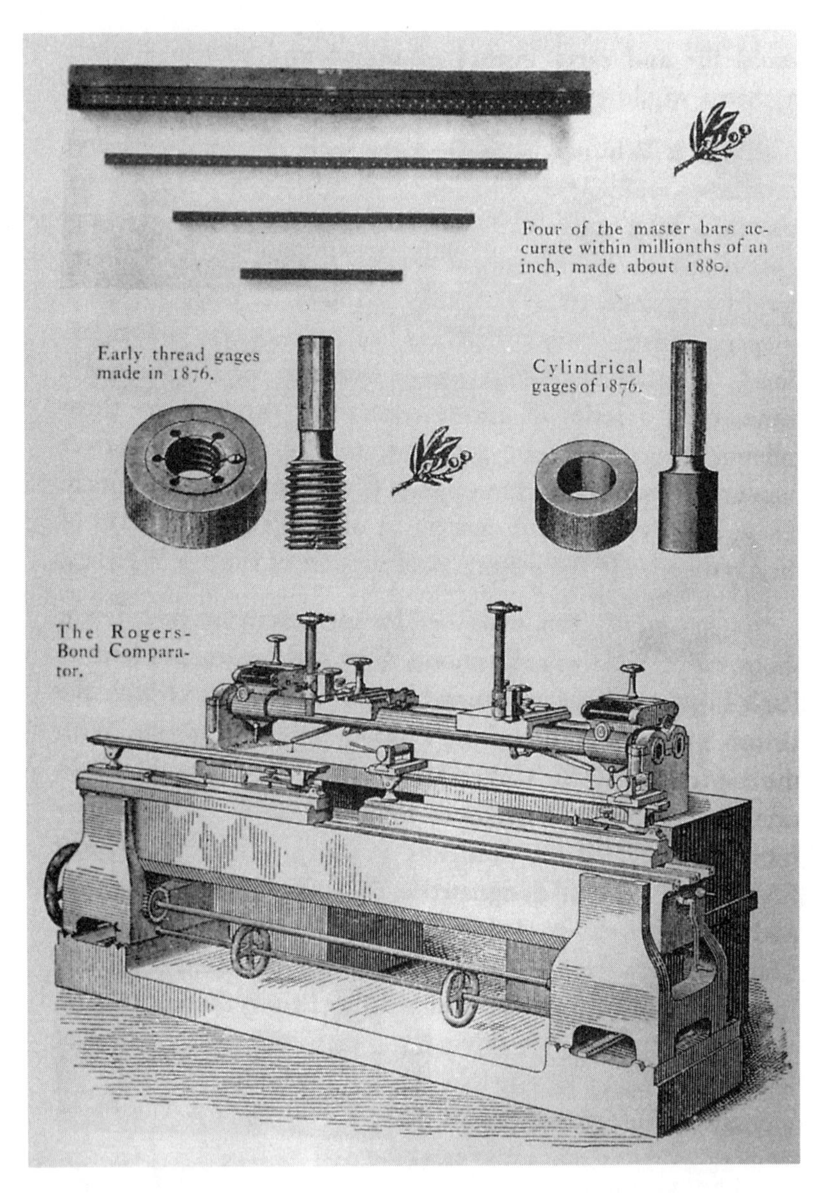

FIGURE 3.3 The Rogers-Bond Comparator, as illustrated in Pratt & Whitney's celebration of industrial accuracy. The figure also includes illustrations of standards of length and gauges, and it is worth noting that a following page depicted the firm's interferometer, probably dating from the 1920s and described as "capable of measuring to fractions of millionths of an inch." Pratt & Whitney, *Accuracy* (1930), 36 (see also 38).

by leading firms. Having secured authoritative standards, Pratt & Whitney produced a new set of production gauges and in 1885 added a "Standard Measuring Machine" to their line of products, "a gauge accurate to hundred thousandth of an inch."[21] In 1887 they published a collection of papers and reports detailing the recent confusion and its resolution. It was titled *Standards of Length and Their Practical Application*, and they made sure a copy got to interested scientists such as Michelson.[22] The book paid attention to the links between standards and industrial practice by including papers on screw threads and applying standards of length to gauge dimensions as well as Rogers's report on units of length. The achievement of reliable standards of length involved a delicate teasing apart of realms of work often regarded as private, and required close associations across industrial and scientific realms. In the same period that the width of railway lines run by different companies began to be unified, and railway time began to be standardized in regions across the United States, efforts were initiated to make the nuts and bolts of the industry interchangeable also. These changes gradually reconfigured commercial competition with a demand for technical uniformity across different companies, both reflecting and facilitating new business practices in transportation and industrial manufacturing. They show that in interchangeable manufactures, a widespread concern with *accuracy* relative to a given standard measure was replacing the role more commonly played by *precision* machining, which relied instead on exactly duplicating matched parts, tools, and gauges.[23]

Timekeeping has played an important role in our historical understanding of the role of science in late nineteenth-century modernization, with astronomical observatories often described as a force for rationalization in response to the firm-bound railway times that had proliferated

21. Training Dept., Pratt & Whitney Machine Tool Division, Colt Industries Inc., "Time Line: Company History." Rev. 10/1/81, in Box 5, LRP&W-UC; Pratt & Whitney, *Accuracy for Seventy Years* (1930), 37. Similar changes are registered in the catalogues of Darling, Brown & Sharpe, which show an increasing variety of rules and micrometers over time. See Cope, ed., *A Brown & Sharpe Catalogue Collection: 1868, 1887, 1899* (1997).

22. The volume in the John Crerar Library of the University of Chicago was given to the library by Michelson. It bears a compliments slip and is inscribed with the signature of F. L. O. Wadsworth, Clark University, dated 1 March 1890. Wadsworth was Michelson's assistant while he was at Clark, and accompanied Michelson to Paris for the determination of the length of the meter in light waves in 1892, an episode we consider below.

23. Uselding, "Measuring Techniques and Manufacturing Practice" (1981). See especially 103–4 and the account of Brown & Sharpe and Pratt & Whitney on 114–21.

throughout the Western world.[24] Selling accurate time, observatory scientists helped push for time zones based on geographical position rather than local sun time on the one hand, or network times that followed the reach of the different railway company lines on the other. We can now see that just as the movement to standardize time reached fruition (with the United States adopting standard railway time zones in 1883), the railways themselves recognized a need for standards in another realm, and observatory scientists in particular were again engaged in solving the problems involved.

Industry needs for accurate distance measures centered on controlling the materials used in constructing standard measures and the protocols needed to compare lengths across standards (national and proprietary), commercial rules, and gauges. The Master Car Builders' Association noted that good machinists commonly discerned differences of "much less than 0.001 inch in the diameter" of pins or their bearings, so master bars accurate to millionths of an inch, the Rogers-Bond comparator and the Standard Measuring Machine of 1885 provided more than enough accuracy.[25] The scientific community saw other opportunities in the debate on length standards. In 1879 the Coast Survey scientist and Johns Hopkins faculty member Charles Sanders Peirce worked in conjunction with Lewis M. Rutherfurd on experiments to use the wavelength of sodium light as a new, natural, standard of length. Rutherfurd had helped pioneer astronomical photography and spectroscopy in the 1860s, and provided significant impetus to grating spectroscopy from 1872 by giving away approximately fifty gratings ruled by a machine of his own design. His work helped establish American expertise in astrophysics, and as we shall see, set off a chain of interrelated instrumental developments that helped focus

24. Bartky, "The Adoption of Standard Time" (1989); Bartky, *Selling the True Time* (2000). Morley had participated in discussions on standard time zones. Morley, "Standards of Time" (1879).

25. The description of current machining practice comes from "Report of the Committee of the Master Car-Builders' Association on Standard Screw Threads. Presented at the Annual Convention held in Philadelphia, June, 1882" (1887 [1882]), 71. Bond noted that machinists often worked within 1/50,000 of an inch without being aware of the fact, and described variations of 1/20,000 of an inch as readily perceptible with the use of a standard caliper gauge. Bond, "Gauge Dimensions" (1887), 152 and 157. Noting a discrepancy between Charles S. Peirce's 1883 finding that Bronze 11 was 22 millionths of an inch shorter than the Imperial Yard, and earlier determinations that the deficit was 88 millionths, Rogers wrote: "Whether this apparent variation in length is real, will doubtless be determined by future investigations, but the deviation will not be of any practical account within the limits which you require in your system of gauges." "Report of Professor Wm. A. Rogers" (1887), 2.

research in physics also. Peirce's proposal to establish a new standard of length using the diffraction grating was one step. Investigators could relate the wavelength of light to existing distance standards by measuring the displacement of the image of a slit by a diffraction grating; Peirce reported that the extreme accuracy of Rutherfurd's grating plates made it practicable to measure a wavelength to one-millionth of its own length. Assuming that the wavelengths of light remain constant, the technique could answer questions about the stability of metal meter rods over time. Peirce also developed a theory of the spurious spectral lines or "ghosts" that resulted from periodic inequalities in the spacing of the grating, which could themselves be traced back to irregularities in the screw of Rutherfurd's ruling engine.[26]

William Rogers helped Johns Hopkins physicists take several steps still further and realize the goal Peirce had set—by using gratings ruled in Baltimore. Rogers himself went from constructing accurate scales for distance measurements, to the similar but more demanding task of designing a ruling engine to move a diamond tip across glass or metal plates to inscribe a grid on standard microscope plates—or rule the thousands of closely spaced lines of diffraction gratings.[27] But he also passed on both his expertise and the Harvard Observatory Rogers-Bond comparator to Henry Rowland in Baltimore, who developed the ruling art still further and claimed it for American physics rather than American astronomy.[28] Thus both Rogers and the ruling engines and comparators he built became central factors in a significant confluence of what historians have described as technological convergence and inventor divergence.

Analyzing the way that common elements are utilized in a variety of contexts by different researchers and entrepreneurs, Nathan Rosenberg and Deborah Jean Warner and Chris Evans have argued that the contrasting dynamics of instrumental convergence and divergence were

26. Peirce, "'On Ghosts in Diffraction Spectra' and 'Comparison of Wave-Lengths with the Metre,'" (1879); Peirce, "Comparing a Wave-Length with a Meter" (1879); Lenzen, "The Contributions of Charles S. Peirce to Metrology" (1965); Warner, "Lewis M. Rutherfurd" (1971), 213–14. On the ruling of diffraction gratings, see also Evans, *Precision Engineering* (1989), chap. 5.

27. Rogers, "The Method Employed by Norbert" (1875–76); Rogers, "A New Diffraction Ruling Engine" (1880); Rogers, "Practical Solution to the Perfect Screw Problem" (1884).

28. Bond, "Standards of Length and Their Subdivision" (1887), 131. The comparator was transferred between 1884 and 1887. Rogers wrote: "I am glad you are to have the instrument but it comes rather hard to part with my old friend." Rogers to Rowland, 19 October [1884?] HAR-JHU, Box 9.

characteristic of both the American machine tool industry and U.S. scientific instrument making in the late nineteenth century. Developing a similar line of argument but focusing on generic instruments in scientific research, Bernward Joerges and Terry Shinn have recently used the concept of research-technologies to characterize the elusive nature of those instruments and instrument designers that repeatedly traverse the boundaries between science and industry.[29] The culture of instrumentation emerging around standards of length and ruling engines in the United States offers a significant example of the phenomena these analysts have identified; and Rowland's and Michelson's approaches to standards of length are no exception. Collectively their work illustrates the important relations among astronomy, precision industry, and physics in the United States, with instrumentation and standards providing a fruitful juncture between different communities. Individually, we will see that Michelson and Rowland followed very different strategies. Indeed, to fully understand Michelson's research—and appreciate the distinctive nature of the instrumental culture he elaborated around refractometry—we need to recognize the extent to which it was formed in creative response to Rowland's earlier work. Precision screws, instruments, spectral maps, experimental values, and disciplinary pleas were all at issue in a rivalry that spanned decades.

Henry Rowland had nurtured a dedication toward experiment throughout his undergraduate degree in civil engineering at Rensselaer Polytechnic, even going so far as constructing his own instruments (winding galvanometers and induction coils), and building a steam engine to power his experiments. Indeed, Rowland had such a strong understanding of the need to cultivate the appropriate material and institutional

29. On convergence and divergence, see Rosenberg, "Technological Change" (1963); Evans and Warner, "Precision Engineering" (1988), 10. On research technologies, see Joerges and Shinn, eds., *Instrumentation* (2001); Staley, "Interstitial Instruments" (2002). Joerges and Shinn have pointed to precision mechanics and optics as providing the first grounds for the emergence of a research technology matrix in the nineteenth century, but they focus on the collective activities of German instrument makers in the period (turning to the United States only in the 1930s). For the moment it is helpful to note that the academic participants in the U.S. network we have been tracing were more securely defined by their disciplinary identities (although these were admittedly rather open) than by their engagement with interstitial instruments alone. Figures like Rogers, Rowland, and Michelson may well have seen their instruments to offer protean research technologies (making possible what I have called instrumental programs of research), but their social role as scientists was largely defined by their discipline. For this reason, rather than thinking of them as research technolog*ists*, Evans and Warner's term *academic mechanicians* offers a more apt recognition of the continuing importance of the cultural status they held as astronomers and physicists.

environment for research in his chosen field that when he returned to the school as instructor in Natural Philosophy in 1872, he insisted that his salary be linked to the budget for apparatus: should Rensselaer not spend several hundred dollars on instruments each year, his pay would increase accordingly. In a telling sequence of events, Rowland's first major research—a study of magnetic permeability—was rejected by the *American Journal of Science*, but won rapid publication in the *Philosophical Magazine* when Rowland sent it to James Clerk Maxwell in Britain. It was easier to win recognition in Europe than in the United States. Rowland impressed his countrymen enough, however, for Daniel Gilman to ask him to become the inaugural professor of physics at Johns Hopkins University. Like Michelson shortly afterward, Rowland toured Europe to learn from the established masters (and bring back the best instruments available) but also did enough to establish his originality in research. While at Helmholtz's laboratory in Berlin he carried out a delicate demonstration that setting a charged disk in revolution produced a magnetic effect, as would the flow of current electricity. Rowland returned in 1876 to found a research laboratory in Baltimore, convinced that he would be able to break new ground by working between mathematical and experimental physics, and also by devoting the resources of the laboratory to redetermining the unit of resistance and the mechanical equivalent of heat. Giving new values for fundamental constants constituted an important metrological task at the intersection of the fields of mechanics, thermodynamics, and electromagnetic theory. If the new energy physics linked these regimes intellectually, determining standards with increasing precision governed their relations in practice—and recent experiments had disclosed worrying discrepancies in different determinations. Rowland's contributions would rely on building the best apparatus possible and zealous attention to the reduction of error.[30]

In the early 1880s Rowland came up with a new design for diffraction gratings. By ruling the dispersive grating on a slightly concave surface it was possible to dispense with the lenses that had hitherto been required to focus the light concerned (thereby diminishing its intensity) and also to render the observed distance between spectral lines proportional to their difference in wavelength. That design innovation significantly enhanced

30. On Rowland, see Mendenhall, "Henry A. Rowland, Commemorative Address" (1902); Sweetnam, *Command of Light* (2000); Sibum, "Exploring the Margins of Precision" (2002).

the ability of gratings to compete with prism spectroscopes. Producing larger gratings than his predecessors would allow Rowland to improve their resolving power and light intensity, while ruling more lines per inch would give greater dispersion. Drawing closely on Rogers's design innovations and practical expertise, Rowland was able to machine a precision screw and construct a ruling engine that ruled a series of gratings of impressive dispersive power and clarity; Rogers then put the results to the test of absolute measurement. It will be no surprise that the activity attracted the attention of industrial firms. Hearing from publications of the *Mechanical Engineer* in 1882 that Rowland had a method of making "perfect screws," Brown & Sharpe wrote: "being in need of such a screw we wish to inquire whether we can have the benefit of your invention and if so in what way. We would like to have a screw about 48 in. long."[31] Rowland managed his invention carefully, making the gratings it ruled widely available, rather than the screw itself. Despite writing an influential article on the screw for the *Encyclopædia Britannica*, the physicist kept enough details of the process secret to protect himself from potential competition.

Learning from Rutherfurd's example, Rowland did everything he could to cultivate grating spectroscopy widely within physics and astronomy—but also to ensure that his research remained central to the field. While the gratings were sold to researchers at cost through the firm of John A. Brashear in Pittsburgh, Rowland also retained the very best gratings, using them to make standard photographic maps of the solar spectrum. Featuring the spectral lines of the sun, his maps were relevant to both astronomers observing stars and physicists analyzing terrestrial elements. They made it possible to read off relative distances between different spectral lines and provided a common frame of reference for researchers in different institutions. Meanwhile, Rowland's colleague Louis Bell focused on using diffraction gratings to provide a standard of length in terms of the wavelength of light. In a paper published in 1887, Bell implemented the approach that Peirce had suggested. Bell noted that the final correction of his measurements had to await the establishment of the new metric prototype at the International Bureau of Weights and Measures,

31. For Rogers's tests of Rowland's gratings, see Rogers, "The Absolute Length of Eight Rowland Gratings" (1885). Under Beale's direction, Darling, Brown & Sharpe sought to build a lathe that could cut a standard screw. They wanted the ability to produce commercial screws within 0.0004 inches in 1 foot, or 0.001 inches in 4 feet. Sharpe, "Joseph R. Brown" (1949), 24. Brown & Sharpe to Rowland, 18 August 1882, HAR-JHU, Box 2. On Rowland, see Sweetnam, *Command of Light* (2000); Kargon, "Henry Rowland" (1986).

and used two standard decimeters that had been graduated and determined by William Rogers. He gave a final result for the D1 line of the sodium spectrum, noting that as far as experimental errors the result should be correct to one part in two hundred and fifty thousand; but the error in the regularity of spacing in the gratings themselves "introduces a complication by no means easy to estimate." His final probable error estimate was that the error did not exceed one in two hundred thousand.[32]

As Charlotte Bigg has shown, such endeavors helped Rowland establish the Johns Hopkins physics department as a standards-based research institution. It provided the key instrument and spectral maps and wavelength tables that could set definitive standards for the pursuit of spectroscopy in other laboratories and observatories.[33] There were good reasons to use Rowland's gratings in particular, for they yielded significantly more spectral lines than earlier instruments, and patterns in the structures observed had been delivered by J. J. Balmer's 1885 formula for one hydrogen series (with the search for more general formulas being led in Europe by Kayser and Runge, Rydberg, Schuster, and others through the 1890s). And it is particularly important to note that both Rowland's instruments and his metrological research strategies integrated the Johns Hopkins laboratory into practical research regimes in a variety of contexts. By 1900, between 250 and 300 of Rowland's gratings had been sold, and most significant physics laboratories possessed at least one.

By the time Michelson resumed his work with refractometers in the mid-1880s, then, the precision measurement of distance and use of light as a standard of length had been identified as an active field of research in the United States. While the immediate industrial needs for standards of length had been settled by intense collaborations between instrument makers and government and academic scientists, spectroscopists had taken the advantage of new gratings to forge close links between light and length. The assistance of his Cleveland colleagues Edward Morley and Charles Mabery surely helped Michelson enter this field with enough

32. Bell, "Absolute Wave-Length of Light" (1887), 181.

33. On the solar-terrestrial comparison, see Sweetnam, *Command of Light* (2000), chap. 4. Bigg has analyzed the importance of wavelengths and Rowland's relation to existing practices in spectroscopy, and Hentschel has offered valuable studies of the research of the Rowland school and spectroscopy in general. Bigg, "Spectroscopic Metrologies" (2003); Hentschel, "Redshift of Solar Fraunhaufer Lines" (1993); Hentschel, "Photographic Mapping, Part 1" (1999); Hentschel, "Photographic Mapping, Part 2" (1999); Hentschel, *Mapping the Spectrum* (2002).

confidence to compare his new instrument with earlier spectroscopes.[34] A concrete spur to take a different approach had also been provided by another representative of the French tradition of optical research. Working in Marseilles, Macé de Lépinay had noted that grating measurements of the wavelength of the D2 sodium line varied considerably, from Fraunhofer's 1823 measurement to Ångström's 1868 determination. In 1886 he implemented an alternative approach that made use of the definition of the kilogram as the mass of a cubic decimeter of water. Macé measured the volume of a 4 cm quartz cube with a refractometer arrangement (he generated "Talbot's fringes" by observing a beam of sodium light, one-half of which traveled through the quartz while the other half traveled freely). His measure of length in terms of the wavelength of sodium could be related to the meter through the hydrostatic mass of the cube as determined by the International Bureau of Weights and Measures, but depended also on an accurate knowledge of the index of refraction of the quartz crystal.[35]

The relations we have observed in this field—linking industry and science through mechanics and light—are similar in kind to contemporaneous developments in electrical standards and energy physics. Historians have recently shown that engagement in the latter two fields was critical to the institutional rise of physics laboratories in the last quarter of the nineteenth century. Scores of new laboratories found an important role in training the skills required for emerging power and lighting industries, but also in providing authoritative standards for industry. Notable battles over the appropriate definition and implementation of units of resistance split national and commercial rivalries in British and German electrical manufactures through the 1870s and 80s.[36] In this chapter I

34. As Darwin Stapleton notes, Michelson's early research had not involved the use of prism or grating spectroscopes. Morley, in contrast, had given two papers on diffraction gratings to the AAAS in 1876, and his skills would have been complemented by those of Charles Mabery, a leading spectroscopist who joined Michelson at Case in 1883. See Stapleton, "Context of Science" (1988), 17.

35. Macé de Lépinay, "Valeur absolue" (1886). For an account of Macé's work, see Bigg, "Behind the Lines" (2002), 20–22.

36. Visiting Paris in 1881 Michelson worked as a juror in the International Electrical Exhibition. The related conference on electrical standards saw a heated dispute over different proposals, with William Thomson advocating an absolute standard of resistance rather than the arbitrary standard favored by Helmholtz and Werner Siemens. On standards, see Schaffer, "Late Victorian Metrology" (1992); Schaffer, "Empires of Physics" (1994); Gooday, "Precision Measurement" (1990); Gooday, *Morals of Measurement* (2004). Their work builds on

want to emphasize two facets of the paramount role that standards played in laboratory physics. One concerns American strengths in industrial mechanics. When Rowland established his laboratory at Johns Hopkins in 1876 he addressed the field of energy physics first, initiating redeterminations of the mechanical equivalent of heat and instituting comparisons of temperature-measuring devices before he turned to spectroscopy in the manner outlined above. Nevertheless, it is particularly revealing that light and length—and astronomy and physics together—were to be so important in the nation that pioneered the manufacture of interchangeable parts as a key facet of the "American system of manufactures."[37]

My second point concerns the diverse ways in which standards-mediated relationships were played out in different contexts. As Graeme Gooday has urged in *The Morals of Measurement*, physicists' recommendations were often only loosely followed by industrial electricians, if at all. Electrical practitioners engaged in the regime of standards production and electrical measurement in a variety of ways, strongly related to their engagement with consumers, with rather different implications for the power relations between academic and industrial disciplines.[38] Observing Michelson at work will offer yet another perspective on physicists' engagement with standards than those documented by Gooday, Simon Schaffer's study of the Cavendish Laboratory's arbitration of practical resistance standards, or the example of Rowland's provision of standard spectral instruments and maps for spectroscopical research. Developing his own laboratory and establishing his own field of expertise, Michelson followed Rowland and his collaborators into optical standards. He first aimed to set still more precise standards of wavelength, and later looked to rule more accurate gratings using new, optically governed instruments. But characteristically, Michelson engaged with these fields in a strategically important but distant way. He provided singular values without shaping the practices others employed. Michelson's arguments for a generic class of instruments were accompanied by strenuous metrological

---

important studies of Kelvin and British telegraphy, and is complemented by Olesko's work on German regimes of measurement. Smith and Wise, *Energy and Empire* (1989); Smith, *The Science of Energy* (1998); Hunt, "Doing Science in a Global Empire" (1997); Olesko, "Precision, Tolerance, and Consensus" (1996).

37. Randall Brooks has shown that the British leaned on U.S. expertise in this field: Pratt & Whitney played an important role in the BAAS Committee on Screw Gauges that met from 1895. Brooks, "Standard Screw Threads. Part II" (1988).

38. See especially Gooday, *Morals of Measurement* (2004), chaps. 1 and 2.

work; but Michelson interacted with contemporary disciplines largely through metrological *values*. For all he did to promote them as *ultimate* and *practical* standards, his techniques and instruments remained largely personal.

Light and Length

Immediately after completing their ether-drift experiment in 1887, Michelson and Morley published a paper titled "A Method of Making the Wave Length of Sodium Light the Actual and Practical Standard of Length."[39] The idea had occurred to Michelson earlier, perhaps working with his 1881 ether-drift apparatus and observing the regular displacement of interference fringes with minute turns of the screw that carried the silvered mirror (in 1887 he noted one complete turn of a screw with one hundred threads per inch traversed nearly one thousand wavelengths).[40] But there was a significant question to be answered. How far would the homogeneity of particular light sources allow a moving mirror to track individual light waves: fractions of an inch, inches, or feet? Setting up his instrument in 1881, Michelson used a sodium source and could follow the fringes some distance. But carrying out the experiment itself using the far less homogeneous light of a lamp, the fringes quickly disappeared when the mirror was moved any distance away from the white light that showed an equal path length (indeed, this gave a way of determining the precise point of equal path length). Between their two repetition experiments, in June 1887 Michelson and Morley conducted preliminary standards experiments with apparatus like that used in Potsdam. With no need to turn the instrument it could easily be made stable on a brick pier, and the screw micrometer movement was transferred from the partially silvered mirror to the mirror on one of the two perpendicular arms. Moving this mirror over a longer distance of travel than previously, Michelson and Morley proceeded to work out how many wavelengths they could count for several different light sources.

Fizeau had observed interference in sodium only up to a path-difference of 50,000 wavelengths. With a smaller density of sodium vapor Michelson and Morley moved the mirror over 2 inches, counting more than

39. Michelson and Morley, "Wave Length of Sodium Light" (1887).

40. Michelson and Morley, "Relative Motion" (1887), 339.

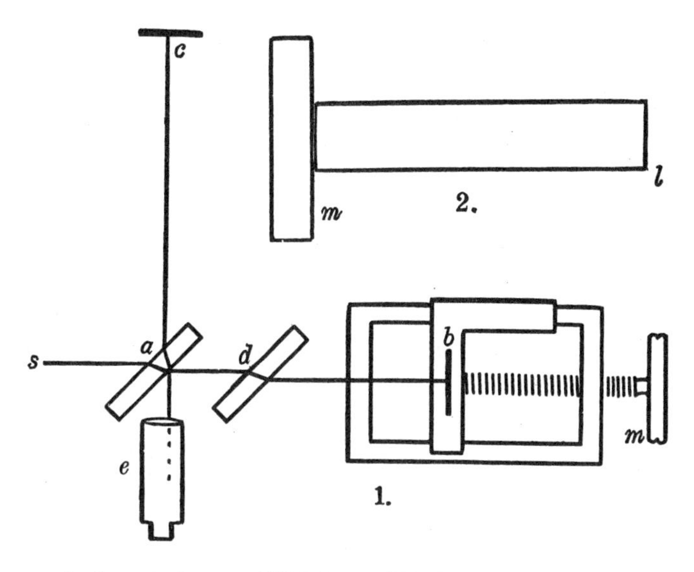

FIGURE 3.4 A schematic diagram of Michelson and Morley's first standards apparatus. The plane mirror *b* is moved by a screw. *lm* in their fig. 2 depicts an intermediate standard decimeter in the form of a prism of glass with one plane end, *l*, and a second, slightly convex end touching the plane *m*. See figs. 3.8 and 3.9 for illustrations of an intermediate standard and the meter apparatus in final form. Michelson and Morley, "Wave Length of Sodium Light" (1887), 428, fig. 1.

200,000 wavelengths. Yet interference patterns alternated with uniform illumination as a result of the superposition of the two series of waves in sodium light, so they tracked the fringes of the light of thallium, lithium, and hydrogen as well. In each case they observed interference for up to fifty to one hundred thousand wavelengths. Such sources could be used to check any determination with sodium. Writing up their work later, Michelson and Morley described how relatively small distances—of a *countable* number of wavelengths—could be used to step off the larger distance of an intermediate standard decimeter (with mirrors at both ends) with an interferential comparer then being employed to step off the whole meter.

But before publishing on standards, Michelson and Morley conducted their ether-drift repetition on 8, 9, 11, and 12 July. This was important, since the standards program rested on the assumption that the wavelength and velocity of light did not vary with location or orientation (or that any variation could be established extremely precisely). Given the null result they found, Michelson's conceit was that the use of light waves would establish an *absolute* standard, dependent only on the invariable properties of light and the ether. Michelson and Morley wrote first of making sodium light the "actual and practical" standard of length, then of establishing light waves as the "ultimate" standard.[41] They wanted to address the first sentiment in Mr. Pratt's call, "to get this matter settled once for all time." The use of light waves would allow a standard of length to be established independent of the international meter held in Sèvres.

In 1883 William Thomson had told the Institute of Civil Engineers in London that the surest way to gain an understanding of the philosophy of measurement would be to "cut off all connection with the earth, and think what we then must do, to make measurements which shall be directly comparable with those we now actually make, in our terrestrial workshops and laboratories." His fantasy involved a scientific traveler setting off through the universe with only his books or knowledge of their principles, and a few simple tools. Thomson noted that in present-day laboratories, students were more likely to measure electrical *resistances* between 100 and 10,000 ohms to an accuracy of 1/10 percent, than they were to be right to one millimeter in a meter in their measurements of *length*. No doubt impressed by Rowland's gratings, Thomson described how his traveler could recover the centimeter using a grating ruled on glass, demonstrating it to his audience with a simple grating and two candles.[42] Michelson would have him use the glass plates and mirrors of a refractometer instead.

From 1887, Michelson and Morley continued the standards program rather than fulfilling the ether-drift protocols they described (which involved repeating the experiment throughout the year). Instead they extended their study of potential light sources in the laboratory and began constructing an interferential comparer and intermediate standards. Our study of Michelson's work with light and length will first establish how his desire to promote a general class of instruments—and need for better appliances—shaped Michelson's relations with instrument makers and collaborators. Then we investigate both the broad intellectual framework Michelson gave his campaign for refractometry and the apotheosis of the

41. Michelson and Morley, "Ultimate Standard of Length" (1889).

42. Thomson, "Electrical Units of Measurement" (1889 [1883]), 107 (for the philosophy of measurement), 77 (for resistance vs. length measures), and 107–10 (on the recovery of the centimeter). This was one of six lectures titled "The Practical Applications of Electricity," and Thomson described ways of reaching a unit of time through measurements of the velocity of light or of the ratio between electrostatic and electromagnetic units (110–19).

method in the determination of the length of the meter in wavelengths of light.

The Refractometer and the Instrument Maker

There were several significant problems of a mechanical and optical nature to be solved in building the apparatus to make any comparison between light and length—but the unusual potential of the instrument was underlined by the important fact that the sensitivity of interferential techniques could themselves be used to check the construction of the instrument components. This circumstance gave Michelson the chance to articulate a new relation between light and mechanics by submitting the latter to the demands of the former. For example, the metal ways on which the carriage holding the mirror moved had to be precisely parallel, the mirrors on each end of the intermediate standards had to be adjusted to measure the appropriate distance within a few waves of the desired length, and the angle between the plane of the mirrors on each end of the standard had to be less than half a second. In each case, the sensitivity of the refractometer enabled the scientist to work to tolerances not normally possible by observing variations in the appearance of the fringes and making appropriate adjustments. Michelson's material demands were stiff, even for the most capable of America's mechanics and opticians. His instrument makers offer ample confirmation of the productive relations we have already observed between astronomers, precision instrument makers, and the premier machine tools makers of the United States. Working in Cleveland, Michelson and Morley engaged the local firm of Warner & Swasey to machine the apparatus for their comparator, which was then sent on to John A. Brashear to complete the optical work for the intermediate standards. While simpler in design than the elaborate cross of the ether-drift apparatus, the comparer proved significantly more difficult to build.[43]

In the 1870s Worcester Warner and Ambrose Swasey had been employed as mechanics in the Pratt & Whitney workshops in Hartford, Connecticut, but left in 1880 to set up independently in the machine tool trade. After opening for business in Chicago they soon moved to Cleveland to be nearer their most important customers. From the onset they involved

43. On Brashear see Brashear, *A Man Who Loved the Stars* (1988 [1924]); Fried, "Masterful Techniques" (1991). On Warner & Swasey, see Stapleton, "Context of Science" (1988); Pershey, "Early Telescopes" (1984); Pershey, "Warner and Swasey" (1990).

themselves in making astronomical instruments and particularly telescope manufacture, which quickly became a high-profile adjunct to their more routine work. Submitting a variety of orders both together and separately, Morley and Michelson were the firm's most frequent customers in the 1880s.[44] They were far from its largest clients, however. In 1887 Warner & Swasey finished their work building what for a decade would remain the world's largest telescope, the 36-inch refractor for the Lick Observatory near San Francisco. Through the 1890s they continued to bid for telescope manufactures, beating out Alvan Clark and Sons to reconstruct the Naval Observatory 26-inch refractor, and then building the mechanics and dome for the 40-inch refractor for Yerkes Observatory of the University of Chicago.

In the meantime they filled many orders for their local customers, including several relating to the Fizeau experiment, from February 1885 to March 1886. In 1887 the Warner & Swasey Sales Book records invoices for 140 hours work for the ether-drift apparatus with its sixteen mirrors set on a stone block.[45] But their most substantial work for Morley and Michelson involved more than 590 hours making the apparatus for the interferential comparer the scientists ordered for their standards project. This was clearly a more demanding job; although Morley requested its completion by August 1887, Warner & Swasey only managed to send it to Brashear for the installation of the optical parts in late winter.[46] The

44. Michelson's first orders related to velocity of light. See Stapleton, "Context of Science" (1988), 15. Pershey's study of the relations between Warner & Swasey's telescope work and their machine tool design leads him to assert: "The relationship of technology and science, via instrument making, does not imply or suggest any cognitive or even methodological connection. Rather, the two cultural activities of science and technology are related only in an economic and political way. They interact and draw on one another through outside agencies and agents. . . . it is not surprising to see very little direct transfer of knowledge from one to the other." Pershey, "Warner and Swasey" (1990), 228–29. Michelson's work in interferometry provides a counterexample, leading as it did to the eventual adoption of interferometric methods of instrument construction in firms such as Brashear, Adam Hilger, and Pratt & Whitney.

45. That included eighty-five hours making "apparatus for carrying a mirror," invoiced on 26 May, and two invoices on 30 June, the major one being for "50 Hrs. making 4 plates with projections[;] this includes time on special Jig Taps & Die[;] 58 lbs. Cast iron [@].03[;] 3 Hrs. Forging on Shifting Device[;] 2 Hrs. M[ach]ine work on same." 26 May and 30 June 1887, Sales Book, W&SC-CWRUA, also cited in Stapleton, "Context of Science" (1988), 17–18.

46. "Sketch of Experimental apparatus for the determining of the length of a wave of Sodium light, Profs. Morley & Michelson," (2 pp.), 23 January 1888; "Experimental Measuring Machine for Prof. Michelson," 2 February 1888; "Measuring Machine for Prof. Michelson," 20 February 1888, Notebook 1886–1888, W&SC-CWRUA. 31 October 1887 (this entry

instrument was finally ready in May 1888, much to Morley's disgust. Blaming the delay on Warner & Swasey, he reported to his father that they would not give them any more work.[47] After Brashear had finished the optical parts, the physicist Charles S. Hastings at Yale tested his work. A letter Brashear wrote to Edward Holden at the Lick Observatory on 5 April 1889—referring in this case to a later version of the apparatus—gives some measure of the extent to which tasks of this nature stretched the Brashear workshop:

> we have completed two of the most difficult pieces of work ever undertaken by us. One was a set of 16 *plane* and parallel plates for Prof. Michelson for his research on the absolute value of a sodium wave length, and its transference to a steel bar in a freezing mixture, (i.e., multiples of the wave length) which is proposed as a new standard of measurement.
>
> The limiting error allowed was *.05 sodium wave length* for every plate.... Hastings wrote us an unsolicited letter in which he said after a critical test he had written Michelson that "it was the most perfect piece of optical work he had ever seen, and that nowhere else in the world could he have obtained it." I give you his words just as I take them from his letter. After Michelson tested them he used two adjectives—Exquisite—Superb—both underscored. So they must be good. It did take a vast amount of time to work and test them finally. You couldn't *look* at them [the fringes] without disturbing them.[48]

Michelson's demands produced long-term changes in the practices of the workshop. Later Brashear wrote to Lord Rayleigh and noted that the

records the initial 593 hours spent making a "meter sub-dividing" machine, including drawings, patterns, forging, and machine work), 5 November 1887, 30 December 1887, Sales Book, W&SC-CWRUA. As quoted in Ibid. Stapleton notes that the instrument makers invariably listed the scientists as "Morley and Michelson," at p. 19. Warner & Swasey's initial plan to purchase the optics from Alvan Clark and Sons was a casualty of their competitive bids for observatory contracts, so they turned to Brashear instead. Around 1900, dissatisfaction with Brashear's business management and cost control led them to form their own optical department. See Pershey, "Early Telescopes" (1984), 311.

47. E. W. Morley to S. B. Morley, 10 May 1888, Morley Papers, Box 2.

48. Brashear to E. S. Holden, 5 April 1889, The Brashear Archives, the Brashear Association, Pittsburgh, PA; copy held in the Holden Archives (Courtesy, Mary Lea Shane Archives of the Lick Observatory, University Library, University of California–Santa Cruz). The second piece of work also owed something to Michelson's advocacy of interferential measurements in the United States: Brashear had produced two prisms for a study of stellar aberration being conducted by G. C. Comstock at the Washburn Observatory, also discussed below. See Susalla, "The Old School in a Progressive Science" (2006).

firm's "prism surfaces are all tested by interference methods and made plane within 1/10 λ. We make so many accurate planes for refractometers that we carry the same accuracy into our prism work."[49]

Michelson's need for precisely plane and parallel ways (the surfaces on which the carriage holding the mirrors rested), however, went beyond instrument makers' capabilities. He wrote: "The accuracy of the ways must be so great that the greatest angle through which the mirror... turns in passing along them is less than one second of an arc. This accuracy cannot be attained by the instrument maker, but the final grinding must be done by the investigator himself."[50] Some indication of the manner in which observing changes in the appearance of the interference fringes enabled Michelson to detect errors and establish proper operating conditions is given by an entry made in Michelson's notebook on 26 May, probably in 1891 or 1892: "In setting up Brashear's refractometer suspected inaccuracy of glasses on account of an apparent (horizontal) parallax. Trouble was traced to want of *parallelism* of the two pl.[ane] par.[allel] glasses. As soon as this was remedied (by normal appearance of fringes) the defect disappeared."[51]

This passage shows also that in his notebooks Michelson associated his instruments with their makers: this is "Brashear's refractometer."[52] Michelson's first publications on velocity of light and ether drift also carefully named the firms he used. Historians and sociologists have shown that despite the extent of the engagement of technicians and artisans such as instrument makers in the design, construction and sometimes performance of experimental work, their visibility in the published records of scientific labor has been strongly subject to both the social relations in which authority is vested in particular periods and to whether experiments

49. Brashear to Rayleigh, 2 August 1895, RA-HAFB.

50. Michelson, *Light Waves* (1902), 41.

51. Albert A. Michelson, Notebook, about 1891–95, Mount Wilson and Palomar Observatories Library, copy held in MC-USNA, X-503.3, 15. He went on to comment: "The 'normal' appearance is that of no rings with white center; for when this is obtained the fringes are symmetrically colored, whereas with a dark center the edges of the central (dark) fringe are salmon and olive. The change due to the light coating of silver is not an exact half wave but about 0.40 0.35 ±.05. Better; the fringes are exactly reversed, and shifted 0.15 λ in the direction corresponding to diminished path." Michelson's notes on adjusting the "compensator" against parallax continue with entries on 28 May, 17 June and 15 October, 16–17.

52. In contrast, Brashear had earlier written to George Ellery Hale, "we have a new apparatus for Michelson which as you remember he calls his refractometer." Brashear to Hale, 10 December 1889, YOA, Yerkes Observatory Directors' Papers Box 1, Folder 1.

work as expected. Steven Shapin has maintained that as a rule in seventeenth-century Britain, Robert Boyle's technicians went unmentioned, remaining invisible unless something went wrong—when the incompetence of their labor could be cited to explain the reasons for failure.[53] Here we can see that as author, Michelson carefully managed the visibility of those on whom his work depended with the aim of building trust in his results, his reputation, and that of his instruments; this called for quite different strategies at different times.

Early in his career it was important for Michelson to promote trust in his results by indicating that he had used reputed instrument makers. His results gained credibility and luster from the well-known excellence of the lenses of Alvan Clark and Sons. By the time he had won his professorship at Case, however, Michelson's reputation had grown—and he was soon to mount a powerful advocacy of a generic class of instruments. Instrument makers disappeared from his publications. Just as Michelson and Morley relied on extending the realm of validity for the results of their ether-drift experiment beyond the bounds of a few days spent in the basement laboratory in Cleveland, they suggested that their achievements were not specifically tied to the particular apparatus they worked with, but to the general class of instrument. While they cited the scientists they drew on, and indicated that they received the support of the Bache Fund, their papers did not mention the firms responsible for making the apparatus.

A related tactic was to emphasize that their results did not rely on their *own* skills, either. In 1889 and 1890 papers, the results of an unnamed observer were included. In the latter case, Michelson listed three sets of readings with means and errors, specifying that one had been taken by "an observer who had no previous practice in this kind of measurement," and that in each case the scale readings had been unknown to the observer.[54] This helped establish results as a straightforward consequence

53. Shapin highlighted an issue in the relations between scientists and artisans that Sorrenson, Morus, and Jackson have shown plays out in different ways in different periods and contexts. See Shapin, "The Invisible Technician" (1989); Shapin, *Social History of Truth* (1994); Sorrenson, "George Graham, Visible Technician" (1999); Morus, *Frankenstein's Children* (1998); Jackson, *Spectrum of Belief* (2000).

54. See Michelson and Morley, "Ultimate Standard of Length" (1889), 185; Michelson, "Measurement by Light-Waves" (1890), 119. The results were a direct measurement of the wavelength of the brilliant green line in the mercury spectrum. Morley obtained a mean of 117.15, with an error in waves of 0.0056, Michelson's mean was 118.50 with an error of 0.0059, and the third observer's mean was 119.05 with a somewhat larger error of 0.0110 waves. The

of the instrumental arrangement, safeguarding them from the perception that they might depend on idiosyncratic skills. (Michelson and Morley also varied the conditions within the instrument for similar reasons, altering all the standards in length and adjustment between measurements, for example.[55]) Recall that when Simon Newcomb revisited Michelson's velocity of light measures in order to critique the result obtained, he needed to treat Michelson's instruments as both particular and personal, specific instruments used in highly specific ways. Michelson's work to promote a new class of instruments involved moving in just the other direction, doing all he could to suggest that refractometers did not rely on unusual skills or expertise.

Michelson's publications emphasized the generic nature of his instruments and observations. His relations with other scientists and several instrument makers over the next two decades (and in the course of founding two further physics laboratories) suggest that Michelson's own deepest needs were for the expertise and experience of capable but dependent *assistants*, people under his private control with a status intermediate between the independent scientists with whom he sometimes worked and nameless inexperienced observers. Michelson's desire for better resources and a more settled faculty led him to move twice, first to Clark University in Worcester, Massachusetts, in 1890, and then in 1893 to the still more generously endowed University of Chicago, where he founded the Ryerson Laboratory of Physics. Michelson used the first relocation to break off his collaboration with Morley and shed his association with Warner & Swasey. He wanted a different kind of collaborator and new instrument makers. By 1892, however, he was also clearly unhappy enough with what the Brashear firm could achieve alone in Pittsburgh to use the more local American Watch and Tool Company to build the apparatus for his determination of the international meter (Brashear provided the optics, but his estimate for the apparatus far exceeded the appropriation Michelson had received for the instrument). The succession of firms suggests that several U.S. instrument makers were capable of filling Michelson's needs, but also that none of the relationships established satisfied Michelson completely.

comparison was repeated in Michelson, "Light Waves and Their Application to Metrology" (1894).

55. Michelson and Morley, "Ultimate Standard of Length" (1889), 185.

An 1889 letter to Henry Rowland gives a highly interesting sketch of the kind of people Michelson wanted to work with. First he asked if Rowland knew of a good man willing to take a fellowship with chances of advancement in a year or two. Then he went on to describe a second wish: "Another thing—I want a private assistant—a man handy with his fingers and a good observer—who will do just what he is told—and who is not too ambitious—If such a phenomenon should present itself I could make it worth his while to consider the position."[56]

Once he moved to Chicago, Michelson found this rare phenomenon. Again the city offered the services of excellent local mechanics in the firm of Octave L. Petitdidier, and to get still more out of the working relationship he established with two Petitdidier employees, Michelson eventually offered Fred and Julius Pearson jobs in the physics department workshop—turning two instrument makers into his private assistants. Fred Pearson worked particularly closely with Michelson, and the Nobel Prize award gave the scientist enough leverage to negotiate for a pay raise for the brothers in late 1907. In several late papers reporting new research on the velocity of light, Michelson noted that he had been "Assisted by F. Pearson," and Pearson coauthored a 1935 paper that published observations completed after Michelson's death in 1931.[57] Skilled collaborators were critical to Michelson, but good hands and observing skills were more important than scientific independence and ambition. Indeed the Fellow he hired in 1889 proved both critically important to Michelson's meter campaign, and too ambitious to remain with Michelson long. F. L. O. Wadsworth had been trained in engineering and mechanics. While at Case he supervised the construction of Michelson's meter apparatus in the shops of the American Watch Tool Company and assisted Michelson in his astronomical and spectroscopical research. When the comparator was damaged on its way to Sèvres, Wadsworth was able to reconstruct it with the help of French instrument makers, an accident that proved the French could also work to Michelson's standards. But

56. Michelson to Rowland, 5 September 1889, copy in MC-USNA, X-494.4

57. Warner, "Octave Leon Petitdidier" (1995). On Michelson's relations with the Pearson brothers, see Livingston, *Master of Light* (1973), 244 (Nobel negotiations), 185–86, 278–81. For the credit accorded and subsequent authorial status of Fred Pearson, see Michelson and Gale, "Earth's Rotation" (1925), 140; Michelson, "Velocity of Light between Mount Wilson and Mount San Antonio" (1927), 1; Michelson, Pease, and Pearson, "Velocity of Light in a Partial Vacuum" (1935).

despite following inducements to join Michelson in Chicago, Wadsworth soon moved sideways to join George Ellery Hale in building up Yerkes Observatory, and then became director of Allegheny Observatory.[58]

Working with new instruments in his laboratory, Michelson most needed people who would be ready to remain nameless and do what they were told. Detailing trials and results in his publications, it was important for Michelson to leave instrument makers unnamed on the one hand, and to explicitly refer to pointedly anonymous and unpracticed observers on the other. Names disappeared from Michelson's publications in this way because of one significant feature of the relations between his research and the broader scientific community. Although in standards of length Michelson had once again entered a field of active concern, characteristically he sought to do so in a distinctive way: he had to convince others of the trustworthiness and value of an instrument few had actually experienced. Sociologists have emphasized the difficulties that scientists face when repetition discloses divergent results, and we have seen these vividly exemplified in Michelson's work on the velocity of light. At this point in his career, however, Michelson faced a different but related problem that is common to highly innovative research. Deploying what were in fact rare skills and using what was uniquely refined apparatus, in the absence of independent repetition by other scientists Michelson's published descriptions of his work (and the protocols they represented) had to make it appear as if the results of such an independent repetition were so sure that it might as well have been undertaken already. An unskilled, unnamed collaborator stood in for the community that did not yet use Michelson's instruments.

"A Plea for Light Waves": Accuracy in the Extreme

Having gained an understanding of how Michelson addressed the material preconditions of refractometry, we are now ready to explore the disciplinary dimensions and intellectual framework of his elaboration of a new

58. Of the meter apparatus, Wadsworth wrote: "As a rule I have found that the expense of obtaining a given instrument is only from 1/3 to 1/2 as much when it is built [...] under my own superintendence as when it is built to order by an established instrument firm. / The Paris interferometer for example could never have been built and set up for the amount appropriated by the International Bureau in any other way. Brashear's estimate on the mechanical parts alone far exceeding the complete amount of the appropriation." Wadsworth to Rayleigh 14 January 1898, RA-HAFB. See also Livingston, *Master of Light* (1973), 176.

instrumental culture—decisively shaped as these were by his responses to Rowland's earlier work. In 1889 Michelson's authority was marked by being given the Rumford Medal of the American Academy of Arts and Sciences, and the honor of presenting the vice presidential address to Section B of the American Association for the Advancement of the Sciences.[59] This was one of only a handful of occasions on which Michelson expressed his sense of the physics discipline, so it is an invaluable guide to the leitmotifs to his research. His title echoed the challenge Henry Rowland's "Plea for Pure Science" had issued six years earlier. Michelson entered the lists with "A Plea for Light Waves," and while certainly less urgent rhetorically, he aimed to go beyond even Rowland's high standards. Rowland's address had focused on institutions, identifying weaknesses in American universities and arguing for the concentration of resources in a few elite institutions. It has attracted scholarly attention principally for Rowland's enemies: mediocre colleges, the corrupting example of consulting professors in the pockets of industry, and money-grubbing industrial firms (with an unstated target in Thomas Edison). But the image of the ideal laboratory Rowland promoted is just as revealing.[60] For Rowland, the perfect laboratory would be large, stocked with perfect equipment, and graced by large revenues, a corps of professors and assistants, and a machine shop. While no such institution yet existed (worldwide), Rowland drew on a pattern that came from close to home. As we might by now expect, his model for the physics laboratory was the astronomical observatory. Housing the simplest and most perfect branch of physics, everyone now knew an observatory needed perfect equipment and recognized that even that would be useless without sufficient income for a staff of assistants to make observations and computations.[61] Rowland thought physics laboratories required resources on the scale now familiar in astronomical

59. Awarding the Rumford Medal, Lovering stated that Michelson might be thanked "not only for what he has established, but also for what he has unsettled." Lovering, "Michelson's Recent Researches on Light" (1889), 468.

60. Rowland, "A Plea for Pure Science" (1883). See Hounshell, "Edison and the Pure Science Ideal" (1980); Dennis, "Accounting for Research" (1987). Dennis argues that in Rowland's vision of the moral economy of pure science, the research university (itself combining disciplinary and pedagogical roles) should be insulated from the marketplace mentality through a sufficient endowment.

61. Rowland, "A Plea for Pure Science" (1883), 246. Agnes Clerke depicted a similar continuum between physics and astronomy, writing of the science of heavenly bodies as "a branch of terrestrial physics, or rather a higher kind of integration of all their results." Clerke, *Popular History of Astronomy* (1887), 461.

observatories; and the transfer of the Rogers-Bond comparator at Harvard Observatory to Johns Hopkins soon after Rowland spoke, clearly indicates that the physics discipline was now taking on roles previously assumed by astronomers in the United States.

Characteristically Michelson's plea was far less institutionally oriented, even as it elaborated an ideal of purity. It was also so resolutely present and future-centered that Michelson did not mention velocity of light or ether drift, despite focusing on the instrument his early experiments had spawned.[62] Michelson chose to define himself as an optical researcher, developing that category largely in opposition to the diverting seductions of electrical research. In comparison to electricity and magnetism (which soaked up much of his teaching), Michelson thought optics appeared "fairly completed." He would combat that perception by focusing on the phenomena of interference and bringing the interferential refractometer into play alongside the microscope, telescope, spectroscope, and camera. The traditional instruments had already achieved practical perfection; in contrast, the interferential refractometer was open to further development and offered a wide range of applications.[63] These included distance measurements and the provision of a new standard of length (which could also furnish a standard of mass through the construction of a cubic decimeter). Using the refractometer to provide an optical test plane also allowed the accurate manufacture of plane surfaces by studying the interference fringes while the surface was being corrected. The refractometer could yield accurate measurements of coefficients of expansion and elasticity and of the density and optical properties of solids, liquids, and gases, as French and German researchers had shown. But Michelson's primary research concerns lay in establishing light as a standard of length and developing the spectroscopic information his instrument had begun to reveal.

Almost as soon as he began his work on standards of length, Michelson had encountered spectroscopy in two distinct ways. He had noted a fine structure hitherto inaccessible to spectroscopes; and experimental work with gratings had established values for the wavelength of the spectral lines in which he was interested. Michelson and Morley merely referred to these in 1887, but by 1889 had begun to draw on them much more

62. Michelson, "A Plea for Light Waves" (1889).

63. Ibid., 67–69. Michelson's teaching is described in McMahon, *The Making of a Profession* (1984), 44–45; Kevles, "Physics and National Power" (1988), 249.

fully—as standards against which to validate their observations of the wavelength of sodium and mercury lines. Their paper, "On the Feasibility of Establishing a Light-Wave as the Ultimate Standard of Length," compared their values with Rowland's grating measurements. Establishing a close agreement, they concluded that: "In view of this final and almost complete confirmation these results may be taken not merely to prove the feasibility of the method [to establish a standard of length], but as an accurate and reliable measurement of the relative wave-lengths of these radiations."[64]

Thus existing spectroscopic measures provided the most important benchmark for Michelson's metrological claims with a novel instrument; but his Plea made it clear that Michelson wanted to go further. He told the AAAS the potential of the refractometer method excelled "by far the best results that can possibly be obtained by the most perfect gratings."[65] Repeating comparisons of a kind he first offered in 1887, he reported that Rowland's new gratings enabled an accuracy of one part in half a million for relative determinations; and Bell's elaborate research on absolute wavelengths claimed but one part in two hundred thousand. Michelson's refractometer, however, would be accurate to one part in two million even for absolute comparisons between the wavelength of light and the meter. In the meantime, refractometer accuracy was already yielding new information. Michelson had shown, for example, that both the red line of hydrogen and the green line of thallium were in fact close doubles. Together with his determination that the ratio of the wavelengths of the two thallium lines was 1.0000212 (note the seventh decimal place), these findings demonstrated the "very considerable superiority" his instrument held over spectroscopes, "at any rate for the class of work in question."[66]

While metrology motivated his entry into the field, Michelson's spectroscopic interests extended to the cause of the phenomena. The remainder of his address was devoted to the discussion of how the temperature, thickness, and density of the source might affect the behavior of spectral lines (in particular causing symmetrical or unsymmetrical broadening), all features that might offer insight into the cause of the spectral lines in the motion of molecules in the source.[67] He finished by noting that perhaps he

64. Michelson and Morley, "Ultimate Standard of Length" (1889), 186.

65. Michelson, "A Plea for Light Waves" (1889), 74.

66. Ibid., 74–75.

67. Ibid., 75–78. On the broadening of spectral lines, see especially Michelson, "Broadening of Spectral Lines" (1895); Rayleigh, "Limit to Interference" (1889). In 1961 Breene wrote

had offered a stronger plea for his instruments and theories than for light waves; but creating the slightest interest would accomplish his purpose. Michelson situated himself as somewhat isolated by his primary concern with light. Like the astronomical concerns that helped shape much of his research, this was science on a hilltop rather than science in the service of practical applications.[68] Yet given the intimate relations we have already traced between the observatory, physicists, and industry, it is important to note that neither Rowland's nor Michelson's calls for purity meant science divorced from industry or application. Both railed against what Michelson famously termed *prostituted physics* (urging his student Frank Jewett not to join an industrial laboratory), but the vehemence of their calls for purity should not obscure the actual links they utilized. Rather, a proper relationship was in order, one based on the scientist's ability to deliver authoritative standards rather than commercial imperatives. (Edison provocatively offered the following take on standards: "I measure everything I do by the size of the silver dollar. If it don't come up to that standard then I know it's no good.")[69]

As he widened his study of refractometry, Michelson developed a formal comparison between traditional instruments and the refractometer that helped demonstrate the properties of his field of research. Speaking on "Measurement by Light-Waves" at the Jefferson Physical Laboratory at Harvard, he outlined an analogy in the fundamental properties of the telescope, microscope, and refractometer by treating the refractometer as a masked lens.[70] The resolution and definition prized in the microscope and telescope rely on the accuracy with which all parts of the lens contribute to make the elementary waves reach the focus in exactly the same

that the early work of Michelson and Rayleigh "still forms the basis for everything we do in this field." Breene, *The Shift and Shape of Spectral Lines* (1961), xi.

68. George Ellery Hale first met Michelson on this occasion. Hale, "Some of Michelson's Researches" (1931), 176. For a more extended discussion of Michelson's important relations to the astronomical community, and the extent to which his vision of physics was shaped by astronomy, see Staley, "Michelson and the Observatory" (forthcoming).

69. Jewett joined what became Bell Telephone Laboratories, and Michelson's view is discussed in Weart, "The Rise of 'Prostituted' Physics" (1976); Reingold, "Physics and Engineering" (1988), 288. For Edison, see Josephson, *Edison: A Biography* (1959), 283. As cited in Livingston, *Master of Light* (1973), 244.

70. The tactic could well have been suggested by his early experience moving from Fizeau's lenses to his own half-silvered mirror. See Michelson, "Measurement by Light-Waves" (1890).

phase. Consider a microscope with an objective consisting of a single perfect lens. Magnification of an image increases its indistinctness, since the image of a luminous point is actually composed of small concentric rings of light. Nevertheless, by increasing the size of the interference fringes in relation to the crosshairs of the eyepiece, such magnification allows a great increase in the accuracy with which the position of the point can be established. By taking only the external annular ring of the lens, or better still two small portions at opposite ends of the diameter, the size of the interference fringes can be increased up to any point without affecting the intensity of the light. The portions need not be curved and may be either plane mirrors or prisms: the microscope (or telescope) has been converted into a refractometer, sacrificing resolution and definition for the sake of accuracy.[71] Michelson published a diagram of five basic groups of refractometers, showing how refractometer arrangements could be regarded as using only two pencils of the entire field of light employed in earlier instruments. The fifth group in his illustration showed that a similar analogy held with the spectroscope: the refractometer made it possible to gain accuracy at the expense of resolution and definition by using only the extreme portions of the surface (fig. 3.5).[72]

Very literally, Michelson's research involved the search for accuracy in the extreme. The refractometer was complementary to existing instruments, but also excelled them in accuracy.

These sections have shown that Michelson worked on several fronts simultaneously in order to consolidate interferential refractometry, and that his efforts continually negotiated a significant tension between generic instruments capable of being used in a wide variety of contexts, and extreme results of singular value. Firstly, working with mechanics, opticians, and a variety of arrangements, Michelson had established a practical mastery of the components of the new form of instrumentation, oriented most fully around observing light fringes over long path-differences in the two arms of his apparatus. Secondly, emphasizing theoretical accuracy and results independent of specific observers or instrument makers had helped consolidate the status of the instrument in general. Thirdly, his discussion of light paths had outlined a theoretical overview in which

71. Ibid., 116–18.

72. Ibid., 120–21. See also Michelson, "Light Waves and Their Application to Metrology" (1894).

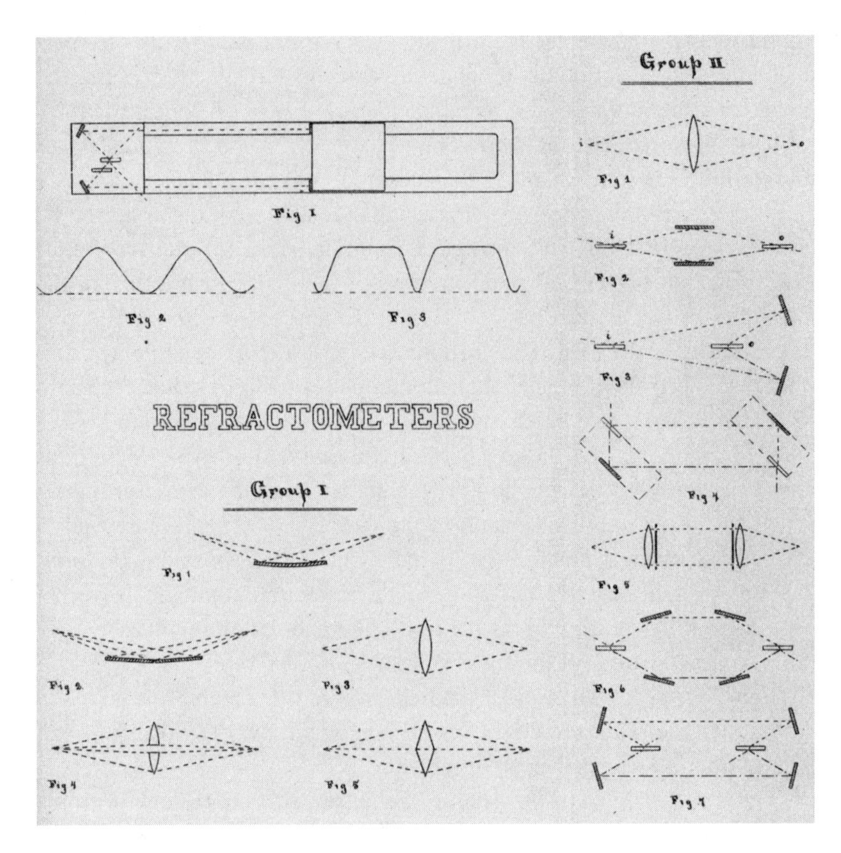

FIGURE 3.5 *(left)*

interferential refractometers took their place in the family of optical instruments as an extreme form in which resolution and definition were subordinate to accuracy. Finally, Michelson needed to establish the specific theoretical understanding that would allow him to infer the complex distribution of light in a source—or spectral line structure—from his observation of variations in fringe intensities.

Soon after publishing his 1881 ether-drift experiment, Michelson had addressed the theoretical basis for the instrument by showing that the refractometer could be treated as being equivalent to a wedge of air between two plane surfaces.[73] Michelson's work on astronomical interferometry led him into his first extended empirical and theoretical study

73. Michelson, "Interference Phenomena in a New Form of Refractometer" (1882).

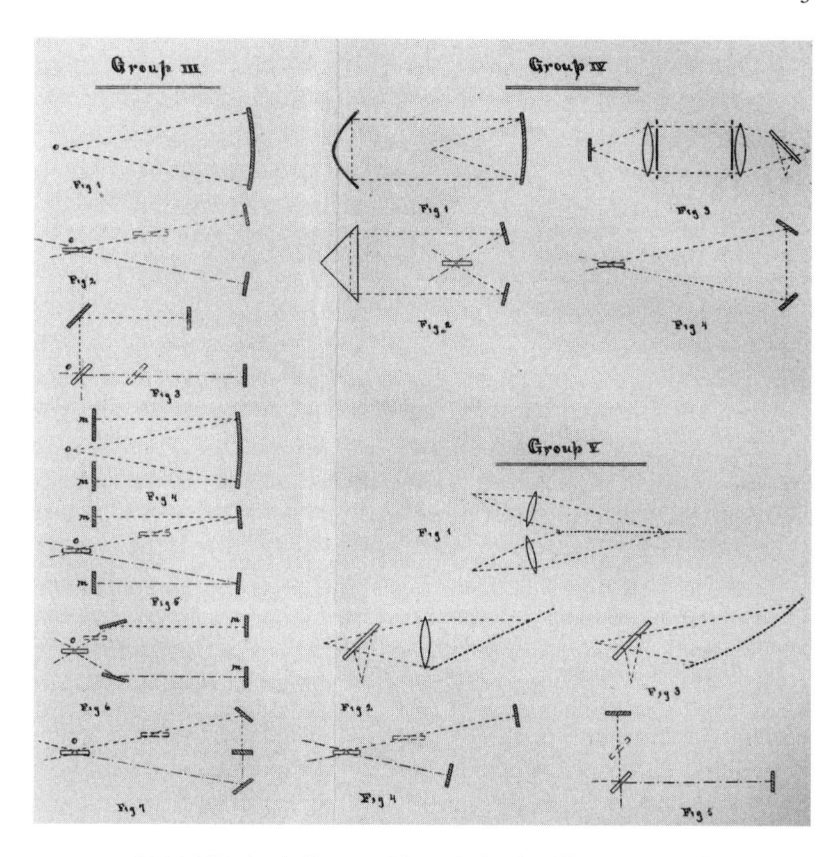

FIGURE 3.5 *(right)* Michelson's diagram of the optical paths of five groups of refractometers (*left* and *right*). Group V (*right*) illustrates instances where the paths of the two pencils are unequal, and Michelson wrote: "Figs 1, 2 and 3 represent various forms of grating spectroscope, of which the last is a concave grating in which the spectral image coincides with the slit. The transition to the corresponding refractometer forms, 4 and 5, is apparent." Michelson, "Measurement by Light-Waves," (1890), 119, plate III.

of the relations between different light sources and the observations he was taking of the changing intensity of fringes—what he described as the "visibility curve" revealed as the instrument traversed large numbers of fringes.[74] Then the considerable variations he observed in searching for a suitable light source for a new standard of length led him to address

74. The method allowed the measurement of the diameter of astronomical objects. Michelson, "Astronomical Measurements" (1890). On Michelson's astronomical work in the 1890s and 1920s, see Staley, "Michelson and the Observatory" (forthcoming); DeVorkin, "Michelson and the Problem of Stellar Diameters" (1975).

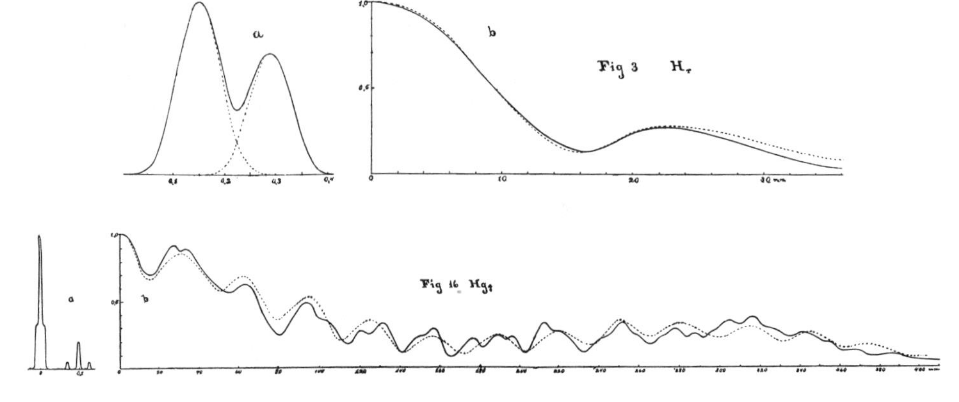

FIGURE 3.6 Two of Michelson's diagrams of the relations between visibility curves and the presumed structure of complex spectral lines. The visibility curve of the red line of the hydrogen spectra is the full curve in Michelson's fig. 3*b*; the dotted curve represents the visibility curve that would result from a double source in which the components have the intensity ratio 7:10, and in each of which the light is distributed according to the exponential law. The full curve in fig. 3*a* shows the distribution of light in the source. Figs. 16*a* and *b* illustrate the presumed structure and visibility curve of the green line in the mercury spectrum, one of the most complex he examined. Michelson, "Spectroscopic Measurements, II" (1892), plates 5 and 6. The diagrams are discussed on 287–89 and 292–93.

spectral line structure specifically. In 1891 and 1892 he published an important two-part paper, "On the Application of Interference Methods to Spectroscopic Measurement," intended to establish the theoretical and empirical framework for an interferential approach to spectroscopy.[75] In the first, theoretical, paper Michelson studied the general form for the visibility curve stemming from the superposition of two equal homogenous streams of light over different path distances, and he solved this for specific cases where the original source of light (either single, double, or multiple) differed in distribution. The measure of visibility involved comparison of the intensity of light at the center of a bright band and the adjoining dark band. In the second paper Michelson compared the theoretical values with empirical estimates of visibility, described in detail the experimental arrangements he had utilized, and related observed visibility curves to the presumed structure of a wide range of spectral sources (fig. 3.6 illustrates two sets of observations and inferred structures). Sean Johnston has pointed to a fascinating contrast between the accuracy with

75. Michelson, "Spectroscopic Measurements, I" (1891); Michelson, "Spectroscopic Measurements, II" (1892).

which Michelson could determine length by counting fringes, and the relative crudeness of his visual observation of the visibility curves of those fringes.[76] Perhaps because his instrument revealed structure invisible to other spectroscopes, Michelson was largely content to establish maxima and minima with precision and infer the general shape of the visibility curve between these points. While he continually pointed to the fact that his instrument revealed hitherto unrecognized structure, it was by establishing a new standard of length that the accuracy of the method would be expressed most fully.

Ultimate Standards: The Meter in Cadmium Wavelengths

Michelson and Morley's preparatory work on length standards caught the attention of Benjamin A. Gould, the U.S. member of the International Bureau of Weights and Measures. He prompted the director of the bureau, J. René Benoît, to contact Michelson and Morley through O. H. Tittman in the United States Coast Survey. Benoît was already familiar with interferential measurements. He had himself used an interferential dilatometer to compare the lengths of meter prototypes (employing techniques Fizeau had pioneered to measure variations in plate thickness), and also supported Macé de Lépinay's 1886 collaboration with the International Bureau to compare the weight of a cubic decimeter of water with the length of the decimeter in terms of the wavelength of sodium. Benoît suggested that Michelson and Morley continue their analysis of the incandescent vapors until they found one from a homogeneous source to serve their purpose and establish a comparison with the international meter held in Sèvres.[77] Michelson was ready to do so by 1892—without Morley. Now he could report that most of the bright lines of the spectrum were complex, but cadmium vapor disclosed one line (red) that was almost ideally simple, and two other simple lines. Using these radiations would permit successive determinations to be made without modifying the apparatus, simply by selecting a different part of the spectrum of light given off by the same source. Gould saw both the fabrication of the apparatus required and the performance of the determination itself to be significant

76. Johnston, "An Unconvincing Transformation?" (2003). See also his more general study, Johnston, *A History of Light and Colour Measurement* (2001).

77. O. H. Tittman, U.S. Coast Survey Office, Washington, DC, to J. René Benoît, 22 October 1890, B.I.P.M. Archives (copy M.Mus.). Also cited in Livingston, *Master of Light* (1973), 137–38.

achievements. When requesting Clark University to release Michelson for his journey to France, he promoted the project in the following terms: "The proposed investigation is a magnificent one; audacious, yet already proved by Professor Michelson to be feasible. The honor inuring to our country by the selection of an American Professor to carry it out, and an American artist for constructing an apparatus requiring such surpassing delicacy, is one which I am confident you will appreciate as highly as I do." In an echo of Michelson's claims for the unvarying veracity of the technique he had developed, he went on: "It is my conviction that the assent of Clark University will not only redound to its high honor, and be gratefully recognized throughout the civilized world, but will constitute an unending title to remembrance, and full appreciation in the history of science. It seems to me a just source of pride that our country should be called on to take the chief part, both scientific and technical, in such an undertaking, and I will not deny that I am considerably elated by it."[78]

Michelson carried out absolute measurements in Sèvres over nine months between 1892 and 1893. The light employed was produced by illuminating cadmium vapor kept at low pressure in a glass globe with an electric current supplied by a storage cell (the vacuum tubes Michelson developed for this purpose were later regarded as superior to arc and spark sources because they enabled close control of pressure). A homogenous radiation (red, green, or blue) was then singled out by using a prism to disperse the beam and a slit to isolate just one spectral line. The light then passed through a semisilvered mirror, being separated into two beams. One was reflected by a reference mirror, the second by mirrors on two intermediate standards placed side by side, also capable of being moved the full length of the meter along parallel ways.[79]

78. B. A. Gould to G. Stanley Hall, 17 January 1892, CUA, also quoted in Livingston, *Master of Light* (1973), 172. Similarly, in 1894 Ogden Rood praised Michelson's "magnificent work . . . a stupendous monument of genius and delicate scientific work, the like of which the world has hardly seen before." He wrote, "Everyone in America will be proud of you, and the people on the other side will begin to open their eyes, and think more of the Republic over the waters. Gibbs and Gould had talked to me a good deal about the affair, but today for the first time I saw the publication." Rood to Michelson, 16 October 1894, MC-USNA, X-49.

79. Wadsworth describes the use made of steel straight edges as guides for the carriages to ensure the requisite accuracy of movement (rather than using planed ways on the bed of the instrument itself), and details the preparation and use of the instrument in Wadsworth, "On the Manufacture of Very Accurate Straight Edges" (1894).

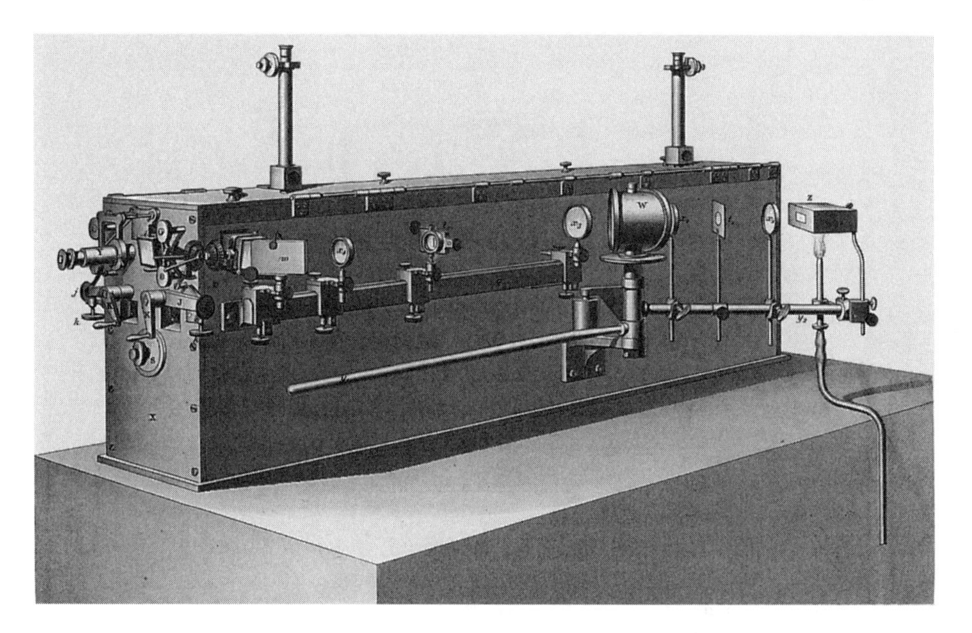

FIGURE 3.7 This image of the meter apparatus in its protective covering shows clearly the light source to the extreme right, with the combination of lenses, prism, slit, and mirrors used to isolate a single spectral line and direct it into the apparatus. Two reading microscopes rise out of the protective box, and the intermediate standards are viewed and manipulated using the telescope and handles in front of the box. Michelson, "Valeur du mètre" (1894), plate 1.

Intermediate standards were required because of the great possibility of miscounting, in any attempt to tally directly more than one and a half million wavelengths over the full distance of the meter. Michelson first counted the number of wavelengths in a small intermediate standard that bore two mirrors, with the back mirror set higher so both could be observed at once. He could set the reference mirror parallel to the front mirror on the standard using white light, and then switch to using the red line in the cadmium spectrum and count the number of fringes passed as the reference mirror was moved the distance required to achieve coincidence of length with the back mirror (1,212.37 wavelengths in approximately half a millimeter). Then he determined the precise relation between the first and a second intermediate standard that was twice as long and placed beside it on the frame. The process was repeated eight times until he knew the number of wavelengths in intermediate standard No. 9, a decimeter long. Moving standard No. 9 back ten times through its own length enabled him to establish the length of the standard meter lying

FIGURE 3.8 An intermediate standard, with mirrors at its front and back (raised) surface and showing the gearing that enabled its manipulation. Michelson, "Valeur du mètre" (1894), fig. 3.

beside the apparatus, reading its fiduciary lines through two microscopes. It was this ability to move from simply counting the number of wavelengths to the standard meter that rendered the measurement of a natural phenomena "absolute." Assuming the wavelength of light was constant allowed Michelson to regard the measure as not only natural and absolute, but ultimate.

Assisted by Benoît and taking measurements over a period of several months (between 22 October 1892 and 3 March 1893), Michelson found the meter to contain 1,553,163.5 wavelengths of the red line of cadmium, 1,966,249.7 of the green line, and 2,083,372.1 of the blue line, all in air at 15°C and at normal atmospheric pressure.[80] Michelson put the achievement this way: "We have thus a means of comparing the fundamental base of the metric system with a natural unit with the same degree of approximation as that which obtains in the comparison of two standard metres. This natural unit depends only on the properties of the vibrating atoms and of the universal ether; it is thus, in all probability, one of the most constant dimensions in all nature."[81]

80. Michelson, "Valeur du mètre" (1894).

81. Michelson, "Comparison of the International Meter" (1893), 560.

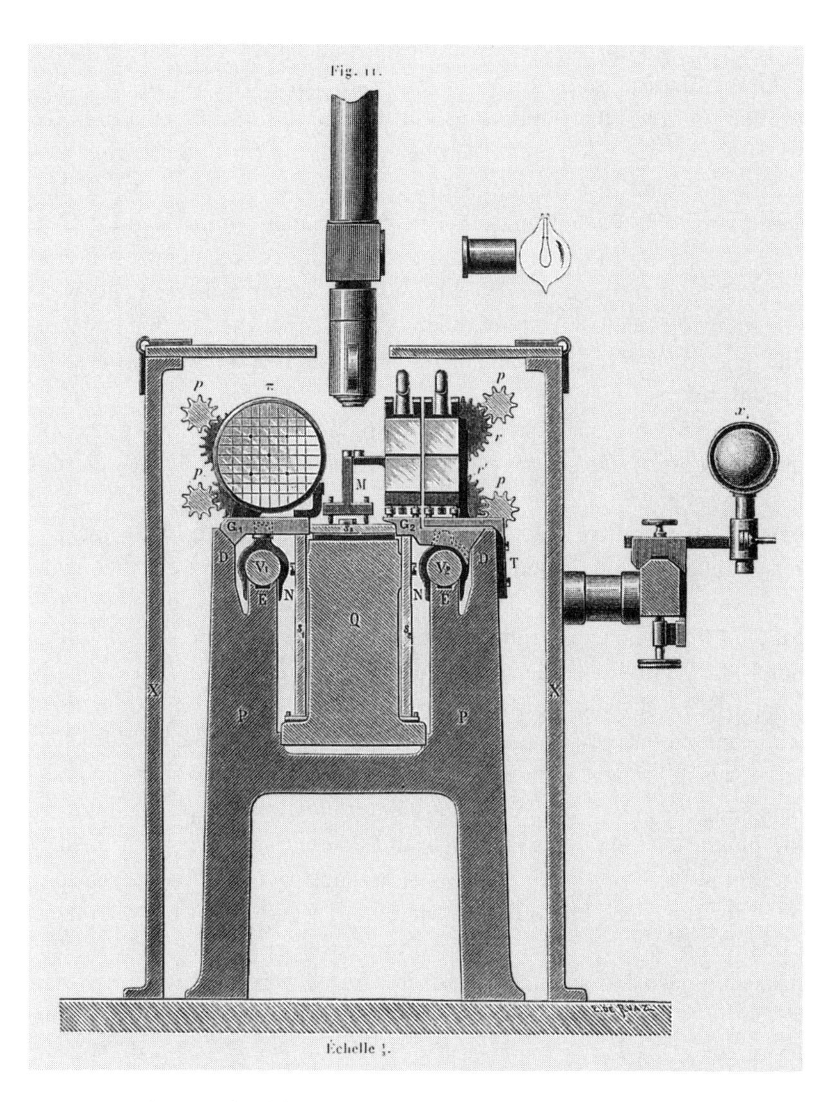

FIGURE 3.9 A front, sectional view of the meter apparatus, showing the circular reference mirror to the left with its inscribed grid, the meter bar in the center of the apparatus (on *M*, below reading microscopes), and the front and back mirrors of two intermediate standards side by side on the right side. Michelson, "Valeur du mètre" (1894), fig. 11.

He had used mirrors and precisely machined screws to move between light as a natural and ultimate measure and the rigid rod that by convention provided the standard for all customary scientific measures. To be sure, the prototype meter had been produced by a painstaking procedure intended to render it as nearly invariable in length as possible. Ironically, while Michelson's ether-drift experiment would lead FitzGerald and Lorentz to argue that all material dimensions might be altered by movement through the ether, his interferometer measurements were regarded as providing the best possibility of testing whether that emblematic rigid rod, the international meter did in fact retain precisely the same dimensions over time.

It is a measure of Michelson's exactitude that his personal notebook includes a list of suggestions for improvements, begun while the determination was in progress on 6 January 1893—and amounting to sixty-nine items. The exhaustive discussion of procedures and results in the final bureau publication also included a section setting out weaknesses that might be addressed in the future. Yet despite this attention to a possible replication, Michelson's experiment stood alone—its outcome a single set of numbers. While his earlier papers on standards of length had explicitly compared results with Rowland's values, Michelson left any comparison with contemporary rivals unstated in 1893.[82] Yet the numbers held a challenge, which Johns Hopkins physicists soon recognized. In 1896 two graduate students noted that Michelson's determinations of spectral lines (at low pressures) were lower than Rowland's measures of arc-spectrum lines (at atmospheric pressure) by 0.208 of an angstrom unit for the red line, 0.173 for the green and 0.186 for the blue. Disagreement between measures by one part in thirty thousand was unsurprising, given the dissimilar techniques and since Rowland's measures were relative rather than absolute. But it was remarkable that the differences varied, and were too great to be thought errors of observation from "either or both of these skillful experimenters." In partial explanation W. J. Humphreys and

82. For the unpublished list, see Albert A. Michelson, Notebook, about 1891–95, Mount Wilson and Palomar Observatories Library, copy held in MC-USNA, X-503.3, 37–39 and 60–64. On the comparison between methods, note that a supplement included an illustration (also published in his paper on spectroscopic methods) contrasting the resolution possible with grating and interferometric methods. Michelson, "Spectroscopic Measurements, II" (1892), plate VIII; Michelson, "Valeur du mètre" (1894), 148.

J. F. Mohler demonstrated that spectral lines were shifted by pressure changes.[83]

Michelson entered a second challenge to grating spectroscopy just a year later, this time based on his claim to discern structure invisible to others. When Pieter Zeeman described his discovery that the spectral lines of substances placed within a strong magnetic field were broadened (or doubled and tripled) in papers that began appearing in late 1896, numerous researchers rushed into the field. Now it was not wavelength but the structure and position of complex spectral lines that was at issue—and Michelson's conclusions proved controversial. To understand the dynamics of the debate that followed we must first know more about the extent to which other scientists used Michelson's results and his techniques.

On the Uses of Interferometers, 1880s–1900

Writing to Rayleigh in the mid-1880s, Michelson noted that he had repeatedly tried to interest his scientific friends in the ether-drift experiment, "without avail."[84] Although optics was likely to remain the preserve of relatively few, Michelson's 1889 address had entered an eloquent plea for colleagues. Developing his major theoretical study of interference phenomena in 1892, Michelson drew on discussions from Rayleigh, Hermann Ebert, and Éleuthère Mascart, and Rayleigh responded with an important discussion of the theoretical limitations of the technique.[85] But the experimental part of his papers on interferential spectroscopy relied solely on research carried out in Michelson's laboratory. While I have noted that Michelson used several strategies to suggest the possibility of *any* scientist working with interferential refractometry, there are a number of interesting features of the work of those who did. First, note that the earlier instrumental traditions on which Michelson drew provided a common background for a number of innovations that relied only marginally on Michelson's experiments, if at all. This is the case for Ebert's 1880s research exploring

83. The pressure change they discovered also addressed differences between solar and terrestrial lines that Lewis Jewell had noted, but it could only account for about 8 percent of the difference between Michelson and Rowland's values. See Humphreys and Mohler, "Pressure" (1896), 114.

84. Michelson to Rayleigh, 6 March 1887, RA-HAFB.

85. Rayleigh, "Interference Bands" (1892).

the nature of different light sources by studying high path differences using a combination of a Newtonian lens and a piece of plane parallel glass. It is true also of adaptations of Jules Jamin's apparatus to measure refractive indexes introduced by Ludwig Zehnder and Ludwig Mach in the early 1890s. Like Michelson before them, both wanted to separate the light paths widely and used a combination of glass plates to do so (though in Zehnder's case these were unsilvered). Neither made any reference to the American's work.[86] Similarly, in 1898 Ernst Pringsheim drew on the earlier work of J. A. Sirks to develop an interference microscope for biological research that modified Jamin's arrangement by using slightly wedge-shaped plates of glass.[87]

These German physicists developed related optical techniques largely independently of both the specific arrangements Michelson had advanced and the broad intellectual stamp he sought to give a new form of instrumentation. Nevertheless, a number of researchers were closely aware of his studies, several saw his results on ether drift to provide a critical contribution to understandings about the relations between ether and matter, and some followed his plea to use interferential methods. Several features stand out. First, theoretical interest began widening at the time Michelson started outlining his broader program in the late 1880s and early 1890s. Second, with isolated exceptions others began to use similar techniques only from the late 1890s onward (a period in which Michelson had begun to use the term interferometer to describe his instruments). Third, they rarely undertook close repetitions of Michelson's earlier work. Fourth, they often had rather close relations with Michelson himself. Together, these factors illustrate subtle tensions between the broader intellectual environment that conferred interest in his results, the more specific research environment that conveyed sufficient resources to make repetition possible, and the fact that the direct communication of specific skills might still be required to facilitate its achievement. They also indicate the extent to which actual economies of performance in the sciences revolve as much around varied elaborations as they do around the methodological ideal of precise replication. In a case where the resources involved are rare,

86. Ebert, "Wellenlänge des Lichtes" (1887); Ebert, "Hohen Interferenzen" (1888); Ebert, "Helligkeitsvertheilung" (1891); Zehnder, "Interferenzrefraktor" (1891); Mach, "Interferenzrefraktor" (1892). The Mach-Zehnder interferometer has been used in aerodynamic research to monitor changes of refractive index (and therefore density) in compressible gas flows.

87. Pringsheim, "Interferenzmikroskop" (1898).

scientists' research strategies are likely to reflect perceptions about the trustworthiness of the work concerned, their assessment of the cost (and benefits) of repetition in terms of time and the cultivation of specific resources, and judgments about the significance of the techniques or results at issue. The test of direct replication is perhaps the most solid foundation for the continued use of a scientists' work, but it is by no means the only one. As we shall see, results, methods, and interpretations survive over time for a wide variety of reasons, but they remain vulnerable to challenge when they depend on rare resources.

As well as providing theoretical counsel that was immensely valuable to Michelson, Lord Rayleigh occasionally drew on Michelson's instrumental innovations. Rayleigh's wide-ranging interests in optical phenomena had found their earliest focus in experimental and theoretical investigations of optical instruments and especially diffraction gratings from the 1870s. Through the 1880s and 90s he contributed to theoretical discussions of both grating spectroscopy and refractometry, as well as using what Oliver Lodge described as "a beautiful interference arrangement of Michelson" in 1888 experiments to analyze the effect of electrical current on the refractive index of liquids. However, in the experiments on the refractive properties of argon and helium that later won him the Nobel Prize, Rayleigh introduced a modification of Jamin's refractometer using a lens and two parallel vertical slits to make comparative measures of the refraction of light passing through two parallel tubes.[88] Soon after the turn of the century he also undertook experiments on double refraction of light that were inspired by the negative results of the Michelson-Morley experiment.[89]

Ether-drift experiments remained important because of the challenge they presented to consistent theories of optical phenomena. George FitzGerald and H. A. Lorentz both independently suggested that the null result could be explained by a contraction in the length of the apparatus

88. Rayleigh, "Argon and Helium" (1896), 201–2. Born and Wolf describe this as superseding the Jamin interferometer. Born and Wolf, *Principles of Optics* (1999), 299–302 and 344.

89. Strutt and Stacey discuss Rayleigh's optical research: Robert John Strutt, *Life of John William Strutt* (1924), 86–98; Stacey, "Rayleigh's Legacy" (1994). For the "beautiful" arrangement, see Rayleigh, "Velocity of Light in an Electrolytic Liquid" (1888); Lodge, "Sketch of the Electrical Papers" (1888), 623. Rayleigh's double refraction experiment is Rayleigh, "Double Refraction" (1902). On the relationship between Michelson and Rayleigh, see Howard, "The Michelson-Rayleigh Correspondence" (1967); Shankland, "Rayleigh and Michelson" (1967).

(or dilation in its width) that was caused in some manner by motion through the ether and that exactly compensated the expected ether drift. FitzGerald postulated such an effect in a letter to *Science* in 1889, and pressed Oliver Lodge insistently enough to lead him to publicize it further in an 1892 address on "Aberration Problems and New Ether Experiments."[90] Lorentz developed a contraction hypothesis in 1892, incorporating it somewhat uneasily in a broader theory of the optics and electrodynamics of moving bodies. Lorentz took Maxwell's electromagnetic field equations as basic, assuming their validity in an absolute ether—an ether at rest. In a major innovation, he introduced charged bodies, positive and negative ions, that mediated the relations between the ether and matter. Given the assumption of a stationary ether, Lorentz had to show why no ether wind was detected as a result of the earth's motion, and developed a set of equations detailing "corresponding states." They explained why optical experiments betrayed no evidence of the earth's motion to first order in the ratio of the velocity of the earth to the velocity of light; and in 1892 his postulation of the length contraction to account for Michelson and Morley's more accurate second order result still seemed somewhat arbitrary.[91]

Experimentalists responded to this continued theoretical interest in ether drift. In Britain, Oliver Lodge used an interferometer to carry out an exploration of the influence of moving matter on the velocity of light in his "whirling machine." The optical instrument was placed between two disks of metal set in rapid motion, and Lodge obtained a negative result.[92] Michelson himself varied the conditions involved by undertaking an ether-drift experiment with apparatus extending up the wall of his Chicago laboratory in 1897, thereby testing whether height affected the experiment to at least a limited extent. Now he described the null result

90. FitzGerald, "The Ether and the Earth's Atmosphere" (1889); Lodge, "Aberration Problems" (1893), 749–50. See Brush, "FitzGerald-Lorentz Contraction" (1967); Bork, "'FitzGerald' Contraction" (1966).

91. Michelson wrote to Rayleigh: "I would like to have your opinion on the rather startling tho perhaps not at all improbable explanation of the negative result we found." Michelson to Rayleigh, 17 May 1893, RA-HAFB. On Lorentz, see Janssen, "Lorentz's Ether Theory and Special Relativity" (1995); Janssen, "Reconsidering a Scientific Revolution" (2002); McCormmach, "Lorentz" (1973).

92. Note especially Lodge's discussion of the care needed to establish a negative result. Lodge, "Ether and Matter" (1897), 149. On Lodge, see Clow, "The Laboratory of Victorian Culture" (1999). On the Whirling Machine experiment, see Hunt, "Experimenting on the Ether" (1986).

as offering potential evidence for the view that "the length of all bodies is altered (equally?) by their motion through the ether."[93] Thus new theoretical interests helped subtly change the framework in which ether-drift experiments were carried out; and Michelson and others elaborated new arrangements rather than undertaking simple repetitions. Writing from Melbourne, in 1898 William Sutherland called for renewed repetitions of the experiment, but Oliver Lodge immediately addressed his claim that lateral shifting of the wave fronts could plague experiments in which the light traveled independent pathways.[94] As we shall see in chapter 5, the most significant experimental concern came from 1900 onward, stimulated by the critical attention the subject received at the International Congress of Physics.

A number of American physicists took up refractometer studies without centering their research efforts in the field. In general they possessed the instrument through collaborating with Michelson (often prior to one or the other moving elsewhere), or as a result of Michelson's kindness. The former is the case for Morley and Rogers's use of an interferometer in measuring the expansion of metals, E. S. Johonnott's use of the interferometer to calculate the distance between water molecules in a fine soap film, and S. W. Stratton's later transfer of an interferometer to the Bureau of Standards when he became director of that institution in 1902. Scientists who received interferometers as gifts include Holden at the Lick Observatory and Rayleigh, whose double refraction experiment was repeated in an improved version by DeWitt Bristol Brace at the University of Nebraska in 1904.[95] Less direct relationships were also reflected in George Carey Comstock's use of an interferometer to discern the effective wavelength of starlight at the Washburn Observatory of the University of Wisconsin in Madison in 1897 and a similar interest in Michelson's techniques in the physics department of that university. The first PhD in physics at Madison was completed by John Cutler Shedd in 1899. He used

93. Michelson, "Relative Motion" (1897), 478.

94. Sutherland, "Relative Motion of the Earth and Aether" (1898); Lodge, "Mr. Sutherland's Objection" (1898).

95. For Morley, see Morley and Rogers, "Expansion of Metals" (1896); Evans and Warner, "Precision Engineering" (1988). Johonnott, "Thickness of the Black Spot in Liquid Films" (1899). Morley was offered but declined the directorship of the new bureau, which among other duties took over the work of the Office of Weights and Measures of the Coast and Geodetic Survey. In 1899 Michelson had asked Stratton to campaign on his behalf for the position, and felt betrayed by Stratton's move. On Brace, see Cahan and Rudd, *Science at the American Frontier* (2000).

Michelson's interferometer in spectroscopical studies of the Zeeman effect, a topic we shall take up shortly.[96]

In the Anglo-Saxon world at least, the instrument had not yet traveled much farther than Michelson had been able to take it, a situation that reflects sociologists and historians emphasis on the difficulties of replication when skills and resources are in play that are rare or difficult to communicate. But in fact a significant school of interferometric techniques and new variations of the instrument was being built up in France. This also seems to have developed at least in some measure through the impetus Michelson gave the subject, and through personal contact with Michelson on his two visits to Paris in 1882 and 1892–93. We noted that in 1886, Macé de Lépinay had undertaken interference measures of the wavelength of sodium and collaborated with the International Bureau of Weights and Measures in relating the volume of a small quartz cube to the definition of the standard kilogram as the mass of a cubic decimeter of water. In a step that indicated the bureau's commitment to Michelson's wavelength standards, they first developed proposals using Michelson's methods to determine the millimeter and centimeter in terms of cadmium wavelengths. Then in 1896 and 1897 P. Chappuis adapted a Michelson interferometer to measure the dimensions of a 5 cm glass cube in cadmium wavelengths.[97]

Just as significantly, at the same time Michelson began his work in Sèvres, Charles Fabry completed his 1892 thesis in Macé's Marseilles laboratory on the "Théorie de la visibilité et de l'orientation des franges d'interférence." That subject had already been discussed in a series of exchanges between Mascart and Rayleigh in the *Philosophical Magazine* in 1889; and in 1896 Fabry developed a new form of interferometer with Alfred Pérot. Their instrument used the multiple reflection of light between two plane parallel, slightly silvered mirrors to produce interference characterized by extremely sharp maxima and minima. While initially designed to measure the thickness of the plates of glass involved, Fabry and Pérot were able to relate wavelength to distance by recognizing that the instrument transmitted only specific wavelengths, depending on the

96. Comstock drew on work in which Karl Schwarzschild developed an adaptation of Michelson's astronomical interferometer to measure double stars, having been asked to give a report on Michelson's procedures by Seeliger. Schwarzschild, "Messung von Doppelsternen" (1896). See Susalla, "The Old School in a Progressive Science" (2006).

97. Comité international des poids et mesures, *Procès-verbaux* (1897), 61–64 (determinations of the millimeter and centimeter) and 66–73 (cube).

distance between the plates. By 1899 they had described two forms of the device. One was adapted to measuring distance in terms of wavelength, and they used this form to determine the dimensions of Macé's quartz cube (using cadmium radiations). In the second form, the distance between the plates was kept constant while the angle at which the light was incident was varied, in order to deliver high-resolution spectroscopic information about its structure.[98]

In summary, we can see that despite considerable interest in ether-drift experiments and the relations between ether and matter, Michelson's results rarely faced direct challenge, and also that with the significant exception of the related work of Marseilles physicists and the International Bureau of Weights and Measures, no body of practitioners was built up refining approaches to the interferometric study of light. To some extent this accords with Michelson's early aim to provide results that would not need repetition, measuring the velocity of light, ether drag, and ether drift. Developing light as a standard of length Michelson established a similar relationship to the community of physicists in general and spectroscopists in particular. While he wrote variously of providing a "practical" and "ultimate" standard, his standard of length had two surpassing features. As the wavelength of a particular spectral line it offered a natural unit depending only on the properties of the vibrating atom and the universal ether, ultimate in the sense that it constituted "in all probability, one of the most constant dimensions in all nature."[99] In principle, any scientist using the same technique in any suitable location could generate such a standard. Nevertheless, the determination used a method of surpassing accuracy. To be sure, by 1900, Chappuis, Fabry, and Pérot were using similar

98. For the exchange between Rayleigh and Mascart, see Rayleigh, "Limit to Interference" (1889); Rayleigh, "Faint Interference-Bands" (1889); Mascart, "Achromatism of Interference" (1889); Rayleigh, "Achromatic Interference-Bands" (1889). The new interferometer was described in Fabry and Pérot, "Nouvelle méthode de spectroscopie intéferentielle" (1898); Fabry and Pérot, "Mesure de petites èpaisseures en valeur absolue" (1896). In a general discussion in 1899 they compared their method of measuring large distances with Michelson's, implying that it would be more convenient. Fabry and Pérot, "Méthode interférentielles" (1899). The concrete metrological study was described in Fabry and Pérot, "Source intense" (1899); Fabry, Macé de Lépinay, and Pérot, "Dimensions d'un cube de quartz de 4 cm de côte" (1899); Fabry, Macé de Lépinay, and Pérot, "Masse du décimètre cube d'eau" (1899). Charlotte Bigg has offered an account of the French tradition and Fabry and Pérot's research in Bigg, "Behind the Lines" (2002), 22–28 and 34–45.

99. Michelson, "Comparison of the International Meter" (1893), 560.

methods in related fields of research. But in each case, the measurements undertaken drew on Michelson's work and values without yet *testing* those values—for they measured a 4 or 5 cm cube in cadmium radiations without either taking the further step of examining what their results meant for the relations between the international units of length and mass, or extending the distance measure twenty-five-fold to redetermine the international meter. As a result, in the absence of direct replication or competitive measures using other techniques, Michelson's relations with the physics community were centered on the strategically important but somewhat isolated position of providing an ultimate standard expressed largely by its metrological values. Spectroscopic research on the Zeeman effect opened up a very different situation.

The Interferometer and the Diffraction Grating

When Pieter Zeeman found in late 1896 that exposing the light source to a magnetic field produced a broadening of the spectral lines emitted, both his Dutch colleague Lorentz and the British mathematical physicist Joseph Larmor quickly showed that the finding could be integrated into their respective ionic theories of matter—given the important assumption that the vibrating ions giving rise to the spectral lines had a charge to mass ratio that was significantly larger than that of hydrogen. Remarkably, this meant that if the charge was of the same magnitude as that of hydrogen, the ion concerned was smaller than hydrogen itself. This extraordinary result and the close confluence of experiment and theory greatly enhanced interest in their approaches to electrodynamics.[100] Experimentalists too rushed to repeat and extend Zeeman's observational work, and Michelson saw the chance to press the "particular adaptation" of the interferometer to discerning differences in spectral structure. In 1897 he published a paper describing several observations at variance with Zeeman's early reports, and especially a doubling rather than broadening of the spectral lines in almost every case he had examined.[101] Zeeman responded enthusiastically, describing Michelson's method as "beautiful." He thought,

100. For a discussion of the subtle interplay between theory and experiment in the early phases of the discovery, see Arabatzis, "The Discovery of the Zeeman Effect" (1992).

101. Michelson, "Radiation in a Magnetic Field" (1897).

"there can certainly be but one opinion as to its particular advantages in such cases," and suggested how differences in their observations could be resolved.[102]

However, a dispute soon erupted—evidently Zeeman's assessment of the advantages of Michelson's method was *not* universally shared. In 1898 Michelson described two sets of four laws to summarize his observations of several spectral lines from eight different substances exposed to magnetic fields of different strengths. The first set had been verified on observations of more than a dozen lines and included the statement that all lines are tripled when radiations emanate in a magnetic field, and, further, that their separation was proportional to the strength of the field, and was approximately the same for all colors and for all substances.[103] This second claim led to a strong challenge from the Dublin spectroscopist, Thomas Preston. Three years earlier Preston's authoritative textbook on optics had expressed his skepticism about interferometrical measures of wavelength. Preston described Michelson and Morley's work with mercury and thallium chloride lines, but also argued that interference methods of determining wavelength were inferior (since they involved the exact determination of some very small length, while the diffraction method lent itself more readily to linear measurement and afforded very pure spectra).[104] Now Preston mounted a major challenge in the pages of *Nature*. His attack reveals the extent to which the reputation of instrument and inventor were related. Preston wrote:

> Investigations demanding special attention are those of Prof. A. A. Michelson, both on account of his reputation as an original investigator and by reason of

102. Michelson observed a doubling of lines when light is emitted along the lines of force. Zeeman initially referred only to a broadening but had since resolved the doublet or triplet and praised Michelson's method for allowing the observation with relatively weak magnetic forces. Zeeman accounted for a more serious variance by suggesting that being unaware of the polarization of the light emitted across the field of force had led Michelson to discern doublets rather than triplets in that direction (with the polarized light being weakened by reflection from the planes of glass). Michelson accepted the explanation. Zeeman, "Doublets and Triplets, II" (1897), 258–59.

103. Michelson, "Radiation in a Magnetic Field" (1898). An instrument Michelson had developed with Stratton aided the analysis of visibility curves. Along the lines of Lord Kelvin's tidal analyzer it was capable of completing Fourier analysis (or the inverse process) mechanically. See Michelson and Stratton, "A New Harmonic Analyser" (1898).

104. Preston, *Theory of Light* (1895), see 152–53, and 238–39 on measurement of wavelengths.

> the nature of the apparatus which he employed. Working with his interferometer, Prof. Michelson concluded some years ago [citing Michelson, "Spectroscopic Measurements, II"] that the spectral lines themselves instead of being, as ordinarily supposed, narrow bands of approximately uniform illumination from edge to edge, are on the contrary in most cases really complexes, some of them being close triplets, and so on. This structure has never yet been observed by means of any ordinary form of spectroscope, and accordingly it has been suggested that it does not exist in the light radiated from the source, but is imposed on the spectral lines by the apparatus used, namely, the interferometer. Be this as it may, the application of this instrument to the study of radiation phenomena in the magnetic field is highly interesting.[105]

Preston was clearly skeptical about the way Michelson had modified his claims in response to others' observations, but focused on the "most surprising" statement that the separation of lines in the triplets produced was independent of both the spectral line and the substance. Since in all other observations the separation of the components varied considerably, Preston questioned the instrument: "That the interferometer has led to such a law as that announced by Prof. Michelson, shows that there is some peculiarity in the instrument not yet taken into account—or else that by chance Prof. Michelson has happened to confine his observations to lines which give approximately the same separation; yet this latter could not easily be done." The challenge was serious, repeated when Michelson replied to point out that his paper had itself noted exceptions to the law (he also discounted diffraction as a possible source of the difficulty). Preston regarded the further structure the interferometer revealed as mainly real, rather than spurious effects imposed by the instrument, but suggested that if Michelson maintained his law, either the cause of the error had to be determined in the instrument, or the interferometer had to be standardized so that it could be employed as a measuring instrument.[106]

105. Preston, "Radiation Phenomena in the Magnetic Field" (1899). Note also his more extended study in Preston, "Radiation Phenomena in the Magnetic Field" (1898). See also Weaire and O'Connor, "Unfulfilled Renown: Thomas Preston" (1987).

106. Preston also rehearsed the kind of change we have seen in Michelson's response to Zeeman and noted that subsequent observations had proved that tripling occurred across the field of force, and that the constituents of the triplets were themselves multiples (Michelson's second set of laws had been devoted to the latter findings, and he joined Cornu in pointing to an unanticipated multiplicity of spectral lines). Preston, "Radiation Phenomena in the Magnetic Field (Letter to the Editor)" (1899). He was responding to Michelson, "Radiation in a Magnetic Field" (1899).

At this point Rayleigh intervened with a supportive comment that nevertheless highlights Michelson's relative isolation. Rayleigh wrote: "The questions raised by Mr Preston can only be fully answered by Prof. Michelson himself; but as one of the few who have used the interferometer in situations involving high interference, I should like to make a remark or two. My opportunity was due to the kindness of Prof. Michelson, who some years ago left in my hands a small instrument of his model." Rayleigh allowed that visibility estimates and the deduction of the structure were "delicate matters," and concluded by describing a modification he had introduced to obviate the need to readjust the traveling mirror after each change of the distance.[107] No doubt it was helpful to remind Preston that Michelson was not totally alone in his use of the instrument, even if only Michelson could speak directly to the point; and Rayleigh's concentration on the practical operation of the instrument also served to highlight its sensitivity. Preston's next letter showed how the dispute was resolved: Michelson had written admitting that the law had probably been generalized from insufficient data. This being the case, Preston's confidence in the instrument was reestablished.[108]

Preston's skepticism extended even to Michelson's standards work—which relied crucially on establishing unusually simple conditions, using a homogeneous light source that rendered interference visible over a long distance. But interferential methods were far more vulnerable in the more complex situations typical of spectroscopical research. Here Michelson had to deliver results coherent with those achieved by other means, or risk the charge they were instrumental artifacts. That so few others used his instrument created a difficult situation. In cases of conflict, Michelson's reputation alone would not suffice—and Preston's question about standardization was particularly pointed, for Michelson's determination of the meter had given absolute wavelengths for cadmium radiations different from those Rowland established. Which instrument was to be trusted, the grating familiar to so many, or Michelson's interferometer?

107. Evidently the ways in Michelson's small instrument were insufficiently true, and Rayleigh described how he had circumvented this problem by using as a reflector a fluid surface that automatically set itself rigorously horizontal after each movement. Rayleigh, "The Interferometer" (1899). Michelson knew the instrument very well, having corresponded with Rayleigh about remedying the defect and having borrowed it to provide demonstrations for a paper he gave in Paris. See Michelson to Rayleigh, 3 October 1892, 11 March and 17 May 1893, RA-HAFB.

108. Preston, "The Interferometer (Letter to the Editor)" (1899).

Michelson had two characteristically instrumental answers to the challenge Preston articulated. Note that responding to Zeeman and Preston, Michelson quickly bowed to *their* results—rather than insisting on his findings and arguing the relations between his observations of visibility curves and inferences about spectral structure. Michelson thereby preserved the ultimate status of the interferometer (its accuracy in the extreme) by deferring to existing spectroscopic practice—without attempting to modify his interferometric techniques.[109] His most important answer to the difficulties Preston dramatized was to offer two new instruments capable of addressing the middle ground he ceded to gratings. The first was his 1898 invention of a new instrument using parallel plates of glass to form a spectroscope of very high resolving power that nevertheless avoided some of the disadvantages of the interferometer. The "echelon" quickly became an important adjunct in high precision spectroscopy, with scientists such as Zeeman and Preston turning to it in order to supplement their work with Rowland gratings (Preston ordered an echelon spectroscope but was unable to use it before he died at age forty in 1900).[110]

Pursuing another tack and working with Stratton in the same year, Michelson built a small ruling engine using interferometric methods to correct the screw (an idea that had occurred to him while measuring the meter in light waves). While the instrument could only rule 2 1/2 inch gratings, it was a prototype for a larger engine already underway. Michelson

109. Completing his PhD at the University of Wisconsin, J. C. Shedd briefly took up that task, but published only sporadically on interference methods after moving to a position at Colorado College in fall 1900. After reviewing the advantages and disadvantages of the micrometric measurement of grating photographs and interferometric visibility curves, Shedd concluded that unless Michelson's echelon spectrograph proved of equal value, in experienced hands the interferometer is "undoubtedly the most powerful instrument of attack." Noting, however, that "under circumstances less favorable than those enjoyed by Professor Michelson, it is difficult to see how the Interferometer Method as he uses it, can be successfully used," Shedd outlined modifications he had made to render the instrument "more available." See Shedd, "Interferometer Study, I" (1899), 7; Shedd, "Interferometer Study, II" (1899). Shedd's work was also published in German: Shedd, "Strahlung in einem magnetischen Felde" (1900); Shedd, "Anwendung des Interferometers" (1901); Shedd, "Formen" (1902).

110. For Preston's intent, see Preston to Rayleigh, 16 March 1899, RA-HAFB. Zeeman describes the echelon as an "exceedingly ingenious and original kind of grating" and wrote of his admiration for both the ingenuity of the idea on which it was based and "the accuracy to which the art of optics has been carried at present." See Zeeman, "Some Observations on the Resolving Power of the Michelson Echelon Spectroscope" (1902); Zeeman, *Researches in Magneto-Optics* (1913), 13–16 on 13 (originality) and 15 (optics).

now wanted to improve on Rowland's ruling engine and gratings. With typical ambition he hoped to rule a 12-inch grating "with almost theoretically accurate rulings."[111] Achieving this would constitute a most effective demonstration of the superior accuracy of interferometric techniques. Michelson devoted much of the rest of his career to the attempt, constructing one small and two large ruling engines with which he did rule several large gratings. Characteristically, however, both the fact that the diamond ruling point itself wears in the process (rendering it difficult to reliably repeat results), and Michelson's attention to experimental refinement rather than routine production, meant he never consistently produced gratings to rival Rowland's.[112] Long after his death, Michelson's two large ruling engines were modified by G. R. Harrison at MIT and by the optical firm Bausch and Lomb to incorporate interferometric control of the ruling process. Michelson's machines did rule a new generation of diffraction gratings—but in other people's hands.[113]

111. Michelson, "The Echelon Spectroscope" (1898), 112. See also the description of the ruling of diffraction gratings and Michelson's method of detecting periodic and run errors interferometrically in Michelson, *Studies in Optics* (1995 [1927]), 99–103, and Michelson, "The Ruling of the Diffraction Grating," ms., no date (probably early 1915), MC-USNA, X-8. From similarities in design, Evans and Warner infer that Michelson may have drawn on the ruling engines of Rogers in designing his own. Evans and Warner, "Precision Engineering" (1988), 5. See Michelson, "Ruling Diffraction Gratings" (1905); Michelson, "A Ten-Inch Diffraction Grating" (1915).

112. For an example of research carried out with one of Michelson's gratings (with a ruled surface 6 1/2 inches long), see Gale and Lemon's paper comparing grating analysis of the principal lines of mercury with measurements taken by echelons, and Michelson's original interferometer studies of 1892. The authors write: "Aside from the very satisfactory results obtained with the grating the most interesting feature of the work was the verification, to such a remarkable degree, of the results obtained by Professor Michelson in 1892. It should be remembered that those results were absolutely the first in this field, that they were obtained by a method requiring such a high degree of personal skill that no one has since used it with success, and that the whole investigation was, in a sense, incidental to the more important task of determining the number of wave-lengths in the meter." Gale and Lemon, "Analysis of the Principal Mercury Lines" (1910), 82.

113. See Harrison, "Diffraction Gratings, I" (1949). Michelson's successful gratings had been ruled on engine #1, and two years before his death he ordered his workshop to find out what was wrong with #2. For a description from within the workshop, see T. J. O'Donnell, "The University of Chicago Physical Sciences Development Shops Applied Telescopic Optics and Interferometry: Albert A. Michelson's Light Wave Measuring Rod: The Cosmic Unit of Metrology. TOD 620, 9th July 1951, Revised 144," Section AA p. 1, in UCL. Evans has described "The Chicago School" and "The Birth of the Modern Age" (with the MIT modifications) in Evans, *Precision Engineering* (1989), 109–10.

Conclusion

Paying close attention to instruments in these two chapters has first generated insight into the origins of Michelson's ether-drift experiment in 1881 and then helped explain why he left the 1887 repetition incomplete, in favor of pursuing an ambitious research program based on rendering light a standard of length and promoting a new class of instruments. Recovering the original context for Michelson's research in this way has also demonstrated just how much both Michelson and Henry Rowland owed their colleagues in the astronomical community, helping us link experimental physics with neighboring disciplines through patronage and materials. Working with light and mechanics, Rowland and Michelson coopted concerns and instruments that moved between precision industry and astronomical observatories on behalf of the emerging physics discipline. As historians like Simon Shaffer have begun to show, the endeavor to set standards with scientific accuracy is so characteristic of the leaders of laboratory physics in the late nineteenth century as to be a defining feature of the period. Further, the specific engagement of Rowland and Michelson in ruling engines and length standards clearly reflects the particular strengths of American manufacturing. But I have also stated here that these scientists pursued standards in fundamentally different ways, with rather different implications for their relations to other researchers. In Michelson's case, despite promoting a general class of instruments, the extreme nature of optical interferometry meant his disciplinary relations were largely mediated by experimental values alone, right through to 1900. This section concludes our reconstruction of the way Michelson defined himself as a physicist by considering three interrelated features of his disciplinary profile: the changing nature of his social relationships; the relations between experiment and theory in Michelson's work; and his sense of the disciplinary transformations facing physics as the century turned.

Michelson's social relations with disciplinary colleagues developed in two major phases, which we can describe by highlighting the contrast they form to those of Albert Einstein—who was born in 1879, the year in which Michelson's first major experiment was completed, grew up in a family of electrical entrepreneurs, and finished four years study of mathematics and physics in the Swiss Federal Polytechnic in Zurich in 1900. We are used to thinking of Einstein as an outsider, at least early in his career. While Einstein's relations with a few fellow students were extremely close, those with his immediate professors were often difficult. As a result, a great deal of his thought was formed in close engagement with the work of distant figures whom he read avidly and critically—especially Ludwig Boltzmann, Poincaré, Lorentz, and Max Planck. In contrast, as we saw in chapter 2, in the 1870s and 80s Michelson quickly formed strong and often lasting relations with people like Commander Sampson, Newcomb, Pickering, and Rayleigh—scientists who became powerful mentors. Further, unlike the young Einstein, Michelson traveled in order to meet members of his discipline, in particular visiting those whose strengths were in the velocity of light and optical research more generally. To be sure, Einstein expressed a great deal of independence by following his family from Munich to Italy as a fifteen year old (rather than finishing his schooling alone in Germany), and then going on to become a student in Switzerland. These early travels undoubtedly contributed to his socially cosmopolitan outlook; but although Einstein began publishing papers in the premier German journal of physics even before he found employment in the Swiss Patent Office in Bern in 1902 (and together with his correspondence after 1906 this made him less of an outsider than usually appears), Einstein did not meet the major figures in his field until he went to his first disciplinary conference in 1909.

In this chapter we have seen that as Michelson became more established he developed a more austere disciplinary profile, in which immediate colleagues, collaborators, and the names of his instrument makers were all somewhat less important than they had been in his early research. This was related to his focus on optics (as a field of less immediate practical and industrial concern than energy physics or electricity and magnetism), to his need to promote generic instruments, and to his singular approach to definitive standards. As the nineteenth century drew to a close, Michelson enjoyed a remarkable reputation. Nevertheless, to most of his contemporaries Michelson's work with interferometry was likely to appear somewhat marginal even if it offered extremely accurate techniques of measurement.

In 1900, for example, the German Paul Drude's important textbook on optics discussed Michelson's interferometer as an instance of high retardation in light path. Drude wrote that the result for a particular light distribution was not pressing, even if the distribution of the intensity and breadth of individual spectral lines could be represented with a certain approximation, which was better than that of the spectroscope or a diffraction grating.[114] Similarly, entries on Light and Spectroscopy in the tenth edition of the *Encyclopædia Britannica* both discussed refractometry. Hale's article on spectroscopy moved through gratings to discuss the echelon spectroscope and interferometers, describing the latter as a most valuable adjunct to the spectroscope and the best-known means of measuring absolute wavelengths.[115] Such articles and textbooks on optics typically included Michelson's interferometry as one of a wide range of more or less useful techniques, without considering the implications of the null result of ether-drift experiments for optical theory. They illustrate a practical sense that Michelson himself clearly expressed in 1896, when he wrote to Johns Hopkins University to beg off an engagement to speak on "The Ether." Having first accepted the task, he now found he had too much work. Michelson was too polite to leave President Gilman completely in the lurch though. He wrote that if his withdrawal "would very seriously embarrass the management and you would approve the change—I might give a talk on 'Applications of Light Waves.'"[116] One interesting feature of this exchange is the implication that Michelson was far more comfortable dealing with the pragmatic features of his experimental program than with discussing more speculative theoretical approaches. Despite his reticence on this occasion, when Michelson expressed his sense of the promise of modern science three years later, it was closely linked to his understanding of the ether.

Characteristically, given the chance to deliver a series of eight lectures to the Lowell Institute in Boston in 1899, Michelson described his own research, speaking on "Light Waves and Their Uses" along lines we are now familiar with.[117] Published as a book in 1902, Michelson's lectures have become best known (indeed, they are now notorious) for their statement of extreme confidence in the essential completion of the foundations of physics, and Michelson's conviction that future discoveries would come in studies to the sixth decimal place. He wrote:

> The more important fundamental laws and facts of physical science have all been discovered, and these are now so firmly established that the possibility of their ever being supplanted in consequence of new discoveries is exceedingly

114. Drude, *Lehrbuch der Optik* (1900), 139–45 on 145. He also described Michelson's echelon (211–15) and referred to his study of the Zeeman effect (411).

115. Knott, "Light" (1902); Hale, "Spectroscopy" (1902), 776–77.

116. Michelson to President Gilman, 4 December 1896, HAR-JHU, Box 7.

117. Michelson, *Light Waves* (1902), 3.

> remote. Nevertheless, it has been found that there are apparent exceptions to most of these laws, and this is particularly true when the observations are pushed to a limit, i.e., whenever the circumstances of experiment are such that extreme cases can be examined. Such examination almost surely leads, not to the overthrow of the law, but to the discovery of other facts and laws whose action produces the apparent exceptions.[118]

In retrospect, it appears remarkable that Michelson's confidence in the stability of fundamental laws had survived such novelties as the discovery of X-rays, and his own participation in two remarkably open fields of research: ether drift and the Zeeman effect. But Michelson's description of the developments that Zeeman's research set in train illustrates perfectly why he was so confident in present foundations. Even that "startling" discovery had rapidly been furnished with a beautiful and simple explanation by Lorentz, Larmor, FitzGerald, and others, based on the longstanding assumption that particles or atoms of matter were associated with an electric charge. Michelson told his audience that such a charged particle, already invoked in explanations of electrolysis, had been termed an *electron*. Then he explained how the effect of a magnetic field on the period of revolution of particles revolving in specific directions would increase or decrease the frequency (and hence shift the spectral line emitted), while leaving the vibrations of electrons revolving parallel to the lines of force of the magnetic field unaltered—thus producing a triplet of lines, of different polarizations. Michelson's narrative went on to show that this explanation of new phenomena using long-known principles had given a new focus to his own experimental research. But in their turn Michelson's interferometric observations indicated still more complexity—including the doubling of the middle line and multiplication of the side lines. So increasing precision—science in the extremes—showed the simple explanation no longer applied; and to date every attempt to deduce some general law covering all known cases had proved unsuccessful.[119] Earlier Michelson had stated that the extraordinary complexity exhibited by spectral lines might dismay the Keplers and Newtons hoping "to unravel the mysteries of this pigmy world," but added: it "certainly increases the interest of the problem."[120] Thus the Zeeman effect showed that even in the pigmy

118. Ibid., 23–24. The reference to the sixth decimal place follows.
119. Ibid., 111–18.
120. Ibid., 82.

world, experiment and theory continually negotiated a delicate interplay between existing knowledge and new effects.

The second major area in which Michelson focused on conflicts between theory and experiment came in his concluding chapter, on the ether. There, Michelson described the considerable difficulty in understanding the contradictory properties that the medium of light must possess and outlined his ether-drift experiment. Its null result showed a difficulty "in the theory itself," which "has not yet been satisfactorily explained." Michelson was clearly doubtful about the contraction hypothesis and conveyed his regret in proving unable to detect the ether by describing the experiment as "historically interesting" because it resulted in the new interferometer, which "more than compensated for the fact that this particular experiment gave a negative result."[121] One feature of Michelson's stance is his clear recognition of disciplinary specializations. Whether or not he thought this went along with distinct social and institutional divisions, Michelson was ready to leave theoretical questions on the nature of the ether to others—like Gibbs or Rayleigh—who he would not hesitate to consult when appropriate.

We will soon see that many of Michelson's contemporaries shared his identification of the Zeeman effect and the status of the ether as significant open questions. Despite pointing to still open questions in this way, at the turn of the century Michelson associated scientific *modernity* with present convergence and a grand, almost utopian hope for unity that was best expressed by the universal ether.[122] In 1888 he had contrasted the present with previous eras by comparing the "barbarous tallow candle" and "semi-barbarous" gas lamp with the wonders of electricity.[123] Then he had conveyed a sense of progress by adorning the things of the past with descriptive adjectives. In 1899 Michelson concluded his lectures by entering into the "sea" of conjectures and hypotheses that surrounded inquiries into the nature of the ultimate particles of ordinary matter—and now it was present science and investigation that won adjectival elaboration:

> Suppose that an ether strain corresponds to an electric charge, an ether displacement to the electric current, these ether vortices to the atoms—if we continue these suppositions, we arrive at what may be one of the grandest generalizations of modern science—of which we are tempted to say that it ought

121. Ibid., 158–59.

122. On the function of such anticipations of completion, see Schaffer, "Utopia Unlimited" (1991).

123. Michelson, "A Plea for Light Waves" (1889), 67.

> to be true even if it is not—namely, that all the phenomena of the physical universe are only different manifestations of the various modes of motions of one all-pervading substance—the ether.
>
> All modern investigation tends toward the elucidation of this problem, and the day seems not far distant when the converging lines from many apparently remote regions of thought will meet on this common ground. Then the nature of the atoms, and the forces called into play in their chemical union; the interactions between these atoms and the non-differentiated ether as manifested in the phenomena of light and electricity; the structures of the molecules and molecular systems of which the atoms are the units; the explanation of cohesion, elasticity, and gravitation—all these will be marshaled into a single compact and consistent body of scientific knowledge.[124]

For Michelson, as practically wedded to a new class of instruments and the uses of light waves as he was, *modern* science and *modern* investigation were clearly best exemplified by knowledge of the ether.

124. Michelson, *Light Waves* (1902), 162–63.

PART II

Physics in 1900

The World's Fair and Congress of Physics, Paris 1900

To date we have followed a single man at work with light and instruments. Now we switch focus radically, turning to his community in the year 1900. We shall not leave Michelson entirely—for his colleagues continued to find uses for both his results and his instruments—but this second part of our study will take advantage of an extravagant exhibition and a remarkable congress to explore how physics appeared on the broadest stage that an era could give. The prospect of marking the turn of the century at an Exposition Universelle in Paris presented rich opportunities. For governments, industrial firms, entrepreneurs, and individual scientists alike, the World's Fair offered an unparalleled theater in which to promote the goods and visions they could offer for other people's futures. And everything promised enterprise on a scale never before realized. Fair organizers budgeted for a projected 60 million visitors through seven months. This was substantially more than the entire population of the host nation (39 million), and dwarfed even the 39 million who had visited the most recent universal exposition in Paris, the 1889 fair with its lasting image in the Eiffel Tower. In 1899 the French Physical Society, led by Alfred Cornu, sent out letters canvassing the possibility of hosting an international conference in conjunction with the exhibition. The enthusiastic response Cornu received from physicists throughout the world encouraged him to set a new goal. The 1900 Congress of Physics would transcend the annual meetings of national science societies or the occasional conferences on specific subjects that had previously brought physicists together. It would fashion a disciplinary vision of physics as a whole. The following two chapters take the opportunity of the Exposition Universelle and the inaugural International Congress of Physics to explore science on display and a discipline at work.

We need some map to guide our tour and study. Chapter 4 begins with the framing perspectives embodied by the entrance to the Paris fair before examining the exhibits of those instrument makers we have already met through our study of Michelson's research, exploring now how they represented themselves in Paris and sought to win customers in Europe. Of course we shall indulge in comparative speculation, throwing more than a sideways glance at the competition American firms experienced—especially when we join Henry Adams, the great American historian who visited the Paris fair in order to study the forces of history. Then on Monday the 6th of August 1900, chapter 5 enters the sessions of the physics congress. There we take good note of keynote lectures and the catalogue physicists gave their subject before examining in detail the numerous ways in which Michelson's work was discussed—and listening in on several papers devoted to the properties of the new ion, the particle many are beginning to call the *electron*.

In visiting the World's Fair and examining the proceedings of the physics congress I am driven by the need to establish critical breadth. Like all investigators, historians' work is informed by more or less explicit principles and organizational structures. We are often guided by single careers and biographies, as in the way I have worked with Michelson to date; or we take advantage of particular events, theories, or research fields to delimit our inquiries and structure our narratives (describing the discovery of the electron or the development of relativity, for example). Similarly, in order to be able to draw secure conclusions about the grasp of particular ideas in different contexts, historians often examine closely the work of particular laboratories and research groups, or specific national and institutional communities. As the following chapters dramatize, this book deliberately moves through and across all these highly valuable forms of organization in order to gain a more distributed footing in that diffuse and variegated entity—a disciplinary community. Examining participation in the Paris World's Fair and international congress will enable us to approach the physics community whole cloth and thereby to develop further two major themes we have already begun to explore.

The first concerns how instruments or results travel. Investigating this question in relation to Michelson's research before 1900, we noted significant features of his work to propagate interferometers as a class of instruments and observed also patterns in the way others used his research—times and places like Germany in the 1890s where Michelson's earlier contributions were hardly recognized, and others like France and the United States, where his work was readily taken up by several researchers. Such observations showed that the fortunes of interferometry and ether drift depended on the existence of particular cultures of instrumentation and traditions of research in specific places. Whether they occurred in several nations or not, local concentrations of expertise conveyed the instrumental resources, research techniques, and intellectual interest required to actively cultivate the tools and insights Michelson offered. Here we examine two of the more formal means that businesses and scientists have often used to propagate their work further—international exhibitions and conferences. In many ways, the appearance of an individual's work in these arenas rendered it representative, often being taken to embody both national achievement and international scope at the same time. Here we will observe the varied ways in which different firms and scientists called on both these representational registers in Paris; and we explore also what the displays of instruments in the exhibition and discussions of results at the 1900 congress can teach us about the interrelations between different traditions of research and national communities of physicists. My aim will not be to describe general links between research styles and nations (though contemporaries did occasionally contrast, for example, British and French styles of mechanics) but to establish how these specific events helped physicists cultivate the wider use of particular resources.

The second major theme concerns perceptions of the character of disciplinary change in 1900. Even our tour of the exhibition will offer some insight into this theme. Although we can only gather a handful of snapshots, it will be helpful to recognize how a few visitors to Paris perceived science in general and physics in particular. These brief glimpses give valuable indications of the faces of physics in broader cultural realms in 1900, and we shall pay particular attention to the senses in which people used terms like *classical* and *modern* in the period before relativity won the status of inaugurating worldview changes. Our study of the Physics congress then offers a detailed exploration of how Michelson's colleagues described the nature of physics at the turn of the century. By examining general addresses by Poincaré and Kelvin, and more focused papers from Lorentz, J. J. Thomson, and others, we explore the broad concepts of change they used and the specific fields on which they focused attention. This will show that by 1900 ether-drift and electron theory had emerged as critical and related fields of inquiry.

CHAPTER FOUR

Science on Display: The World's Fair

Scenes from Paris

The machine has taken over control of the world, declared the socialist minister of trade Alexandre Millerand at the opening of the exposition on April 14, 1900. Even more important than the recent retreat of death through the progress of medicine is the service science provides to the material and moral greatness of states. For Millerand, science held the key to a solidarity that should link individuals and nations alike. Summoning the effects of social-welfare institutions on the one hand and the promise of new technologies on the other, he spoke of the search to fuse individual groups into a unified entity and thereby counter the weaknesses inherent in the individual:

> Ideas, feelings fuse and cross all over the world, just like the thin filaments on which human thoughts are transported at lightning speed. The more the international relations resulting from the diversity of human needs and the ease of exchange become intertwined, the more reason we will have to hope and trust that the day will come when the world realizes that peace and the honorable struggles of human labor reap greater benefits than rivalry.[1]

1. Kraemer, *XIX. Jahrhundert* (1900), 18–19. This chapter owes a great debt to my involvement in two museum exhibitions and the work of my colleagues Robert Brain and

FIGURE 4.1 René Binet's *Porte Monumentale* depicted as it was illuminated at night, from Picard, ed., *Rapport général*, vol. 2: *4. ptie.* (1903), facing p. 12.

The exhibition would surely put diversity and exchange and peace and rivalry on show. As president of the republic, Emile Loubert opened its proceedings under the auspices of the hope that a new stage in the slow evolution of work toward happiness and man toward humanity would soon be completed. Millerand and Loubert may well have found their opening themes—and the message that work will liberate mankind from the bondage of darkness—on their way into the exposition itself. Visitors entered through a monumental gate with a symbol of Paris at its summit and one of electricity at its feet.

The 30-meter-high Porte Monumentale was designed by the architect Rene Binet, who drew the inspiration for its filigreed form from the intricate natural fretwork and spiky globes of radiolaria shells depicted in an Ernst Haeckel book on the artfulness of nature. Two minarets drew

Simon Schaffer: Bennett et al., *Empires of Physics* (1993); Brain, *Going to the Fair* (1993); Bennett et al., *1900: The New Age* (1994); Schaffer, "Time Machines" (2006). Mandell has given an overview of the 1900 exhibition, and Geppert of European exhibitions. Mandell, *Paris 1900* (1967); Geppert, "Welttheater" (2002).

FIGURE 4.2 Salammbô, the electric deity, as depicted in *Encyclopédie du siècle. L'Exposition de Paris de 1900* (1900), 185. Whipple Library, University of Cambridge.

visitors toward a central arch surmounted by a bejeweled representation of the Lady of Paris (modeled on Sarah Bernhardt by Paul Moreau-Vauthier). She was clothed in a robe of sapphire blue—modern dress. While some applauded the courage to depart from the customary antique plaster goddesses, many commentators thought the assertive, almost masculine stance of the statue verged on vulgarity. Below her, a larger-than-life frieze showed an army of workers from different trades hastening toward the common goal, all bearing the products of their industry and dressed in common working garb. Again, observers noted the modernism of the frieze. The group channeled incomers toward the great archway and under a huge open cupola. There, naturalism was deserted as visitors passed between stern and mysterious representations of the Egyptian priestess Salammbô, personifying *electricity*. The *Encyclopédie du siècle* described her as the queen, even the divinity of the present day. Depicting that divinity it was once again necessary to summon the remote impassivity of the hieratic forms mankind had left behind. Both helpful and terrible, nothing was more dangerous than her ire, should one forget the precautions she required. Standing on electromagnets, attributes of her power, the electric deity was clothed in strange jewels. Her metal carapace was reminiscent of the worrying machines of laboratories, with their brilliant coppers and remarkable blown glass: apparatus that could not be approached without apprehension because it concealed lightning. Remote and ambiguous, the calm expression of the goddess conveyed the indifferent, latent force of an uncontested sovereign.[2] Those visitors arriving in the evening would have felt her beneficent side, for the gate was illuminated by 3,116 light bulbs, with twelve arc lamps on the dome and minarets supplemented by eight spotlights and sixteen reflectors: a fiery lattice of emerald green and blue that glimmered with violet flashes.[3]

Contemplating the symbolic form of electricity and bathed in her fey light, visitors completed the business of entrance by handing in their ticket and passing through one of thirty-six turnstiles designed to allow 60,000 people to pass through each hour. On one peak day in October,

2. *Encyclopédie du siècle. L'Exposition de Paris de 1900* (1900), 188–90. Michelson published his own elaboration of Haeckel's work in Michelson, "Form Analysis" (1906).

3. On the entrance, see Muthesius, "Eingang" (1900); Schivelbusch, *Licht, Schein und Wahn* (1992), 17, as cited in Kretschmer, *Geschichte der Weltausstellungen* (1999), 144; Morand, *1900* (1931), 65. The reader will notice a preponderance of German sources in my discussions of Paris 1900. In part, this reflects the unusual degree of German interest in (and dominance of) the event, discussed below.

652,000 entered the exposition, and by the time the affair was over in mid-November, over 50 million people had experienced its offerings.[4] One hundred and twenty hectares of exhibits awaited in Paris, with nearly as much space devoted to railway, transport vehicles, horticulture, and sport at a second site 6 miles away in Vincennes. Intellectually the 83,047 individual exhibits were organized in a new division of groups. For the first time, Works of Art moved above the engineering disciplines to take second place behind Education and Instruction at the peak of the hierarchy. Eighteen different groups were themselves subdivided into 121 different categories. The new hierarchy showed that the ideals of engineering and industrial display had relaxed their hold, but the profusion of categories made any organizing principle hard to discern.[5] The earlier exhibitions of the French Third Republic had been dominated by the primary aim of promoting technology and commerce by standardization, democratization, and extending the free market. Now, with a general waning of enthusiasm for rationalization and mass production, art and cultural traditions were valorized. As commissioner of the exhibition, Alfred Picard exhorted exhibitors to display side by side the continuum from raw materials through manufacturing processes to finished products. Describing the exposition as the "balance-sheet of the nineteenth century," he ensured that the prospective dreams inspired by the year 1900 would be set off by the official theme of a century in retrospect. Organizers were directed to prepare exhibits highlighting the historical development of their field in the last century. In the face of decades of economic growth and rapid change, the task was often felt to be so daunting that it was sheer folly to attempt it.[6]

4. The total number of visitors was 50,860,801 at an average of 250,000 per day, and exhibition revenues were 126,318,168 francs for a profit of 7,000,000 francs. It ran between 15 April and 12 November 1900, with forty-three foreign states and twenty French colonies participating. A survey of statistics of the exhibition was published in "Statistique de l'Exposition" (1900). The 1900 total was not exceeded until 51 million visited the New York World's Fair in 1964–65.

5. After Education and Works of Art came The Appliances, Instruments and Accessories of Science and Art, and the technical fields of Machinery, Electricity, and Civil Engineering and Transportation. Then followed branches of trade and commerce such as Agriculture, Forestry, Mining and Metallurgy, and Textiles and Clothing. Finally, National Economy (Hygiene and Social Welfare), Colonization, and Defense completed the major groupings.

6. In the case of the arts this retrospective imperative had an unexpected effect. The newly built Grand Palais housed two different exhibitions, one of which reviewed the last decade. Curated by a jury dominated by the conservative Académie des Beaux-Arts, on the one hand this display of recent art featured works that were traditionalist in conception. On the other

The Nuts and Bolts of U.S. Display

While most visitors to Paris came to alternately study its lessons and be distracted by its gaudy entertainments, thousands of firms, institutions, and individuals came to show and sell. The abstract possibility of winning new customers was one incentive. The certain reality that visitors would weigh your wares against the best the world could advance was a second inducement. And for those who submitted their work to formal competition in different classes, there were concrete prizes on offer. Specialist international juries met to evaluate the instruments, machines, and goods on show, distributing awards that might grace the catalogues and advertisements of successful makers.[7] John A. Brashear, Brown & Sharpe, Thomas Edison, Pratt & Whitney, and Warner & Swasey all made the journey from the United States. Joseph Brown and Lucian Sharpe knew well that such travel offered advantages going far beyond simply finding new markets. Having exhibited in fairs since 1862, at the Paris Exposition in August 1867 they saw the "screw caliper" Jean Laurent Palmer had patented in 1848. Palmer used graduations on both a stationary barrel and revolving thimble, and Brown & Sharpe saw a way of adapting this to micrometer gauges for sheet metal production. They subsequently began a successful line of what they marketed as "pocket sheet metal gauges" until they introduced the term "micrometer caliper" in their 1877 catalogue.[8]

Class 15 in Group 3 showcased Instruments of Precision, Coins and Medals, and was housed in the Main Building on the Champs de Mars. The director of the U.S. Department of Liberal Arts and Chemical Industries was Alexander Capehart. He noted proudly that the U.S. display was distinguished by presenting the only complete series of astronomical instruments shown in a single section of the exposition. He described it as a result of the "energy and generosity" of the firms and

hand, the centennial display was curated by Roger Marx and surprised visitors by highlighting impressionist paintings as representative of great French art. See Chandler, "Culmination" (1987). On classification, history, and Picard, see Bennett et al., *1900: The New Age* (1994), 21–23.

7. Clemens Riefler, for example, rushed a reference to the Paris 1900 Grand Prix they won into their advertisement of mathematical instruments in the September to December 1900 issues of the premier German physics journal, *Annalen der Physik*.

8. Sharpe, "Development of the Micrometer Gauge" (1892); Uselding, "Measuring Techniques and Manufacturing Practice" (1981), 117–18.

FIGURE 4.3 The 1887 catalogue of Darling, Brown & Sharpe depicted the awards won by the two different branches of the company at the 1878 Paris Exposition with this figure. In Cope, ed., *A Brown & Sharpe Catalogue Collection* (1997), 129. In contrast, the title page of Brown & Sharpe's 1899 *Catalogue No. 101* simply listed the leading awards received to date, 135. Courtesy Astragal Press.

individuals involved, but it was also a reprise of displays the same firms had shown at the Chicago World's Columbian Exposition in 1893.[9] The Warner & Swasey exhibit featured an equatorial telescope with an 11-inch objective, as well as several smaller telescopes, binoculars, and a sextant. (And Capehart pointed out that all the circles on their instruments had been graduated on a specially constructed dividing engine, with a maximum error in the automatic gradation less than 1 second of an arc.)

9. Capehart [Director], "Report" (1901), on 64–65. Looking ahead to the Columbian exhibition, Brashear wrote to Hale: "Regarding an exhibit for the Columbian Exposition I am not yet decided I do want to make a good exhibit if my means will allow me to. A friend has offered me one thousand dollars to purchase raw materials, and I would I think furnish specimens of the leading characteristics of my work. Say a 15 in object glass, some smaller ones, a 10 or 12 in reflector, some smaller ones, prisms, diffraction grating. Quartz and Rock salt lenses & prisms, a Laboratory spectroscope, Helioscope, micrometers, small concave spectroscope etc. I am sure I could make a nice exhibit of it if I can only raise the funds, and I have not much doubt that I could sell everything by the time the exhibit was over. / If I do exhibit it will likely be in conjunction with Warner & Swasey as they have talked over the matter very seriously several times." Brashear to Hale, 30 March 1891, YOA, Yerkes Observatory Directors' Papers, Box 1, Folder 1. Later he wrote: "I almost regret now that I undertook this fair business, as it is going to be quite costly, and I seriously doubt its commercial value, in fact that is a secondary matter to me, as our work is so widely known that we do not need any advertisement in that line, but it would almost be a shame not to have an exhibit in this line of a truly American character." Brashear to Hale, 29 March 1893, YOA, Yerkes Observatory Directors' Papers, Box 3, Folder 9.

Charles Hastings contributed isochromatic glasses, and Henry Rowland showed diffraction gratings. John A. Brashear completed the group with a whole range of instruments that included spectroscopes, photographic telescopes, visual objectives (of 11 and 6 inches in diameter), plane parallel sextant mirrors, and a collection of prisms and oculars. In addition, the firm exhibited several instruments that we now know rather well: refractometer plates (plane and parallel) and diffraction gratings with rulings 15,000 lines to the inch. Clearly Americans' showings highlighted the strength of their astronomical community; and instrument makers brought the results of their work with Michelson and Rowland to the fair. Michelson's own instruments were also on show in Paris—exhibited in a University of Chicago display. Somewhat reluctantly, Michelson had agreed to the shipment of an interferometer, an echelon spectroscope, and a harmonic analyzer. Attending the exhibition, the president of the university, William Harper, noticed that the equipment had not yet been unpacked. On his return to Chicago he instructed Michelson's younger colleague Robert Millikan—on his way to the Congress of Physics—to put the instruments on view. The University of Chicago won a grand prize.[10]

A second major feature of the U.S. displays in Class 15 was Brown & Sharpe's exhibition of accurate measuring instruments, standard gauges, micrometers and verniers, and index and surface plates. Capehart's formal report listed twenty-seven U.S. exhibitors, with a large number showing adding machines and optical instruments. Of this contingent Brashear and Warner & Swasey both received grand prizes for their astronomical exhibits and instruments (Columbia Phonograph Company and the American Graphophone Company received the same distinction for phonographs). Hastings and Brown & Sharpe received gold medals, and Edison also won a gold medal for his phonographs. Honors went to individuals as well as firms: Jason B. MacDowel and William B. Hartwell won gold medals as collaborators for their part in the instruments Brashear displayed, while G. Fecker and W. E. Reed received similar honors for their work with Warner & Swasey. In all, of the 357 awards distributed to 483 exhibits in the class, American firms won twenty-one prizes or honorable

10. On Michelson's instruments, see Jaffe, *Michelson and the Speed of Light* (1960), 130. Michelson responded to that success with a letter to Harper: "I thank you sincerely for your own share in the work and appreciate the wisdom of your course in the matter of sending on the apparatus in the face of many obstacles—my own objections included."

mentions; and eight collaborators were distinguished.[11] Despite the undoubted satisfaction of these honors, we shall see below that one remarkable achievement overshadowed the U.S. performance. The American juror John Rees joined with Britain's C. V. Boys to suggest that the German exhibit as a whole deserved a special grand prize—a distinction the rules and regulations did not allow.[12]

More significant U.S. attention was devoted to Group 4 Class 22, Machine Tools for Wood and Metal. Preparations had begun several years earlier. They were facilitated by a final staff of thirteen under Francis E. Drake's control as director of the Department of Machinery and Electricity. Many more machine tools firms had applied for space than companies in other classes, largely as a result of acknowledged American preeminence in the field (many firms had already developed European markets, or desired to extend their markets). Drake held organizational meetings in New York and Chicago. There, participating firms accepted his suggestion to establish a small display headquarters in the Champ de Mars, and agreed to foot the bill to build an annex to the official exhibition in Vincennes, where more extensive displays of machines at work could be set up. Besides the additional space, the primary advantage of the more distant site was the commercial possibilities it might engender: Vincennes housed the major transport exhibitions. As Drake wrote, "in the estimation of your director the railroad, automobile, and bicycle industries in Europe are among the largest purchasers of machine tools." Out there, in a building modeled on a typical machine shop but festooned with pennants, the 167 U.S. firms exhibiting would be away from the eyes of the merely curious, able to cultivate the interest of aficionados.

Brown & Sharpe did not exhibit everything they built, but showed several sizes of the principal types of machine, noting that all came from regular stock without being finished for exhibition purposes. Twelve of their twenty-one machines were belted so they could be seen in motion, and eight of these were in action producing work. One automatic screw machine produced three collar buttons from a brass rod every minute.

11. Capehart listed twenty U.S. awards. According to the official reckoning, 142 of 159 French exhibits won awards, while 78 of 111 German, 24 of 26 Russian, 21 of 33 United States, 16 of 28 British, and 15 of 22 Italian exhibits also won. Of sixty-one Grand Prizes the lion's share went to France (21) and Germany (16). See Capehart [Director], "Report" (1901), on 29–30, 41 and 44; Picard, ed., *Exposition universelle, Pièces annexes* (1903), 634 and 747.

12. Rees, "Instruments of Precision" (1901), 180.

Thirty-five thousand were given away to interested visitors. Drake noted that the machine Brown & Sharpe displayed for determining the diameters and length of standard gauges could measure as closely as 1/10,000 of a millimeter and proved "of much interest to those familiar with such matters."[13]

Pratt & Whitney showed what Drake described as "a most complete and interesting exhibit." Of course their standards instruments were prominent among the many lathes and machines. They showed a 600-mm standard measuring machine; standard cylindrical size gauges and templates; standard drop-forge steel caliper gauges; international standard screw-thread gauges (6 to 50 millimeters diameter); standard reamers, for machine shop and locomotive work; standard taper pins; and international standard taps and dies. Like Brown & Sharpe they won a grand prize for their efforts. Warner & Swasey gained a gold medal for some of their "well-known high-grade machine tools." In total, ninety-seven U.S. firms won awards (rivaling the 107 awards that 191 French exhibits in the class received, and far exceeding the twenty and seventeen awards won by thirty-six and twenty-one British and German exhibitors, respectively).[14] Perhaps the greatest sensation surrounded the exhibit of the Bethlehem Steel Company, set up near Pratt & Whitney's display. Impatient to get things going early, the firm's workmen had dug a foundation themselves, making sure their heavy lathe would have a stable footing. Dwight Merrick ran it *fast*. Operating at twice to three times the normal speed, the hot cutting tool glowed a dull red and turned out blue chips of the metal being worked that were so thick they had to be broken up by hammer. They hung from a railing in looping spirals, and visitors took them as souvenirs. This was the "Taylor-White high speed cutting process" in action, working metal at rates experts wouldn't believe until they had seen it for themselves. The fame Frederick Winslow Taylor won in Paris helped provide

13. Drake, "Report" (1901), 125. The firm exhibited in four different locations. Their precision instruments in Class 15 were shown in the liberal arts (Group 3) section of the Main Building on the Champ de Mars. A small selection of machine tools was exhibited farther down the great hall in the Machinery section, while the bulk of their wares were shown in Vincennes. Finally, hair clippers were displayed in the building for diverse industries on the Esplanade des Invalides.

14. Drake's report offered an extended description of the main feature of the Warner & Swasey exhibit, a hollow hexagon turret lathe, which was a new machine for turning iron and steel bars up to 2 inches in diameter and 24 feet long. Ibid., 153 and 157–59. The statistics come from Picard, ed., *Exposition universelle, Pièces annexes* (1903), 643 and 750.

a platform for his widespread promotion of the scientific management of tools—and men.[15]

The United States had also been allocated 5,000 square feet in the *salon d'honneur* on the Champ de Mars. In keeping with the official theme of the exhibition, Drake set it apart for a retrospective "museum of science" using original apparatus or models to display the part American genius had played in mechanical and electrical fields. The assistant director, James S. Anthony, curated the exhibit, but Drake wanted to ensure the facade would display the requisite taste. Containing so many objects of particular interest, the entire display called for individuality. Indeed, he had something special in mind—but had trouble achieving it. Drake reports it was extremely difficult "to convey to the decorators the idea that the exhibits which were to be housed would admit of classic treatment, although the pure classic would be too severe to produce a proper effect." The architectural style with which Drake conjured was mostly associated with a venerated past and high culture. In many contexts, like that of secondary education, "classical" schooling was explicitly contrasted to more recent (and less prestigious) forms of training that focused on "modern" languages and business. So could you treat the products of mechanics and electricity—so often symbols of modernity—in a classic style? We will explore the full significance of fin de siècle invocations of the concepts Drake sought to combine in chapter 7. For now, note the cognitive difficulty of mixing classical forms and new industries, but, remembering the monumental gate, observe also that Paris saw surprising juxtapositions that suggest how open and fluid the connotations of terms like *classical* and *modern* were in 1900. A successful compromise meeting Drake's full approval was finally achieved (fig. 4.4).[16]

As elegant as the facade was and as interesting as the small collection of instruments proved, the German visitor Albert Neuburger found their labels highly unsatisfactory. Neuberger wrote an extended description of physics and chemistry at the exhibition for a review of science and culture in the nineteenth century. He reported that objects in the U.S. display were described in a barely legible scribble on small cards hung from the wall, often with their backs to the viewer. What was purportedly the first

15. See Kanigel, *One Best Way* (1997), 341–46; Nelson, *Frederick W. Taylor* (1980), 86–91.

16. Drake, "Report" (1901), 124–26 and 136. More generally, the architecture of the exhibition was often described as backward looking. Rather than keeping faith with the functionalism of iron and glass, it had become an El Dorado for stuccowork and Academy architects of the classical school. See Kretschmer, *Geschichte der Weltausstellungen* (1999), 144.

FIGURE 4.4 The "classic" facade of the historical exhibition on electricity in the U.S. government display. Kraemer, *XIX. Jahrhundert*, vol. 4 (1900), 231.

telephone from Alexander Graham Bell was on show, and Neuburger noted that the rival claims of the German inventor Philipp Reiss were simply hushed up.[17]

In contrast to the creditable and representative showing put up in machine tools, even their organizers found U.S. exhibits in the electrical groups far less successful. Although General Electric and Westinghouse participated, most firms had declined to attend. The typical excuse was a preoccupation with present orders and the discouraging difficulties that freights and duties placed in the way of competing in European markets. Many had licensing or financial agreements with European firms, which also prevented the kind of participation Drake desired. So he put Warren E. Weinsheimer in charge of organizing a collective exhibit "of various small, though not unimportant electrical and mechanical specialties."[18]

17. Neuburger, "Chemie und Physik" (1900), 231–232. Notable displays included A. B. Chaudton's sender and Edison's receiver of the first telegraphic dispatch to encircle the earth directly without being repeated (in 1896), and recent instruments capable of sending script and photographs by telegraph.

18. Drake, "Report" (1901), 132–33.

Preeminent Cooperation: The German Society of Mechanics and Optics

Collectivity was not always used to disguise weakness. German participation in the great international exhibitions in Europe had been curtailed throughout the last quarter of the century. In response to the anti-German sentiment that followed the defeat she inflicted on the French in 1871, the new nation declined to take up the invitation France issued to the world to attend the Paris exhibitions of 1878 or 1889. But as early as 1892 the French campaign to hold a fifth, turn-of-the-century *exposition universelle* was kick-started by the news that Germany was planning an international exhibition in 1896 and, even more worryingly, that she was considering holding one in Berlin in 1900.[19] French politicos thought the chance to set the scene on home ground was too important to cede the occasion to a rival. Cosmopolitanism and the ideology of free trade had distinguished the first International Exhibition in the Crystal Palace in 1851. By now, commentators like Julius Lessing thought that the broad enchantment of those early days had given way to "cool calculation." Decades of protectionism and the representative nature of national delegations had seen a new emphasis emerging. Lessing noted, "while no one in 1851 held high the proposition that the government has nothing to do with this occasion, now one has come to see the government of each individual country as the real entrepreneur."[20] Thus, when the German nation took part in the Chicago world's fair in 1893 it did so with something to prove. By the time preparations were made for 1900 in Paris, the stakes had been ratcheted up yet another notch.[21]

An organizational change gave the 1900 show an unusually cosmopolitan aspect. In comparison to earlier exhibitions, the display halls were now oriented around objects rather than nations, giving precedence to the specific class of exhibits as the primary organizational principle. But for anyone who still wished to evaluate the performance of different peoples, the segmentation in fact facilitated a more rigorous and valid comparison of national achievements. Now one could readily judge like against like:

19. Chandler, "Culmination" (1987).

20. Lessing, "Weltaustellungen" (1900); as translated in Brain, *Going to the Fair* (1993), 161.

21. German participation in the U.S. exhibitions is studied in Fuchs, "Deutsche Reich" (1999).

the agricultural machinery of French, British, and American manufacturers stood side by side, for example. And each nation still had an opportunity to display a unified and representative image of their land and wares, if in less space, with the erection of palaces along the Rue de Nations on the bank of the Seine.[22] Famously, the German Kaiser delicately complimented the host nation by filling the German Palace with his favorite artwork, paintings of the French masters. But it was also known that he took personal interest in far more than the fine arts and even whispered that he had visited Paris incognito several times in order to supervise all the preparations.[23] In several key fields, German contributions to the exposition showcased national strengths by putting on cooperative, collective displays—in the main grounds of the exhibition. Their efforts stood out in the classes devoted to chemical industry, and optics and mechanics.

Perhaps the chemistry exhibits offered the most extreme showing of distinctively national endeavor from German industry, with 800 square meters of displays set apart from the surrounding space in cabinets of a uniform style, and featuring labels that described the instruments and products on display without even identifying the firms that produced them. Otto Witt gave his assessment:

> It has been unanimously recognized by all concerned circles, both domestic and foreign, that the collective exhibition of the German chemical industry was by far the greatest, far exceeding all the other outstanding objects in the chemical section of the Centenary Exhibition in Paris in its completeness and unity as well as its scientific depth, and that it was one of the most interesting and significant displays in the exhibition as a whole.[24]

Admittedly, as editor of the official catalogue of the German empire, *Geheimer Regierungsrat*, and professor of chemistry at the Technische Hochschule in Berlin, Witt was an interested party. He noted that the names and business activities of the firms represented were only communicated in the accompanying catalogue, which offered a historical and statistical survey of the industry (and was freely available); but even then particular objects were directly linked to individual firms only in the case

22. Otto Witt (who also edited the official catalogue of the German empire) compared the organizational principles of different exhibitions in Witt, *Chemische Industrie* (1902), 2–4.

23. Morand, *1900* (1931), 79.

24. Witt, *Chemische Industrie* (1902), 2–4.

FIGURE 4.5 The display of the Normal-Aichungs-Kommission of the German Reich, in the uniform cases that characterized the German display of scientific instruments as a whole. Kraemer, *XIX. Jahrhundert*, vol. 4 (1900), 225.

of Apparatus and Equipment for Laboratories and Chemical Factories (one of eight classes of exhibits).[25] Thus the chemical display and its catalogue embodied the knowledge that the diverse firms participating understood themselves as representatives of the entire German industry, which was to be regarded as one whole. Witt thought the exhibit conveyed an impression of great distinction and refinement, even if the submersion of individual interests had gone too far to recommend repeating the tactic.

25. Witt commented that in this case the diversity of objects made it necessary to identify individual firms. It is worth noting that the historical retrospective the French displayed—recreating Lavoisier's laboratory—was regarded as far more engaging than the German display of 222 preparations of historical significance. In 1889 a French display of human work throughout the ages had contrasted a rendering of an alchemist's cell (with German script on the walls) with a reconstruction of Lavoisier's laboratory. The lessons were not lost on German chemists, who presented a similar contrast in the 1904 St. Louis exhibition, rebuilding Liebig's laboratory instead of Lavoisier's. The Deutsches Museum took up the practice of reconstructing the laboratory workplace in the early twentieth century. Vaupel, "Chemie für die Massen" (2000).

Representing the precision scientific instrument-making industry, the German Society of Mechanicians and Opticians also departed from customary exhibition practices by presenting a unified, coordinated display of the German industry as a whole. Led by Alfred Westphal, a committee had begun preparations four years earlier by first calling leading firms to submit their best products for consideration and then selecting a still more exclusive group for display. In this way firms cooperated to present their most important instruments only, avoiding duplication and allowing a representative coverage of the scope of their craft. Neuburger thought it no surprise that the result contained the best and most worthy instruments that had been produced, and that Germany had again drawn the admiration of the entire world: no other nation could measure up to the value and quality of the instruments displayed.[26] Over 700 marks worth by more than one hundred firms, and like the chemical displays they were shown in unified display cabinets. The instrument makers too produced a joint catalogue, the *Special Catalogue of the German Collective Exhibition for Mechanics and Optics*, printed in the new font designed for the *Official Catalogue of the Exhibition of the German Empire in 1900*.[27] While the chemists had made do with German and French editions, both these catalogues were published in German, French, and English. The *Official Catalogue* suggested that the best indication of the development of the intellectual life of the German nation was the volume of books it published. With about 24,000 new publications yearly, each year Germany produced one copy of a book to every three Germans.[28] The instrument makers' catalogue proclaimed their own productivity in a proud introduction surveying the transformation of German instrument making. In the early nineteenth century their craft had been highly derivative and dependent on skills gained through travel and work in Britain and France. At

26. Neuburger, "Chemie und Physik" (1900), 223. Brachner's study of German instrument making draws heavily on the 1900 catalogue discussed below; Shinn has described this collectivity as an illustration of the rise of the new social role of "research technologists." See Brachner, "German Nineteenth-Century Scientific Instruments" (1985); Shinn, "The Research Technology Matrix" (2001).

27. Deutschland. Reichskommissar für die Weltausstellung in Paris 1900 [ed. Otto N. Witt], *Catalogue officiel* (1900); Deutschland. Reichskommissar für die Weltausstellung in Paris 1900 [ed. Otto N. Witt], *Official Catalogue* (1900); Deutschland. Reichskommissar für die Weltausstellung in Paris 1900 [ed. Otto N. Witt], *Amtlicher Katalog* (1900). There were 6,000 copies in French, 4,000 in English, and 3,000 in German.

28. Deutschland. Reichskommissar für die Weltausstellung in Paris 1900 [ed. Otto N. Witt], *Official Catalogue* (1900), 51–52.

the dawn of the twentieth, it could be represented as world leading—with the direction of travel reversed.[29] Abbe's glassworks in Jena and Siemens & Halske in Berlin were major examples of this thesis, and the German catalogue went on to offer two principal (and related) reasons for the rise of their industry in general. The first was state support, late in coming in Germany but from 1887 given a particularly effective focus through the work of the Physikalisch-Technische Reichsanstalt (PTR) in Berlin-Charlottenburg. That institute had been founded to provide technical aid to industry through the development of standards and testing services, and intellectual guidance by allowing scientific research not possible either within single firms or the universities. It exemplified the second major grounds for German success, the fostering of an "intimate cooperation between men of research and men of practice," which had also been furthered by the foundation of the German Society of Mechanics and Optics in 1879 and its journal *Zeitschrift für Instrumentenkunde*.[30]

This message was amplified in the body of the catalogue. There the three pages listing the optical measuring instruments of Carl Zeiss opened with a paragraph stating that the firm had operated a special workshop devoted to optical measuring instruments under the leadership of Dr. C. Pulfrich since 1892. It pointed out that in many cases the instruments were the end result of many years' practical experience, and that most owed their conceptual and practical development to the workshop itself. Emphasizing both the technical and scientific applications of Zeiss instruments, the exhibit catalogue also indicates that in Germany, unlike the United States, precision refractometry moved between French traditions and continental elaborations of the instruments, without making direct reference to Michelson's contributions. Along with a wide range of refractometers, the firm displayed different length measuring instruments and three kinds of interference apparatus: a new interference measuring apparatus with a platform designed by Pulfrich, a four-plate interference refractometer of Ludwig Mach's design, and two forms of Jamin's interference refractometer (fig. 4.6).

The effect of the *Special Catalogue* was electric. In Britain, *Nature*'s staff columnist quoted extensively from its opening passages in order to

29. *Weltausstellung in Paris 1900: Sonderkatalog der deutschen Kollektivausstellung für Mechanik und Optik* (1900), 1–10. The introduction was repeated in the official catalogue, where it was attributed to Westphal.

30. Ibid., 2.

carries the plate and thereby places each successive line under the cross-lines of a laterally mounted microscope. In instruments of this kind, as usually constructed, it is necessary each time to read the position of the micrometer head with respect to a fixed index or vernier. This process of alternate adjustment and reading is not only very fatiguing but conductive to inaccuracy. The chief novelty of this instrument consists in the fact that each adjustment, instead of being read on the micrometer head, is registered by a printing recorder. For this purpose the micrometer head is provided with lines and figures extending from 1 to 100. Another similar head rides loosely on the same spindle side by side with the fixed head and is connected with the latter by a tooth gearing in such a manner as to travel one division of the scale while the fixed head describes one complete revolution. A pointer is situated below and between these two disks. While turning round, these disks run on two felt rollers saturated with colour for giving the required impressions. A tape passes below both disks which is pressed against the latter by the depression of a button situated on the left side of the instrument, whereby an impression of the lines, numbers and the index is produced on the tape. The pressure on the button displaces at the same time the tape automatically through a certain distance. While measuring, the observer needs therefore not remove his eyes from the microscope, it being sufficient to press the button after each observation. The readings are subsequently furnished by the tape marks. The loose disk indicates the entire revolutions, hundredths are indicated by the movement of the fixed disk, and tenths divisions can easily be estimated from the distance of the index from the adjacent division lines. The screw has a pitch of about 0.33 mm. It is very carefully cut so as to exclude the necessity of providing special screw corrections. The distances separating two lines on the spectrum can therefore be measured directly and accurately within 0.00033 mm.—The track of the slide embraces about 15 cm. For the convenience of the observer the microscope is mounted in a slanting position, and the plate-carrier is correspondingly inclined. While measuring it is often desirable to append to the positions of the various lines notes respecting their intensity, definition, &c.; for which purpose the apparatus is provided with four other buttons so as to imprint 1, 2, 3 or 4 points upon the tape next to the registered measurement. By variously combining these four points it will be seen that fifteen conventional notations are available.

This instrument is the property of Mr. Hauswald, of Magdeburg-Neustadt.

10. Carl Zeiss, Optical Works, Jena.

(See also Sections II, Vc, Vd, Ve and Vf.)

Optical Measuring Instruments.

A separate department has been established since 1892 for the construction of optical measuring instruments. This department is under the supervision of Dr. C. Pulfrich. The instruments made in this section—spectrometers and refractometers, spectroscopes, goniometers and reading microscopes, interferential appliances, telemeters, &c.—are mainly designed for use in physical, chemical and mineralogical laboratories and are available for many scientific and technical uses. The majority of these instruments are original in conception and design and have emanated from the firm's own requirements. They have been constructed successively by various members of the scientific staff either for the immediate needs of the workshops or for subsidiary experimental investigations. All of these instruments have therefore been subjected to practical tests, often for years, in the original workshops, and the majority of them have therefore through continued improvement reached their final stage.

The following instruments are exhibited:

Abbe's Spectrometer, a large model for spectrometric determinations and a smaller instrument for teaching and practicing in college laboratories, including auxiliary and accessory appliances, prisms, hollow prisms, heating apparatus, &c.

FIGURE 4.6a *(left)*

Section Vb. 109

Two New Refractometers, both based upon the method of prismatic deviation. One, having a variable refracting angle, is particularly adapted for the examination of highly refracting fluids. The other is a differential refractometer, constructed for determining the difference of refraction of two fluids.

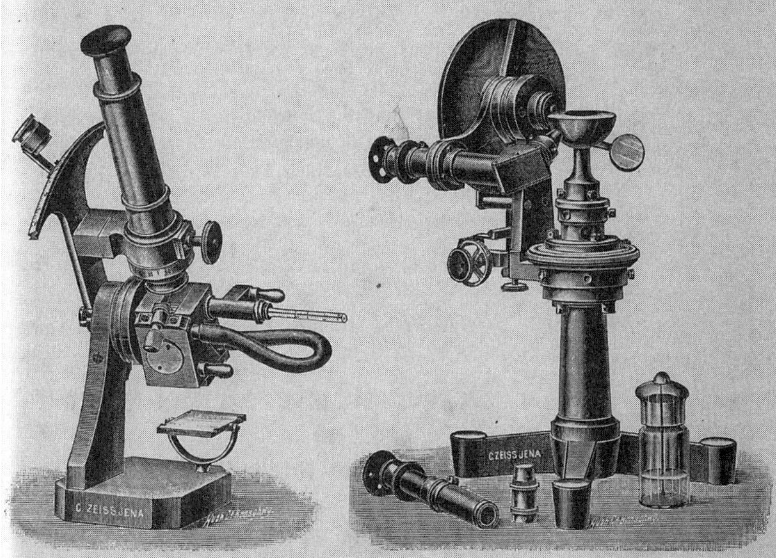

Fig. 1.
Abbe's refractometer with heating apparatus.

Fig. 2.
Crystal refractometer (new model).

Various refractometers based upon the observation of the critical angle of total reflection, e. g.:

Pulfrich's Refractometer (new model).

Refractometer for Colleges.

Abbe's Refractometer. Two models, one of which is fitted with prism adapted for being heated (Fig. 1).

Refractometer for Special Technical Purposes (butter, milk-fat and lard refractometers).

Crystal Refractometer (new model), eminently suited for the examination of small and defective surfaces (Fig. 2).

Laboratory Comparison Spectroscope with wave-length scale (Fig. 3).

New Fluid Prism of high dispersion.

FIGURE 4.6a *(right)*

110 Sections Vb, Vc.

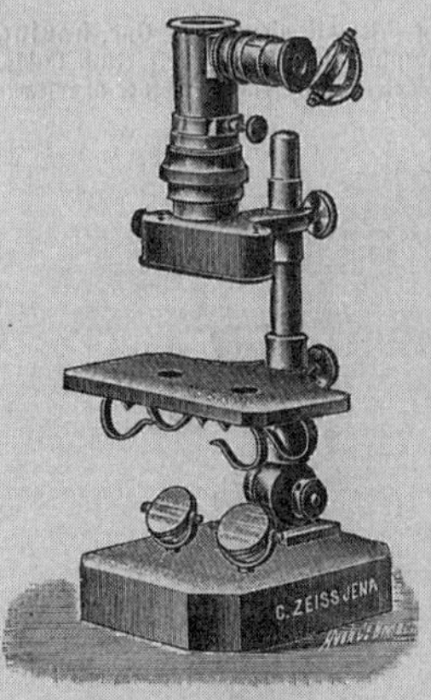

Fig. 3.
Laboratory comparison spectroscope.

New Crystal Goniometer with various improvements in the collimator, by which the path of the rays is rendered completely adjustable from the luminant to the eye of the observer. A special feature of this instrument is its adaptability for observing diffuse reflection as a means of examining small and uneven surfaces.

Various Appliances for Longitudinal Measurement (pachymeter, two forms of comparators, a spherometer and a focometer).

Abbe-Fizeau's Dilatometer.

Pulfrich's New Interferential Apparatus and stage.

L. Mach's Quadruple Plate Interferential Refractometer.

Interferential Refractometer, constructed on Jamin's principle.

New Stereoscopic Telemeter.

Descriptions of these various instruments and price-lists may be had free on application. The majority of these have been published in French, English and German.

C. Microscopes and their Auxiliaries.

1. Gustav Halle, Rixdorf near Berlin, 53 Hermannstr.

Maker of Scientific and Technical Instruments of Precision.

(See also Sections Vb, Vg and X.)

1. **Hand-microscope for demonstrating** the anatomy and physiology of plants, made almost entirely of aluminium and weighing only 370 g, focussing by rack and pinion, provided with two stage diaphragms.

2. **Entomological Stand-microscope,** with adjustable tube-length and rack and pinion adjustment for working-distances of 30 mm to 90 mm; adapted for opaque objects; very light and adapted for use as a hand-microscope. Stage with Cardanian joint and two objectives.

2. E. Hartnack, Potsdam, 39 Waisenstr.

Optical Works.

This firm was formerly established in Paris.

1. **The Microscopes** exhibited are in general design and construction of the form originated by this firm some 50 years ago, but the movements, especially the coarse and fine adjustments,

FIGURE 4.6b The discussion of optical measuring instruments of the Zeiss firm in the *Special Catalogue* of the German scientific instrument makers *(a* and *b)*. *Special Catalogue* (1900), 108–10. Courtesy Museum of the History of Science, Oxford.

endorse the claim made there that Germany now led the world in scientific instruments. Then he went on: "As to the excellence of this joint exhibition, it is difficult to speak too strongly; rumour says that some, at least, of the judges wished to award it a Grand Prix among the nations. Had the rules of the Exhibition allowed it, such an award would have met with the universal approval of all physicists who have visited Paris."[31] The rumor was accurate: British and American jurors had sought to bestow the distinction, but whether it would have found universal approval is another matter, as we shall soon see. *Nature*'s columnist went on to describe the organization of the catalogue, comparing the annual exports and expenditures it listed with English equivalents, and picking out noteworthy attractions. That brought him back to the point:

> But a detailed description of the catalogue would take too much space, and would indeed be of no great value to a reader; the book itself will prove to a physicist a well of information; the exhibit, however, must be seen in its entirety if we wish to realise what our German cousins have done.
>
> Not that the sight is one which brings great pleasure to an Englishman, and if he moves on to examine the English exhibit his thoughts cannot fail to be very grave.[32]

The displays of British firms looked like small portions of the shop windows of leading opticians. As much apparatus as possible was crammed into as small a space as possible, without any evidence of organization or the attempt to either instruct learners or attract sales. In contrast, the German exhibition had been attended by a courteous and knowledgeable representative, Dr. Robert Drosten. He was able to provide access to the instruments, offer detailed information on their construction and operation, and inform visitors how to purchase them. The excellent displays of the Cambridge Scientific Instrument Company had at least been accompanied by a set of cards on the wall, but Drosten could give the serious visitor a catalogue in their own language to be taken away for reference.

The praise and lament from *Nature* brought a flood of responses from writers echoing its views, expanding the list of German virtues to include their catalogue of chemical exhibits and recommending that both books

31. "Instruments of Precision at the Paris Exhibition" (1900), 61.
32. Ibid.

be provided to public libraries attached to scientific institutions.[33] The response on the other side of the Atlantic was similar. In his official report as the American juror for the class, John Rees wrote praising the organization and management of the German exhibit and recommending that in the future the United States prepare a catalogue "similar to this most excellent book." He thought its introduction should be required reading: "it would be well for all American mechanicians and opticians to read and thoroughly digest, for it shows by what processes the German mechanicians and opticians have been able to go so far ahead of other nations."[34] Having already made sure it was reprinted in *Science*, he appended both the introduction and the three versions of the catalogue to his report. Within a year the French Industry of Precision Instrument Makers had also responded. They took the extra step of producing their own catalogue in reply, intended to demonstrate conclusively the vigor of their industry. While paying the compliment of imitation, a searing introduction—by Alfred Cornu—explicitly rebutted what Cornu regarded as the extravagant German claims to dominance.[35]

The Forces of History

Such challenges and counterchallenges show that many experts were deeply concerned with promoting national interests, on a stage that was both commercially and culturally significant. Lessons of this kind were seldom lost even on less technically proficient visitors, many of whom went to study just this energetic dimension. Henry Adams, for example, the scion of a great American political family and celebrated historian of America and medieval France, spent six intense months "haunting" Paris. Adams secured the services of the Smithsonian astronomer and physicist Samuel Pierpont Langley to guide him through the displays—a fact that renders his observations particularly interesting to us. Langley's career had followed what we can recognize was a characteristic pattern: before becoming superintendent of the Smithsonian in 1887, Langley had

33. See letters from E. T. Warner and H. Davidge, C. V. Boys, and R. Meldola, *Nature* 63 (1900): 107, 156, and 179 respectively.

34. Rees [Juror], "Instruments of Precision" (1901). He had also published the introduction in Rees, "German Scientific Apparatus" (1900).

35. *L'industrie française des instruments de précision* (1901). Cornu's introduction is on v–xii.

been an assistant at the Harvard Observatory, taught mathematics at the Naval Academy, and directed Allegheny Observatory, focusing much of his research on the study of the sun's radiation (Langley developed the "bolometer" to measure the energy transmitted in different parts of the spectrum). Langley's present concerns were with the engines required to pioneer heavier-than-air flight, and Adams found him able to strip the exhibition to its bones—knowing what, how, and why to study. Set down in Adams's autobiography, the results of their examination of the exhibition formed a turning point in the historian's cultural education; and they will help conclude our own excursion to the fair. Already stunned by the power of new science and technology on display in Chicago in 1893, Adams thought the electrical dynamos—with the quiet hum of their ceaseless revolution—had shifted mankind's guiding power from the cross and virgin of the Middle Ages to a new and no less mysterious agent of change. He was intent on the study of forces; and Langley helped Adams cull much of the fine arts display and move past most of the industrial exhibits. His narrative describes them standing before the automobiles, now capable of a horrific 100 kilometers an hour, and then pausing in front of the dynamos.[36]

But to reach them from the rows of automobiles shown in Vincennes, Langley and Adams had to ride the new Métro, or catch an electric tram, or travel by boat down the Seine the 6 miles toward the Eiffel Tower. From its feet they would look down the Champs de Mars, flanked by the pavilions of the main halls on both sides. A visitor could enter them and wend his/her way through countless exhibits. Mining and Forestry, and then Textiles and Clothing ran down the left-hand side. Arts and Sciences, Education, Engineering and Transport, and Chemical Industry were to the right. Or one could take the promenade between the buildings, walking toward the Water Palace with its elaborate rococo fountains powered by electricity and gorgeously lit by electric light. Behind it stood the extraordinary Electricity Palace and Machinery Halls; and there many of the giant steam engines and dynamos on display were also in commission. The enormous Helios dynamo inspired an almost universal awe. The basement below the Electrical Palace even put on show the switchboards that delivered both alternating and direct current to exhibits throughout Paris. Powered by its own displays, the exhibition dramatized the clean separation that interchangeable forces enabled between the engine house, the

36. Adams, *The Education of Henry Adams* (1918), chap. 25.

FIGURE 4.7 A night view of the Palace of Electricity with the spectacular water fountain before it. Kraemer, *XIX. Jahrhundert*, vol. 4 (1900), facing 148.

dynamo, and the countless different arenas in which electricity could be put to use. It was electricity that had enabled so many to run their machines on site and fulfill Picard's call to display products and processes together. Without electricity the shafting required to transmit motive power to each machine would have rendered the scheme impractical. Thus the exhibition was a showcase for the distributed power of new technological systems, and on a scale never before approached. It dwarfed more specialized exhibitions dedicated to electricity in Paris (1881), Munich (1882), and Frankfurt am Main (1891), and even among the universal exhibitions only Chicago had come close. The network realized in 1900 gave a new meaning to Paris as the "city of light," and it would help visiting civic leaders reimagine their own cities.[37]

Struggling to relate both the discoveries and economies of force, Adams greeted the dynamo as a familiar. But the gap between its turn

37. On the role of electricity in realizing the aims of the exhibition, see Great Britain. Royal Commission for the Paris Exposition 1900, *Official Catalogue* (1900), ix. For a description of the Helios dynamo, Naumann, *Ausstellungsbriefe* (1909), 109–12. On the creation of electrotechnological systems, Hughes, *Networks of Power* (1983). Nye has described the cultural dynamics of electrification in the United States in Nye, *Electrifying America* (1990).

and the machine that powered it, between steam and electricity, still fractured the historian's grasp of his own age; and Langley too had to explain "how little he knew about electricity, or force of any kind." For Adams, the absolute fiat that electricity demanded—the way it pressed its audience toward faith—became a symbol of the still more mysterious forces of the new physics as a whole.

> Frozen air, or the electric furnace, had some scale of measurement, no doubt, if somebody could invent a thermometer adequate to the purpose; but X-rays had played no part whatever in man's consciousness, and the atom itself had figured only as a fiction of thought. In these seven years man had translated himself into a new universe which had no common scale of measurement with the old. He had entered a supersensual world in which he could measure nothing except by chance collisions of movements imperceptible to his senses, perhaps even imperceptible to his instruments, but perceptible to each other, and so to some known ray at the end of the scale. Langley seemed prepared for anything, even for an indeterminable number of universes interfused—physics stark mad in metaphysics.[38]

Adams thought Langley prepared, but ultimately the physicist was still unable to help the historian comprehend the present:

> for [Langley] constantly repeated that the new forces were anarchical, and especially that he was not responsible for the new rays, that were little short of parricidal in their wicked spirit towards science. His own rays, with which he had doubled the solar spectrum, were altogether harmless and beneficent; but Radium denied its God—or, what was to Langley the same thing, denied the truths of his Science. The force was wholly new.[39]

After ten years pursuing the lessons of science for humanity, Adams "found himself lying in the Gallery of Machines in the Great Exposition of 1900, his historical neck broken by the sudden irruption of forces totally new."[40]

Of course Henry Adams got up, brushed himself down, and continued the analytic work—and his awed response to the epochal dimensions of

38. Adams, *The Education of Henry Adams* (1918), 381–82.
39. Ibid., 381.
40. Ibid., 382.

inhaltlich die beredteste Sprache von allen Teilen der Ausstellung.

Kraft-centrale.

Bei der Ausstellung des Jahres 1900 ist das Prinzip der vollständigen Konzentration aller Kraftmaschinen streng durchgeführt. Es ist dies nur dadurch ermöglicht worden, daß für den Antrieb der Arbeitsmaschinen, die sich an den verschiedensten Stellen der Ausstellungshallen befinden, durchweg Elektromotoren zur Anwendung kommen. Die ausgestellten Kraftmaschinen sind infolgedessen fast ausschließlich Antriebsmaschinen für Elektricitäts-Generatoren und die Kraftmaschinenhalle ist eigentlich nichts anderes als eine gewaltige elektrische Centralstation. Die Ausstellungsleitung hat von vornherein für einen möglichst vollständigen Ausbau dieser Centrale Sorge getragen, indem sie die Kraftmaschinen nicht bloß als Ausstellungsobjekte ansah, sondern zugleich als einen wesentlichen Bestandteil des Ausstellungsbetriebes; zu diesem Zweck hat sie mit allen bedeutenderen Maschinenfabriken des Kontinents Verträge abgeschlossen, die dahin zielten, Dampfkessel, Dampfmaschinen und Dynamomaschinen gegen entsprechende Vergütigung unmittelbar in den Dienst des Ausstellungsbetriebes zu stellen. Die Folge dieser Bestimmung war allerdings, daß sich die Aussteller bezw. die Lieferanten beschränkenden Bedingungen in Bezug auf die Bauart der Maschinen fügen mußten und ihnen nur relativ geringe Freiheit übrig blieb, eigenartige Konstruktionen zur Darstellung zu bringen.

Um von einer Centrale aus die gesamte Beleuchtung und Kraftversorgung zu bewirken, war es natürlich von vornherein notwendig, Einheitlichkeit in Bezug auf die Eigenschaften des elektrischen Stromes zu erzielen. So war vorgeschrieben worden, daß elektrischer Strom als Gleichstrom nur mit einer Spannung von 250 oder 500 Volt erzeugt werden, daß Drehstrom bei 50 Perioden eine Spannung von 2200, 3000 oder 5000 Volt haben sollte. Andererseits war im Interesse der allgemeinen Betriebssicherheit bestimmt worden, daß der erzeugte Dampf die Kessel durchweg mit einer Pressung von 11 Kilogramm verlassen sollte, um aus dem gemeinsamen Dampfleitungsnetz mit ca. 10 Atmosphären an die einzelnen Dampfmaschinen abgegeben zu werden.

Anker-Segment der großen Helios-Dynamo-Maschine.

FIGURE 4.8 The Helios dynamo was so imposing it was easier to capture in words than photographically. As depicted in Kraemer, *XIX. Jahrhundert*, vol. 4 (1900), 78.

fin de siècle change was complemented by more immediate lessons to be drawn from the exhibition. Indeed, when his brother Brook wrote comparing human energies and describing an economical formula that incorporated the relative geographical and geological value of Euro-Asia and America, Henry offered a view from Paris:

> As far as I have guessed the results of this Exposition, the Germans alone show very marked development of energy. Their machinery seems to have impressed people much.... Looking forward 50 years more, I should say that the

FIGURE 4.9 The Siemens & Halske display in the German section of the Electrical Palace. *Encyclopédie du Siècle. L'Exposition de Paris de 1900* (1900). Whipple Library, University of Cambridge.

> superiority in electrical energy is going to decide the next level of competition. That superiority depends, in its turn, on geography, geology and race-energy. All these elements have somewhere exact numerical values, and the value of your theory depends on getting the values for these unknown quantities.[41]

Then he turned to the geopolitical questions raised by the China crisis, with European powers jockeying for position in Asia:

> It is very clear that nobody means to fight. I regard the Chinese question as settled in principle. Russia and Germany will necessarily take the north as their sphere, England, with our support, will retire southwards. The difficult problem to me is France, and I want much to win France over, but the German Jews now own France, as far as I can see, in very simple fee, and they will never let us into that market. They are pure Euro-Asian, and mean to hold France tight.[42]

41. Henry Adams to Brook Adams, 7 October 1900, as cited in Brain, *Going to the Fair* (1993), 166–68.
42. Ibid.

Several months earlier the resolution of the Dreyfus affair in favor of the Jewish soldier controversially imprisoned for treason had helped avert a potential boycott of the world's fair by other nations; but more than Dreyfus's freedom would be required to resolve the tensions underlying the period. It was common for observers in 1900 to bind electrical energy to race energy. Whether with hope or foreboding it was typical to link technological progress with understanding of peoples, to sheet industrial competition into geological and geographical gifts and needs—and on a national scale. In 1930 the French author Paul Morand looked back on the turning point of 1900. His reminiscences demonstrate how widespread and lasting impressions of German dominance were: "All Paris rushed to the section of German optics to see the German instruments of precision. And the stuffs of Grefeld and Elberfeld! And the new chemical industries! And the laboratories!" Morand recited a long list of attractions (stud farms, German uniforms, Berlin china ...) and noted that besides the German dynamos, generators, and the crane dominating an entire gallery, "the machines of other countries looked like toys."[43] But the competitive realities of international commerce were registered in absences as well as participation in the galleries of Paris. The electro-technology firm run by Albert Einstein's uncle Jakob and father Hermann had exhibited dynamos, arc lamps, and Jakob's patented electrical meter in the 1891 electrical fair in Frankfurt. After failing to gain significant capital and losing a critical bid to provide Munich with electric lighting to Schuckert of Nuremberg, in 1894 the brothers dissolved Jakob Einstein & Cie. and sought new business opportunities in Pavia and Milan. Rather than continue boarding to finish secondary school in Munich, Albert followed his family in 1896 and remained a sensitive observer of his father's business trials as he studied in Zurich. The Einsteins did not display in Paris. In contrast, by 1900 Schuckert was trading as the Elektrizitäts-Aktiengesellschaft and had a working capital of 60 million marks, employing 7,400 hands and 1,130 officials. Their main exhibit displayed dynamos, transformers, motors, and a switchboard, and in the electrical palace they showed a historical collection of dynamos, arc lamps, and measuring instruments, with a graphic tableau depicting the development of their factory.[44]

43. Morand, *1900* (1931), 78–79. On the machines of different nations, see also Naumann, *Ausstellungsbriefe* (1909), 109–15.

44. On the Einstein firm, see Pyenson, *The Young Einstein* (1985), chap. 2. For Schuckert in Paris, see Deutschland. Reichskommissar für die Weltausstellung in Paris 1900 [ed. Otto N. Witt], *Official Catalogue* (1900), 73, 173, 175–76.

Our excursion through the Paris exhibition has found electricity illuminating its entrance, allowing a new distribution of power from the central plant of an extraordinary palace, and, for Adams, symbolizing the mysterious new forces of a physics that had reached toward a new universe, stepping beyond the scale of common measurement. In exhibition halls and instrument displays we have seen the power of collective visions and noted the value accorded catalogues and books. For many observers a common lesson emerged from their visits to the fair—discerning a direction of travel in Germany's favor. The next chapter will explore these themes in a different setting, calibrating Adams's somewhat breathless vision of wholly new forces against the collective endeavor of the physics community itself.

CHAPTER FIVE

A Discipline at Work: The International Congress of Physics

Samuel Pierpont Langley had not come to Paris in order to chaperone Adams. Rather, like nearly a thousand others, he had accepted the invitation of the French Physical Society to join the work of the International Congress of Physics. The society had begun preparations in January 1899 by creating an organizing committee headed by Alfred Cornu, with Louis-Paul Cailletet as deputy, Lucien Poincaré and Charles-Édouard Guillaume as French and foreign secretaries, and a host of prominent French physicists among its membership.[1] In July the committee sent a letter to physicists throughout France and abroad announcing their intention to hold a meeting in conjunction with the Paris exposition. Specialist conferences, particularly on electricity, had made physicists aware of the benefits of international collaboration—but there had never been a *general* international conference in physics. Without specifying its final form, the committee promised the congress would extend to reports and

1. Cailletet was best known for being the first to liquefy many gases, including oxygen, hydrogen, nitrogen, and air. The physicist Lucien Poincaré came from a family with an extraordinary academic and public profile. He was cousin to the mathematician Henri and brother to the politician Raymond, who moved from finance and education to hold the prime ministership of France from 1912 to 1917, 1920 to 1924, and from 1926 to 1928. In 1902 Lucien became inspector-general of public instruction, and he published popular studies of modern physics. The Swiss physicist Guillaume was a member of the International Bureau of Weights and Measures in Sèvres.

discussion on a limited number of subjects decided in advance, such as the definition and determinations of units (the traditional province of international gatherings), but also the bibliography of physics and national laboratories. In addition there would be lectures on new work and visits to the exposition, laboratories, and workshops. Soliciting remarks and proposals, the committee asked those interested in participating to contact one of the secretaries.[2]

Interest was extraordinary. Receiving nearly a thousand responses prompted the committee to embark on still more ambitious plans. Each year a haphazard collection of papers floated to the surface in the annual meetings of the French, British, German, or American associations of scientists. The latest concerns of different individuals jostled for space, while presidential addresses gave elbow room to a few prominent figures. By instead commissioning suitable authors to write reports on specific topics, the 1900 congress could offer a more systematic lens on the discipline. The committee believed there would be great interest in covering the broadest span possible, noting that while oral discussion did not suit every topic, it would be extremely useful to read critical papers—succinct and precise—on the current state of science on all important subjects. They wrote soliciting reviews of research in different fields from carefully chosen authors. By the time the official invitation was published in July 1900, sixty-one physicists had agreed to offer papers, and the list of their names and report titles was surely the most concrete enticement to register and send the 20 franc fee for the conference. Most of the reports would be available beforehand to those who expressed interest in specific subjects (all translated into French), and there were more in the pipeline. After the conference they would be collected into a set of proceedings edited by Guillaume and Poincaré and distributed to every member for the cost of postage.[3]

In fact, physicists were late to the game of general, international conferences, and they were certainly just one of many groups that responded to the opportunity the centenary exposition offered. One hundred and

2. For early announcements in *Science*, see "Scientific Notes and News" (1899): 301; Guillaume, "International Congress" (1899); and in *Physikalische Zeitschrift*, "Internationaler Kongress für Physik zu Paris vom 6. bis 12. August 1900" (1899).

3. This was conveyed in the pages of *Physikalische Zeitschrift*, "Tagesereignisse" (1900). The second, official invitation described the work of the conference as communication concerning new questions, visits to the exhibition, laboratories and workshops in Paris, and reports and discussions on recent progress, before listing the reports.

twenty-seven conferences were held in Paris that year, on topics ranging from concerns as expansive as the History of Religion (meeting over five days) to Basque Studies (two days) and Acetylene (an entire week). Mathematicians held their second international conference, while Medicine met for its thirteenth (with 6,000 attending), and the International Congress on Electricity attracted 917 participants.[4] Nevertheless, considering the historical development of the physics discipline, the scope of the committee's organizational ambition was unprecedented, and the scale of physicists' participation truly remarkable. Historians and sociologists have counted the number of academic physicists in 1900 and compared the state of the discipline across different nations in that year. They found that despite different growth rates, for a brief period around 1900 numbers across the three European giants of Britain, France, and Germany happened to be roughly comparable with one hundred or so physicists in each country, while the United States already supported nearly twice as many with 215; and they estimated the total population of working physicists to be between 1,000 and 1,500, worldwide. Well over half as many came to Paris: 789 men and two women registered as participants, and the exhibition statisticians recorded the number attending at 836.[5] Among physicists then, the 1900 exhibition had prompted a new kind of conference. That meeting was greeted with great enthusiasm, and it would result

4. An overview of the conference statistics is provided in Picard, ed., *Exposition universelle*, vol. 6 (1903), 19–22. The mathematical congress was held concurrently with the physicists' (with Henri Poincaré as president and Joseph Larmor presiding over the section on Mechanics and Mathematical Physics). It was reported to have suffered from the warm weather and fear of exhibition crowds and exhibition extortions. Although it attracted only about 225 of the 1,000 or so expected participants, David Hilbert's address drew notice and was to help orient the mathematics of the coming century. "The International Congress of Mathematicians" (1900), 418.

5. On the physics discipline, see Forman, Heilbron, and Weart, "Physics *circa* 1900" (1975). Their results are summarized on 3–10 and tables I and II; each of the big four supported about the same number of physicists per million of total population, with the same expenditure per milliard of total income. For a more recent overview, see Kragh, *Quantum Generations* (1999), chap. 2. The congress participants were listed by name, position, and address in Poincaré and Guillaume, eds., *Travaux* (1901), 129–69. The two women were Marie Curie and Isabelle Stone, who in 1897 became the first woman to gain a PhD in physics in the United States (from the University of Chicago). She spent a year in postdoctoral studies in Berlin, and was also one of two women at the founding of the American Physical Society in 1899 (with Marcia Keith of Holyoke College). Stone went on to a professorship at Sweet Briar. Rossiter, *Women Scientists in America* (1982), 27 and 89. The congress audience was described as being about 1,000 (in its publications) and 900 in Simon, "Physikerkongresse" (1900). It was listed as 836 in Picard, ed., *Exposition universelle*, vol. 6 (1903), 22.

FIGURE 5.1 Physicists at the International Congress of Physics; Cornu is the taller central figure with a fob watch. Courtesy AIP Emilio Segrè Visual Archives, Brittle Books Collection. This photograph is from the front of the AIP library copy of the Congress *Travaux* (1901) and appears to be unique.

in a new kind of book. Publicizing the results of the endeavor in an article in *Nature*, Guillaume averred that the congress reports provided "the most complete representation of any science at a given epoch yet made."[6] Noting the scale and scope of the inaugural International Congress of Physics it should puzzle us now, firstly, that physicists no longer remember the event, and second, that it has never attracted serious scholarly attention.

What do the inaugural International Congress and its three volumes of reports reveal about physics in 1900? As we did entering the exhibition, we shall begin with opening addresses and framing perspectives before exploring how Michelson's work and the electron fared in the more intimate give and take of specific reports. First consider briefly some procedural matters and the entertainment program.[7] After Cornu presided over the opening in the Palais des Congrès, the congress met in the rooms

6. Guillaume, "International Physical Congress" (1900), 426.
7. See Poincaré and Guillaume, eds., *Travaux* (1901).

of the French Physical Society at l'Hôtel de la Société d'Encouragement.[8] Its sessions were held on the mornings of a five-day week, with keynote lectures from Henri Poincaré on experimental and mathematical physics and Lord Kelvin on the ether and matter held on Tuesday afternoon and Wednesday morning. Wednesday afternoon saw a demonstration from Henri Becquerel and Pierre and Marie Curie appropriately titled "Radioactive Bodies." On Thursday Cornu drew on his long-standing interest to give a talk on the speed of light. On Friday and Saturday the physicists visited the exposition. There they took in an ascent of the Eiffel Tower, an evening visit to Deloncle's Optical Palace, and a visit to the electrical installation of the moving pavement that enabled visitors to travel easily through the exhibition grounds.[9] On Sunday, after the formal closing of the conference, its members visited the laboratories of the Sorbonne. There, and in the École Polytechnique, both French and foreign physicists had installed a number of experiments in progress, best viewed in small groups. The light entertainment of the congress reflected the specifically optical strengths of French physics; but writing for the new German journal *Physikalische Zeitschrift*, Hermann Theodor Simon said he thought that guided visits to the exhibition and official entertainments like the reception with the French president on Friday afternoon were superfluous. It would have been better to leave the time free for individuals to make what they could of the rich offerings in Paris. The journal Simon helped edit ran a series of articles on both facets of the Paris experience, examining major exhibits and reviewing more than half the reports given to the Congress of Physics.[10]

8. In addition to the organizing committee and Professor Gariel and Lord Kelvin as principal delegates, several ministers of war and of education were gathered on the stage as delegates of national governments.

9. Guillaume noted that the demonstrations of radioactivity were particularly effective because so few physicists had witnessed the rare phenomena previously. Guillaume, "International Physical Congress" (1900), 427–28. Deloncle's Optical Palace housed what was without doubt the most extraordinary instrument on show in Paris—a refracting telescope with a lens of 1.25 meters in diameter and a 60 meter focal length, requiring a giant siderostat to reflect light down its horizontal tube. Briefly the world's largest telescope, it was never rebuilt outside Paris in the dark skies that would have allowed it a working life.

10. Simon, "Physikerkongresse" (1900), 572. Simon edited the journal with Eduard Riecke. It had been founded in 1899 to complement existing journals by providing an organ for disciplinary news, general articles, and the rapid publication of research results. They reviewed forty-six of the eighty congress reports.

Alfred Cornu must have felt a sense of pride and anticipation looking out at the dense crowd of physicists gathered in the opulent surroundings of the Palais des Congrès at 3 p.m. on Monday the 6th of August. Speaking in a time skittish with contest, his gracious welcoming speech carefully juxtaposed the troubled world through which his audience moved with the purity of their science. Cornu promised that the associative endeavor in which they were engaged was "a work of progress, a tight bond for the concord of nations, since in the terrain of pure science rivalries, elsewhere so cruel, are resolved through a generous competition [*émulation généreuse*] always profitable to humanity."[11] The Dreyfus Affair had been defused by a presidential pardon that satisfied few; Paris had been racked by a long series of strikes. European powers were still riven by the Boer War and disturbed by echoes of the China crisis; and the exhibits that surrounded the Palais surely gave eloquent testimony to national posturing and sharp commercial contest. But pure science offered different terrain. Noting the honor of being graced by so many official delegates (ministers of war and education among them), and the great debt owed to the organizing committee, Cornu's most earnest thanks went to the savants responsible for the reports to come. These were the principal work of the conference and its defining characteristic. They provided the basis for his claim that the proceedings would form a true scientific monument, one that "ensures our Congress a durable memory, an important role in the history of the progress of natural philosophy at the end of the nineteenth century."[12] Cornu promised that the authorized voices of the speakers would venture into the heights of difficult speculation, where the geometer, the poet, and the philosopher united in the task of de-robing nature to reveal her precious secrets.[13] Individual reports would be devoted to specific current issues, gaining perspective on work at the coalface, piece

11. Cornu's speech is in "Procès-verbaux" (1901), 5–8 on 5–6. I have translated "*émulation généreuse*" as generous competition, but the imitative connotations of the word *émulation* have special resonance given Cornu's competitive response to the display of wares from the German Society of Mechanicians and Opticians, and their representation of the history and present strengths of the instrument-making communities of Germany, Britain, and France.

12. Ibid., 7.

13. Ibid. The rhetoric had a concrete referent: one of the most celebrated pieces of art nouveau statuary on show in 1900 was titled *Nature Unveiling Herself before Science*. Carolyn Merchant has described it as illustrating an important long-term change: "After the Scientific Revolution, *Natura* no longer complains that her garments of modesty are being torn by the wrongful thrusts of man." Instead, Louis-Ernest Barrias portrays her "coyly removing her

by piece. The signal honor of offering an overview of the discipline went to Henri Poincaré.

Keynotes and Categories: The Library of Science and the Catalogue of Physics

Poincaré was a student of both Cornu and the mathematician Charles Hermite at the École Polytechnique—and published his first scientific paper before he graduated in 1875. His second degree came from the school of mines, and his first job was as a mining inspector (in which capacity he had to investigate and report on a fatal explosion in the Magny pit). At the same time as he built his academic career, being appointed to the Sorbonne in 1881 (and later holding a position at the École Polytechnique as well), Poincaré worked in the Ministry of Public Service supervising railway development. Although known firstly for his pioneering mathematical work in celestial mechanics, the theory of automorphic functions, and algebraic topology, by 1900 Poincaré had become one of France's most respected and broadly honored scientists with a prominent place in the scientific bureaucracy of his nation. That year he contributed to international congresses of mathematics, physics, and philosophy, and if Cornu had any single person in mind describing the qualities of the speculative heights, it was very likely Poincaré. Within two years he would bring together many of his addresses in the popular and sometimes even poetic book "Science and Hypothesis," in which an account of new geometries provided a platform for a philosophically rich discussion of the methods and results of science. Addressing the Congress of Physics, Poincaré began by posing an epistemological question that was also rapidly becoming a disciplinary issue. "Experiment," he said, "is the sole source of truth. It alone can teach us something new; it alone can give us certainty. These are two points that cannot be questioned. But then, if experiment is all, what place is left for mathematical physics?"[14]

---

own veil and exposing herself to science. From an active teacher and parent, she has become a mindless, submissive body." Merchant, *The Death of Nature* (1989), 189–91.

14. Poincaré, "Physique expérimentale et la physique mathématique" (1900), 1. The lecture was subsequently published in English as chapters 9 and 10 of Poincaré, *Science and Hypothesis* (1905), quotation at 140. In the foreword, Joseph Larmor noted the delicate play of suggestion and allusion characteristic of Poincaré's writing and highlighted the advantage

He could have asked what place is left for mathematical physicists. While local differences gave very specific complexions to the picture, in each of France, Britain, and especially Germany this question was an urgent one for increasing numbers of (mostly) young men who chose to specialize in the tools of mathematical resources and sought to work alongside or next door to the established academicians. As Michelson's career and defense of optics has illustrated so clearly, from the 1870s the first generation of professional physicists had won their positions as experimentalists, and much of their teaching serviced the rapidly expanding electrical and heating industries. Now their students fashioned new disciplinary roles in a search for places. But Poincaré posed his question first as an epistemological one, and answered that the reason mathematical physics existed was the need to *generalize* from limited facts in order to be able to *use* our observations. Then he developed a comparison to explain the role of mathematical physics in the economy of the scientific machine. Science, he said, is a library that must go on increasing indefinitely. With limited funds for purchases the librarian must strain every nerve to prevent waste. Experimental physics has to make the purchases and alone can enrich the library. The duty of mathematical physics is to draw up the catalogue—if well done the library is none the richer, but the reader can utilize its riches. And finally, by showing the librarian the gaps in his collection mathematical physics can help him make a judicious use of severely limited funds.[15]

Poincaré's choice of metaphor should be taken seriously. Our study of material culture and instruments in Paris has already illustrated the extraordinary practical and symbolic importance of single books—two collective catalogues. Now consider for a moment the symbolic power of books en masse and the disciplinary significance of editorial direction. In the late nineteenth century, no group was more aware of the cultural value of libraries than Germans. One room of the German palace in Paris was devoted to printing and books, another to reading and writing, and a third reproduced the library of Frederick the Great. Nothing represented the intellectual life of the nation better than shelves of books. Seven years

that accrued from scientists becoming their own epistemologists, at xii. Recent studies of Poincaré include Galison, *Einstein's Clocks* (2003); Zahar, *Poincaré's Philosophy* (2001).

15. Poincaré, "Physique expérimentale et la physique mathématique" (1900), 4; Poincaré, *Science and Hypothesis* (1905), 144–45.

earlier, Poincaré's German colleague and mathematical rival Felix Klein had chosen a library to display the power and productivity of his discipline, in a display on the German University at the Chicago World's Fair. Klein transported to Chicago copies of every mathematical dissertation written in a German university since 1750, about five hundred textbooks by German authors, the complete sets of seven German journals, and the proceedings and publications of the scientific academies of Berlin, Göttingen, Leipzig, and Munich. Thus Klein created in Chicago an image of the Reading Room that was the central site and intellectual workplace of his Göttingen fiefdom.[16] He also ensured that over half the papers in the Mathematical Congress held in Chicago were given by German mathematicians (though most were read in their absence), and helped staff new departments of mathematics in the United States with his followers.[17] Of course Klein was not content just to fill the shelves; helping determine what made its way there offered an important way of promoting disciplinary reform. A tireless critic of narrow conceptions of mathematics, Klein soon embarked on a whole array of efforts designed to increase the intellectual and institutional engagement of mathematicians with applied fields, and especially engineering. In the 1890s he founded the massive six-volume *Encyclopedia of the Mathematical Sciences*, using the position of editor to present his integrated view of the role of mathematics in an age of specialization and traveling Europe and crossing disciplinary boundaries to solicit articles on the full range of mathematics and its applications. The first volume appeared in 1898, the last three volumes dealt with physical topics, and the project took thirty years to complete.[18] So the sheer expanse of library shelves might express national and disciplinary

16. On Klein's concept of the Reading Room, see Frewer, "Mathematische Lesezimmer" (1979).

17. On the Mathematical Congress, see Parshall and Rowe, "Embedded in the Culture" (1993). Having just seen his students staff the department of mathematics in Chicago, Klein described America as ripe for scientific colonization but thought Germany could learn from its open, flexible education system with significant involvement from industry. Felix Klein, Bericht über die Reise nach Chicago zwecks Teilnahme am mathematischen Congresse, Klein Nachlass IC, NSUB Göttingen, quoted in Parshall and Rowe, "Embedded in the Culture" (1993), 45. For different interpretations of Klein's work within the mathematics and engineering discipline, see Manegold, *Universität, Technische Hochschule und Industrie* (1970); Pyenson, *Neohumanism* (1983); Rowe, "Essay Review" (1985).

18. *Encyklopädie der mathematischen Wissenschaften mit Einschluss ihrer Anwendungen* (1898–1935). For a highly interesting study of Klein's aims within the mathematics discipline including a discussion of the encyclopedia project, see Rowe, "Klein, Hilbert, and the Göttingen Mathematical Tradition" (1989), 204–10.

vigor, and determining how they should be used was still more critical for disciplinary progress. This was true both within and across disciplines. On the broadest scale, Klein and Poincaré had both participated in an international committee and several international conferences steering efforts for an international catalogue of scientific literature.[19] More immediately, their concerns highlight the critical intellectual and disciplinary significance of the relations between mathematics and physics at the turn of the century.[20]

Poincaré's address focused on the importance of the *catalogue* for the library of science. How did the French committee and the proceedings editors direct their discipline and point to gaps in its holdings? The congress and its reports were organized in seven sections: general questions and units and measures together composed the first section; the second section was devoted to mechanical and molecular physics; the third to optics and thermodynamics; the fourth to electricity and magnetism; the fifth to "Magneto-optics, cathode rays, uranium and etc."; the sixth to cosmic physics; and the final section to biological physics. In this ordering we can read the imprint of disciplinary priorities and an implicit history.

The premier place given metrology should be no surprise given its importance for the relations among physics, business, and government. Our study of Michelson and Rowland has emphasized the highly productive and multifaceted nature of their research on a range of standards, as they established careers and founded new laboratories. Many of the most important meetings and committees associated with previous international exhibitions had taken as their sole subject the negotiation of standards that could be agreed and put into practice internationally. International standards in weights and measures were established by committee in association with the 1867 Universal Exposition in Paris, and various international meetings addressed the issue of establishing a system of electrical units from 1881. The questions involved were of far more than

19. Ironically, the attempt to introduce worldwide bibliographic practices was hampered by German resistance to going behind the titles to establish a comprehensive subject index that would deal more fully with content. See the reports in *Nature* and *Science*, "International Catalogue of Scientific Literature" (1899) and "The Third International Conference on a Catalogue of Scientific Literature" (1900). Catalogues did not meet universal approval, however; Nebel laments the official catalogue guide to the 1900 congress in Nebel, "Die Weltausstellung und ihre Kataloge" (1900).

20. For a recent overview of the varied terms in which mathematical or theoretical physics was pursued circa 1900, see Seth, "Crisis and the Construction of Modern Theoretical Physics" (2007).

merely pragmatic significance too; Peter Galison has recently argued that Poincaré's work on such committees helped shape his epistemological stress on the role of freely chosen *conventions* in the conceptual development of science.[21] In 1900 the measurement section of the congress took its main task to be to determine the present state of metrology as a whole (on the basis of a detailed review of the history of standards development). Close attention was given to metrological and legal definitions of standards, the legal status of the electrical units in different countries, and to improving several definitions. The metrological strengths of Johns Hopkins now resulted in Joseph Ames presenting the report on the mechanical equivalent of heat—in a group of papers that shows the organizing committee distributing reports in metrical thermodynamics between Britain, the International Bureau of Weights and Measures, and the United States.[22] Similarly, the eighty reports across seven sections of the congress were delivered by members of thirteen different nations, surely reflecting an attempt to draw on the relative research strengths of different segments of the international community. But the committee turned to local researchers more often than others, and the distribution was certainly not proportional to the size of different communities. French physicists were responsible for thirty-six reports, while Germany contributed ten, Britain six, the Netherlands and Russia four, Italy and Austria-Hungary three, and a host of smaller nations contributed two, as did the United States. We will observe the consequences of particular choices across nations and research traditions when we investigate how Michelson and the electron were treated in Paris.

The deliberations of the first section led to the few concrete resolutions that were to emerge from the congress. On the one hand, specific

21. Schaffer, "Empires of Physics" (1994); Schaffer, "Metrology, Metrication, and Victorian Values" (1997); Galison, *Einstein's Clocks* (2003).

22. In addition to Ames's paper, reports were given by Chappuis (of the International Bureau), E. H. Griffiths (Sidney Sussex College, Cambridge), and Carl Barus (Brown University) on precision thermometry, the specific heat of water, and pyrometry. Their work enabled the congress to recommend the adoption of the mechanical cgs units of the joule and the erg for the expression of calorimetric quantities. Summary reports of the meetings of the Metrology section are given in "Procès-verbaux sommaries" (1900), 10–14; "Procès-verbaux" (1901), 12–18. The resolution is in "Procès-verbaux" (1901), 55. The papers were: Chappuis, "L'échelle thermométrique normale" (1900); Barus, "Progrès de la pyrométrie" (1900); Ames, "L'équivalent mécanique de las chaleur" (1900); Griffiths, "Chaleur spécifique de l'eau" (1900).

recommendations were made concerning units for pressure and calorimetric experiments; the subdivision of the spectrum into infrared, visible, and ultraviolet regions; and use of the term *density*. But the congress also promoted a far more general institutional need on the other hand. At its final meeting as a whole, congress members voted unanimously to support the resolution that in view of the immense advantages for science and industry stemming from national laboratories of technical physics analogous to the Physikalisch-Technische Reichsanstalt (PTR) in Charlottenburg, it was a matter of urgency for those nations that did not already possess such a laboratory to create one.[23] Institutionally, German physics offered a model worthy of imitation, and in this the lessons of the exhibition and the congress went hand in hand. Congress members might have seen the arguments for this proposition in Pellat's report on national laboratories of physics and technology, embodied in the German exhibits and catalogues at the exhibition, and in the numerous contributions to the congress from physicists associated with the PTR. Perhaps the most prominent example of the latter was a set of reports on the theory of radiation, blackbody radiation, and gas emissions by Wilhelm Wien, Otto Lummer, and Ernst Pringsheim. In its broadest aims their research was designed to address industrial needs in the heating and lighting industries—with the competition between gas and electrical lighting vividly represented in the displays of the 1900 exhibition.[24] In their congress papers, they subjected Max Planck's formulation of Wien's radiation law to searching criticism in an early stage of the dialogue that was to push Planck toward a successful law of blackbody radiation several months later (thereby opening up what was later to become known as quantum theory).[25]

23. "Procès-verbaux" (1901), 55.

24. Cahan's valuable study describes the general goals of the PTR as well as addressing the scientific and industrial aims that drove its research on blackbody radiation in particular. See Cahan, *An Institute for an Empire* (1989), 143–57.

25. Pellat, "Laboratoires nationaux physico-techniques" (1900); Wien, "Théoriques du rayonnement" (1900); Lummer, "Rayonnement des corps noirs" (1900); Pringsheim, "L'émissions des gaz" (1900). Pellat's address attested for the necessity of both fundamental metrological research and more routine instrument testing, but articulated a clear distinction between each. Britain had already begun steps to found a National Physics Laboratory (with its director pointing to the success of the German instrument exhibits in Paris to help justify the institution). In 1901 the French government created a Laboratoire d'Essais to undertake routine testing while simultaneously increasing the budget to the International Bureau, now

The broad categories that the second, third, and fourth sections represented indicate principal subject divisions in turn-of-the-century physics, and then and now might well have provided the titles of a series of undergraduate courses, or textbooks: Mechanics and Molecular Physics, Optics and Thermodynamics, Electricity and Magnetism. Their ordering also reflects, if only crudely, the temporal development of a discipline that had taken mechanics for its foundation following Newton's work, established distinctive approaches to thermodynamics in the nineteenth century and that had recently seen enormous strides in electromagnetic theory.[26] For many, the links between the different sections might have been as remarkable as the diverse phenomena and fields of expertise they represented. In the third section on optics and thermodynamics, for example, Heinrich Rubens presented a report on his research extending the spectrum into the infrared, demonstrating experimentally the connection between long light waves and electrical waves.[27] Guillaume found this work similar to Righi's treatment of Hertz's electric waves in the fourth section on electricity and magnetism, and suggested that the line of demarcation between these sections was becoming more and more difficult to draw.[28] He noted, too, the increasing convergence of experimental measurements of the velocity of light with the values obtained for the propagation of electric waves determined either by direct experimental measurement or by the ratio of the electromagnetic and electrostatic units. Maxwell's theory of the identity of luminous and electrical oscillations was strongly emphasized by this convergence.[29] We have seen Michelson conveying a similar

---

substantially freed from that task. The United States opened its National Bureau of Standards in 1902. Michelson coveted the directorship and was incensed when it went to his Chicago colleague S. W. Stratton (after Morley declined the position). See Glazebrook, "Aims of the National Physical Laboratory" (1901), 344; Pyatt, *National Physical Laboratory* (1983), 1–35; Bigg, "Behind the Lines" (2002), 24–28; Livingston, *Master of Light* (1973), 208–9.

26. To provide just one example, Cajori divides his 1899 history of physics into centuries. His treatment of the nineteenth century has sections on Light, Heat, Electricity and Magnetism, and Sound; that of the seventeenth and eighteen centuries begins with a section on Mechanics. He describes the nineteenth century as having overthrown the leading theories of the previous century to largely build anew "on the older foundations laid during the seventeenth century." Cajori emphasizes recent unification and intimate relations between "wide realms of physics, once thought to be perfectly distinct," but thinks it will be left to the future to "reveal fully the mysteries of the structure of matter and the ether. Are not both of them dynamical systems, subject to the laws of motion, of momentum, and energy?" Cajori, *A History of Physics* (1899), 137–38.

27. Rubens, "Spectre infra-rouge" (1900).

28. Righi, "Ondes hertziennes" (1900).

29. Guillaume, "International Physical Congress" (1900), 427.

message in his paean to the unifying power of the ether, but as we shall see, Poincaré's keynote address was to critique the status of the ether and problematize the relations between mechanisms and unity. In general the diverse reports of the 1900 congress gave little breathing space to broad theoretical programs for unity, focusing participants on a more fine-grained exploration of empirical connections across different sections.

However much particular phenomena might dissolve the borders between the intellectual divisions of the catalogue, those borders were clearly suspended on reaching the fifth section—which rather than any gap in the collection represented a profusion of new phenomena currently challenging categorization. This was variously titled with the open list of the most dramatic discoveries and recent areas of concern, "magneto-optics, cathode rays, radioactivity and etc.," or with a general concept that linked so many of them—"electro-optics and *ionization* [emphasis added]."[30] The editors had no doubt that its subjects would fall naturally into one of the preceding sections within a few years, but as Guillaume wrote in his report to *Nature*, "at the present time they are still so undefined, they open up such new horizons, that it appeared well to collect them in a special section. The idea proved very fruitful, for the section was largely attended, and the discussions at it proved very fascinating."[31] Following this open category, the final two sections reflected the application of physical tools and perspectives to different fields: to the physical environment of the earth and sun in the sixth section on cosmic physics, and to biology in the final section.

Thus the catalogue that physicists gave their subject at the turn of the century shows them moving toward a unity through the empirical continuity of the spectrum on the one hand and recognizing a fluidity and openness in encountering new phenomena on the other. More general directive perspectives were offered most clearly in the keynote and plenary addresses of Henri Poincaré and Lord Kelvin. Poincaré's task was to deliver a preface, a general overview, and as well as tackling the epistemological challenges facing physics he offered a critical perspective on the central conceptual difficulties posed by recent theory. The bulk of his address considered the problematic of rapid change—a preoccupation of

30. The conference program offered the long list in various forms, while the third volume of the published proceedings was subtitled "Électro-optique et ionisation.—Applications.—Physique cosmique.—Physique biologique."

31. Guillaume, "International Physical Congress" (1900), 427–28.

many who gave "turn of the century" addresses in physics. For Poincaré, rapid change had left foundations uncertain and the new radiations had opened up a new world. Was science therefore bankrupt? The answer came in a more subtle understanding of historical change. Poincaré described scientific progress as a continual evolution between the achievement of apparent simplicity and the recognition of new complexity. It was not image or ontology that provided secure foundations but the true relations expressed in hard-won equations, and especially in general principles. Rather than explanatory mechanisms, the true goal of physics was unity, and Poincaré devoted particular attention to developing a critical stance toward the hypotheses guiding physics. He warned his audience of the need to root out the most dangerous form of hypotheses, tacit assumptions, and celebrated the role of mathematical precision in that task. Then he showed the purchase his epistemological concerns gave on recent progress, in particular by engaging in an extended discussion of electrodynamics and Lorentz's theory (considering also Larmor's specific stances on several questions). Poincaré raised the prospect that the ether would be abandoned as an abstract entity, asserted his confidence that only relative motions would ever be observed, described Lorentz's electrodynamics as riven by hypotheses in its explanation of the absence of ether drift to the second order, and critiqued Lorentz's readiness to see the mechanical principle of action and reaction distributed across ether and matter rather than matter alone. Thus, while many other fields came under his purview, Poincaré's most sustained examination of recent physics focused on conveying why he thought that despite its extraordinary success in unifying different fields, Lorentz's theory would require real modification.[32]

Poincaré had been given the unusual license to step above the catalogue and the specific sections of the congress. Lord Kelvin claimed the right to impart critical direction. Kelvin had given his last lectures at the University of Glasgow a year earlier, forty-four years after encouraging students of natural philosophy to join him in his experimental research and thereby bringing practical laboratory training into an undergraduate education in physics for the first time. It was surely appropriate that

32. The discussion of mechanisms, ether, and Lorentz and Larmor's theories dominates the second half of his address. Poincaré, "Physique expérimentale et la physique mathématique" (1900), 19–26; Poincaré, *Science and Hypothesis* (1905), 161–77. On Poincaré's work in electrodynamics more generally, see especially Darrigol, "Poincaré's Criticism" (1995); Darrigol, "Einstein-Poincaré" (2004).

the congress voted him the position of honorary president at its opening meeting and gave him the honor of a plenary lecture to begin the considerations of the third section in particular. From 1847 his critical engagement with the work of Joule and Carnot had helped define the field of thermodynamics, articulating the great unifying concepts of the conservation and dissipation of energy and promulgating a new science of energy with reforming zeal. Rather than simply delivering the report he had written for the congress, Kelvin was inspired by Poincaré's earlier lecture to embark on what Guillaume described as a "brilliant improvisation on the constitution of the ether."[33] This had been a preoccupation for decades, and one in which Kelvin possessed a sharp critical sense that was driven in part by concerns he knew had been abandoned by many about him, and in part by a clear awareness of problematic issues.

Kelvin had already given one "turn of the century" address before coming to Paris, a lecture to the Royal Institution in April in which he identified two dark clouds on the horizon of the dynamical theory of light and heat. In his formal report for the congress, Kelvin took up only the first of those difficulties but pursued it in more detail.[34] His major concern was with the still unsolved question of how the earth could move through an elastic solid—such as he supposed the luminiferous ether to be. Kelvin's insistence on treating the ether as an elastic solid had become idiosyncratic. First Maxwell and more recently Lorentz and Larmor had developed an approach to electromagnetic theory that Kelvin persisted in regarding as unphysical. They were content to represent the ether through a set of equations (that satisfied the general mechanical principle of least action) without subjecting that medium to the stronger restraints embodied in the mechanical theory of light. He was well aware, however, that his present study remained an exploration of abstract mathematical dynamics, without physical application, and without moving into the terrain of electromagnetic theory.[35] On the floor in Paris, Kelvin did move into that terrain, offering a prospective sketch of a fuller theory that incorporated atoms of electricity—which, following Larmor, he called electrons

33. Guillaume, "International Physical Congress" (1900), 426. On Kelvin and thermodynamics, see especially Smith and Wise, *Energy and Empire* (1989); Smith, *The Science of Energy* (1998).

34. The general address was: Kelvin, "Nineteenth Century Clouds" (1970 [1900]). The congress paper was published in French and English versions. Kelvin, "Mouvement d'un solide élastique" (1900); Kelvin, "Motion Produced in an Infinite Elastic Solid" (1900).

35. Kelvin, "Mouvement d'un solide élastique" (1900), 1.

("at present").[36] Nevertheless, the terms in which Kelvin identified a particularly stubborn problem were probably more important to his contemporaries than his sketchy attempts to provide a mechanical theory of electromagnetism. Kelvin allowed that he still had to reconcile his hypothesis "with the result that ether in the earth's atmosphere is motionless relatively to the earth, seemingly proved by an admirable experiment designed by Michelson, and carried out with most searching care to secure a trustworthy result, by himself and Morley." Kelvin could see no flaw in either the idea or the execution of the experiment, but he did recognize a possibility of escaping from the conclusion through the "brilliant suggestion made independently by FitzGerald and by Lorentz of Leyden, to the effect that the motion of ether through matter may slightly alter its linear dimensions."[37]

Kelvin and Poincaré may well have appeared to their contemporaries as representatives of the old and new physics, and they dramatize well the extent of change within the discipline. Yet before we assimilate this too quickly to our present understanding of the transformation from classical to modern physics, we should note several features of their addresses. First, it is apparent that despite their different stances toward the past and future, both pointed their contemporaries in the same direction, focusing on present difficulties in electrodynamics. Further, taking up quantum theory in chapter 7, I will claim that there are good reasons not to describe Kelvin as a "classical" physicist, despite his allegiance to mechanical explanations. Examining the second cloud Kelvin identified on the horizon of the dynamical theory of light and heat will show him warning physicists *not* to accept at least one principle we now associate closely with classical physics—the equipartition theorem. Our study there will highlight surprising tensions in when and why *specific* theories like Planck's radiation law or Lorentz's electrodynamics first came to be regarded as classical or modern. But just as importantly, our present understanding of that conceptual contrast clearly involves very general features of the physics discipline and its history. So my final observation is to note the absence of the term *classical* in these addresses. It is particularly significant that despite his concern with characterizing the nature of change and even given his advocacy of unity over mechanical foundations, Poincaré made no

36. Ibid., 20. He could have followed Poincaré too. Poincaré, "Physique expérimentale et la physique mathématique" (1900), 17–18; Poincaré, *Science and Hypothesis* (1905), 164–65.

37. Kelvin, "Mouvement d'un solide élastique" (1900), 19.

reference to "classical" theory or a "classical" era in his presentation in Paris. We will set this omission in the broader context of Poincaré's thought and the work of his contemporaries in chapter 7.

Interferometric Standards of Length and the Velocity of Light in Paris, 1900

The most general level at which the congress offered the physics discipline intellectual leadership was marked, then, by the need to comprehend the implications of rapid change and critical perspectives on recent work in electrodynamics. We will now explore the way its proceedings represented current research, and pursue our study of the way scientific instruments or results travel, by opening up specific reports to examine first the changing status of Michelson's work, and then the fortunes of a new particle, the electron. The references to the implications of the Michelson-Morley ether-drift experiment that marked Poincaré and Kelvin's keynote addresses were far from the only occasions on which Michelson's work was discussed in Paris, but the two principal contexts in which it was raised saw rather diverse treatments of his results. While congress reports showed his determination of the meter in terms of the wavelength of cadmium to be foundational to an expanding field of interferential metrology, his velocity of light measures were discussed far more critically.

Reflecting the fundamental status of length measurements, the first paper to follow Henri Poincaré's overview in the congress volumes was an account of standards of length from J.-René Benoît, director of the International Bureau of Weights and Measures (and someone we met through his role in Michelson's meter campaign). After comparing the standards in different nations and discussing the prototype international meter, Benoît turned to the permanence of standard measures and the possibility of natural units. Here he discussed interference measures, stating that Michelson's improvement of interferential techniques and discovery of new monochromatic sources of light had together offered the first solution to the problems facing the attempt to render light a natural standard, capable of controlling the permanence of the fundamental prototype. Benoît was of course very familiar with Michelson's work, but had no need to describe it in detail. Pointing readers to Macé de Lépinay's congress report on interferential metrology, Benoît wished only to add that the determination had achieved an accuracy in the region of 1 in 1 million, and that repeating it with the benefit of the experience gained and further improvements in apparatus might enable an accuracy approaching that with which the meter itself could be defined. He mentioned too, that Fabry and Pérot had developed an instrument that would "provide another means of solving the same problem, following a similar general procedure."[38]

Benoît's focus on the future and his reference to Fabry and Pérot's technique are both revealing, indicating that the Bureau of Weights and Measures was not content to let its stewardship of length measures rest with maintaining the present international prototype alone. While Michelson's instrument had gained a great deal of credibility from his determination in Sèvres, that collaboration had also enabled the bureau to increase its involvement in research rather than testing. Now the 1900 Congress of Physics offered a highly visible way to promote the vitality of the International Bureau both within France and internationally. Macé de Lépinay's report indicated that the collaboration between the bureau and Marseilles scientists had helped focus renewed attention on the metrological possibilities of interferential methods. After offering a preliminary theoretical account of interference measurements, Macé described the "memorable" researches of Michelson and Benoît at length (without mentioning the rival determinations from diffraction gratings).[39] Then he outlined more briefly three methods that had offered a metrological determination of the end-to-end length of a solid body in terms of wavelengths (rather than between line measures). These were determinations of the volume of a quartz cube employing his own use of Talbot's fringes, Fabry and Pérot's techniques with multiple reflection between lightly silvered panes of glass, and a version of Michelson's apparatus that the bureau scientist Chappuis had used in 1897.[40] Interestingly, Macé compared the results of these measurements by reducing them to the mass of a cubic decimeter of distilled water, but had earlier noted that the wavelength measures they offered could be expressed in metric units thanks

38. Benoît, "Détermination des longueurs en métrologie" (1900), 69–70.

39. Macé de Lépinay, "Déterminations métrologiques par les méthodes interférentielles" (1900), 121–27. Macé explicitly referred to the collaboration, thereby heightening the validity of what could otherwise be regarded as a single result. In contrast Michelson typically referred to Benoît without naming him. Michelson, *Light Waves* (1902), 98.

40. Macé de Lépinay, "Déterminations métrologiques par les méthodes interférentielles" (1900), 127–30.

to Michelson and Benoît's absolute determination. Macé's report shows that Michelson's work played a foundational role for a range of interferometric measures by rendering them absolute, linked to metric measures. We have some indication of how other physicists evaluated the complex of studies Macé described too: reviewing the report for the *Physikalische Zeitschrift*, Johannes Stark referred readers directly to Michelson's publications rather than attempting what could only be a brief overview, but did take the trouble to outline Fabry and Pérot's method because he considered it both valuable and still little known.[41] As we have seen, if Michelson's interferometric techniques were known in Germany by 1900, they were seldom utilized. Drude judged them marginal to optical spectroscopy; German physicists had largely neglected Michelson's instruments; and they were absent from the Zeiss displays of refractometers in 1900.

The relations between French optical research and the interests of the International Bureau of Weights and Measures helped ensure the centrality of Michelson's research to metrology, and thereby guaranteed its prominence in the proceedings of the International Congress. Yet Michelson's measures of the velocity of light were in danger of disappearing from view, despite the fact that Alfred Cornu's address on that subject gained significant attention, garnished as it was by displays of Fizeau and Foucault's original apparatus.[42] Cornu had measured the velocity of light himself, several years before Michelson and Newcomb entered the field. His report showed that two decades later the president of the French Physical Society was not ready to accept the values the Americans had obtained without raising serious questions about both the revolving mirror method they employed and their error estimates.

We saw in chapter 2 that Michelson and Newcomb carefully evaluated the relative merits of different experimental arrangements as they designed their apparatus; and Newcomb's correspondence showed that once the experiment was underway, reconciling his values with earlier results was critical. His final publication indicates that by the time he finished his work Newcomb had a different view. In a section titled "Comparison of results," Newcomb listed results in two groups. The first group included Michelson's 1879 and 1882 values and two values from his own

41. Stark, "Macé de Lépinay, Massbestimmungen" (1900).

42. Simon, for example, commented on the historical interest of Cornu's presentation. Simon, "Physikerkongresse" (1900), 572.

experiments: one a limited selection of results "supposed to be nearly free from constant errors," the second including all runs with diverse weightings for different series of results. Then came a second group of five results, which "may be added," Newcomb wrote, "for reference." They included results obtained by Foucault, Cornu in 1874 and 1878, a revision of Cornu's second determination by Listing, and Young and Forbes's result. Despite the *implication* that his final figure would reflect the experience of all researchers, the significance of Newcomb's comparison would have been lost on any careful reader. Giving each determination equal weight, the mean of the four U.S. experiments is 299,858.25, while the mean of all determinations is 299,745 km/s. Rather than either of these values, the final value Newcomb offered as "the most probable result" was his own best result of 299,860 with a probable error estimate of ±30 km/s (somewhat higher than the error estimate he stated for his own runs alone). Without any formal discussion of appropriate weightings, Newcomb apparently gave his own "best run" results a higher weight than any others; one might even conclude that other results and especially those of non-U.S. researchers were included *only* for reference.[43]

As one might expect, Cornu evaluated this complex of results very differently (and it is worth noting that there had been no new experimental determinations since the early 1880s). His congress report presented a critique of methods that was motivated by the impossibility of reconciling (his own) toothed-wheel measurements with what he commented were currently regarded as the most precise results (Newcomb's), at least without multiplying Newcomb's probable error estimate tenfold. Cornu described both Michelson's and Newcomb's arrangements, but regarded Newcomb's more elaborate experiment as more authoritative. In his detailed discussion of results and methods, Cornu drew critical attention to the behavior of light in the neighborhood of, and reflected from, a rapidly revolving mirror. In particular, he questioned whether the ordinary laws of reflection applied to a revolving mirror or to a light ray whose axis moved with a velocity comparable to that of light; and he questioned whether the entrainment of light waves in the vortex of air close to the

43. Newcomb, "Measures" (1882), 202. A year earlier Todd had given Cornu's and Michelson's first determinations a weight of 1, while he weighted the second determinations of Cornu (rediscussed by Helmert) at 25 and Michelson at 100. Todd, "Solar Parallax" (1880). In contrast, Gibbs used Newcomb's value, and Preston described Newcomb's determination as "perhaps the most reliable and complete." Gibbs, "Review" (1886); Preston, *Theory of Light* (1895), 502.

mirror would affect the displacement. Thus, although Cornu accepted that Michelson's and Newcomb's error estimates were appropriate to the numerical range of their results, he argued the operation of some systematic error had to be considered a serious possibility. Cornu's probable value of the speed of light was the arithmetic mean of determinations from the two methods: 300,130 ± 270 km/s.[44]

The 1900 congress offered physicists the chance to set past achievements against present standards, as a community. Its keynote addresses and reports show that Michelson's contemporaries placed new scrutiny on his velocity of light measures, on the status of electrodynamic theory and the ether in the light of the null result of the Michelson-Morley experiment, and on the metrological achievements of interferential techniques. In each instance, rather than stating a definitive consensus, these reports offered a snapshot in time. They reveal and in some cases heighten—without finally settling—a long-standing complex of issues in the assessment of methods, results, and their implications. Briefly tracing the way these particular reports were picked up in subsequent years, and thereby helped shape debate on the issues they identified, shows that their effect was felt in different ways. Cornu's report on the velocity of light quickly got both Lorentz's and Michelson's attention. In 1901 Lorentz considered and dismissed Cornu's criticisms concerning the behavior of light; a year later Michelson offered new ways of reconciling firstly the results and secondly the methodological differences at issue between the followers of Foucault and Fizeau. Michelson engaged in a more inclusive weighting exercise than either Newcomb or Cornu and suggested a new experimental arrangement using a combination of a revolving mirror and grating to interrupt the light.[45] Yet in this instance the exchanges stimulated by Cornu's report remained largely abstract. Despite his speculative foray

44. For the toothed wheel method, Cornu used a value that he had obtained in 1878 (300,400 ± 300 km/s), and he accepted Newcomb's value for the revolving mirror method. Cornu, "Vitesse de la lumière" (1900), 225 and 246 (for his conclusions) and 230–46 (for his discussion of the revolving mirror results and critique of the arrangements). Also in 1900, Drude gave the average of determinations from Cornu, Michelson, and Newcomb, while Schuster later described Cornu's congress paper as giving a very clear discussion of the relative merits of the different methods and accepted Cornu's 1900 figure. See Drude, *Lehrbuch der Optik* (1900), 112; Schuster, *Theory of Optics* (1904), 37.

45. See Lorentz, "Méthode du miroir tournant" (1901); Michelson, "Velocity of Light" (1902). He noted that a considerable diversity of opinion would attend the attribution of relative weights to various determinations, but also that there was no doubt that the weights to be assigned varied enormously. Then he focused on the authoritative status of the later values of Cornu, himself, and Newcomb to offer another most probable value for the velocity of

into methods, for example, Michelson was not to undertake new experimental determinations until the 1920s.

In contrast, congress discussions of both ether drift and interferential metrology were soon elaborated into more extensive debates and significant new research (which we explore in greater detail in following chapters). The renewed focus on ether drift in Paris directly prompted new contributions on both theoretical and experimental fronts. Poincaré himself sharpened his presentation of the conflict between Lorentz's recourse to an absolute frame of reference and the principle of relativity in both his popular writings and technical papers on electrodynamics; his criticism prompted Lorentz to gradually strengthen his theory by integrating the contraction hypothesis ever more fully into the fabric of his electrodynamics. Kelvin's views led Edward Morley and his student Dayton Miller to revisit the ether-drift experiment. Kelvin might have seen no flaw, but in conversation after his address he urged the two Cleveland physicists to repeat it with more powerful apparatus. In 1902 and 1903 Morley and Miller carried out experiments with wood and steel interferometers. Now they framed the experiment as a test of the Lorentz-Fitzgerald hypothesis concerning the relations between ether and matter. In particular they aimed to establish whether different materials were affected by the hypothetical contraction in a similar way, and whether magnetic attraction on the iron parts of the instrument had affected earlier observations.[46] In 1905 the effect of height and optical exposure were also tested by working in a temporary hut constructed on a Cleveland hilltop (at an altitude of about 285 meters), with glass windows rather than opaque walls of brick or stone. These experiments did not reveal an effect of the magnitude expected on the hypothesis of a stationary ether; but Miller did note small positive effects that stimulated him to return to ether drift in the 1920s.[47]

light as 299,890 ± 60 km/s. (Neglecting earlier work Michelson gave equal weight to Listing's value of Cornu's results, his own second determination, and Newcomb's result from all observations.) Michelson's paper showed that he had also paid close attention to congress reports on the measurement of the ratio between electromagnetic and electrostatic units, and of the velocity of Hertzian waves, describing the increasing accuracy of a group of measurements that reflected a wide network of theoretical issues and experimental results.

46. Miller, "Ether-Drift Experiment" (1933), 208.

47. Loyd Swenson has shown that Miller subsequently undertook an extended series of experiments on a mountaintop and throughout the course of the year in the 1920s. By this time he framed his experiments as a test of the theory of relativity, noting, "The tests of the theory of relativity, made at the solar eclipse of 1919, were widely accepted as confirming the

If anything, congress reports on interferential metrology had sidestepped potential debate by according Michelson's cadmium measures fundamental status without detailing their relations to competing determinations with grating spectroscopes. With their position strengthened by the strong support of the International Bureau, in 1902 Fabry and Pérot entered specific challenges to grating wavelength standards. In doing so they enfolded Michelson's absolute values into an explicit debate on methods that was soon to become central to the spectroscopical community.

Thus our study of the exhibition and the congress shows that in 1900 Michelson's reputation and the use of his work in the international community rested on a broad platform of related endeavors. In the exhibition halls, his instruments were put on show by a university looking for international recognition. The refractometers he had designed were displayed by a U.S. instrument maker seeking customers, and, less visibly, the precision their manufacture demanded was also embodied in the many prisms that Brashear put on show. Yet competing displays by the German firm Carl Zeiss suggest that precision optical measuring devices were increasingly important in scientific and technical work, but also indicate that such instruments drew on diverse sources and might not require particular reference to Michelson. In the Congress of Physics, use of Michelson's work rested in part on discussion of the implications of his ether-drift experiment, but also on diverse evaluations of the weighting to accord different methods both in velocity of light and wavelength metrology. His contributions to these fields fared very differently in Paris, but in each case the long-term fortunes of experiments first carried out in the 1880s and 1890s would have to depend either on comparative exercises in weighting or on the more secure sanction of experimental repetitions and redeterminations—and in 1900 these were still in the offing. Now consider how congress reports registered the pulse-beat of a new discovery.

---

theory. Since the Theory of Relativity postulates an exact null effect from the ether-drift experiment which had never been obtained in fact, the writer felt impelled to repeat the experiment in order to secure a definitive result." The positive result Miller announced drew some attention, but by then offered too little and too late to convince most physicists to doubt the conclusions that had already been drawn on the basis of the less systematic experiments previously undertaken, or a theory that was then regarded to have several independent sources of support. See ibid., 217; Swenson, *Ethereal Aether* (1972).

The "Electron" in Paris, 1900

"Electro-optics and ionization." The use of the word ionization to link together the diverse subjects of section five of the congress and its reports characterizes well the tentative sense of unity physicists felt in regard to "Magneto-optics, cathode rays, radio-activity and etc." Here are some of the names given to the charged bodies giving rise to spectral lines in magneto-optics (the Zeeman effect), the emissions inside the cathode ray tube, and those that issued from radioactive elements: ions, electrons, lightions, cathode rays, corpuscles, electrions, Becquerel rays, beta-rays. Almost as many names as there were serious investigators of the subject. In 1900 the concept that all agreed linked them was that of the ion. They were charged particles, specifically, negatively charged particles. The direction of their deflection when passing through an electric field had shown that. What else they shared, and their implications for physics in general, were at issue.

Looking back on this period, it has long been customary to attribute the discovery of the electron to J. J. Thomson in a series of papers published between 1897 and 1899, in which he established that the particle emitted from the cathode had a charge to mass ratio significantly higher than that of hydrogen, and on this basis argued for the existence of a subatomic particle. However, recent historical studies have raised important questions both about the contributions of other physicists and the concept of discovery more generally.[48] They can be posed quite sharply by noting, first, that the concept of the electron predated Thomson's work. In 1891 George Johnstone Stoney used that name to denote the elementary quantity of electricity passing in electrolysis, and Joseph Larmor adopted it to describe the positive and negative ions central to his electrodynamic theory from 1894. Second, Pieter Zeeman's 1896 discovery was quickly interpreted as evidence for ions with a mass smaller than that of hydrogen, giving new impetus to the development of Lorentz's and Larmor's existing theories. As we saw in chapter 3, Michelson was one of many who connected Lorentz's interpretation of the Zeeman effect to

48. The work of Falconer, Feffer, and Arabatzis has offered a subtle portrait of Thomson's research, and the question has recently been put on a still broader footing by the many excellent contributions to the edited volume, *Histories of the Electron*, and Arabatzis's recent book. See Falconer, "Corpuscles, Electrons" (1987); Feffer, "Discovery of the Electron" (1989); Arabatzis, "Rethinking the 'Discovery' of the Electron" (1996); Buchwald and Warwick, eds., *Histories of the Electron* (2001); Arabatzis, *Representing Electrons* (2006).

the electrons of electrolysis. Third, other researchers in Germany—Emile Wiechert and Walter Kaufmann—had measured the mass to charge ratio of cathode rays before or simultaneous to Thomson's first publication in 1897. And fourth, Thomson pointedly used the term *corpuscle* rather than identifying cathode rays with *electrons*. Following Michelson has scarcely prepared us for the quick density of these crisscrossing claims to significance. Like many scientists, most of Michelson's work trod the more stable ground of redeterminations and replications of measurements and experiments already carried out by others—working to pull the honor of a definitive measurement rather than *discovery* out of a thicket of competitive experiments. It is in his advocacy of a class of instruments that we have seen Michelson writing of invention and gradually reaching for the terminological novelty of the "interferometer." Discussion of the theoretical lessons to be drawn from the ether-drift experiment developed over a still longer period. While all reckoned early that the experiment had shaken theory, it was only much later—following lines of questioning like those Poincaré voiced in Paris—that its results were generally taken as constituting a kind of inverse discovery for the extent to which they threw the existence of the ether into doubt.

Recent work has drawn attention to different traditions about whether the discovery of the electron was multiple or singular. While pointing to the importance of Thomson's institutional basis and students at the Cavendish Laboratory, this scholarship has been unable to give a complete answer as to why Thomson eventually came to be attributed with the discovery so widely.[49] Without attempting to resolve that question fully, our study of the 1900 congress can clarify some of the issues involved. Discovery claims always have to meet a communal test. No matter how clearly we can distinguish the intent and ambition of individual scientists in one or another private moment or published paper, their views must be accepted by others—and widely accepted—to count as knowledge. The congress sought to be internationally representative and to provide a reliable guide to the current state of science. For that reason, studying treatments of the new entity in Paris will offer an excellent means of exploring the complex processes that mediated individual perspectives, specific research traditions (in different nations), and the views of the physics community more broadly. Notice first that the range of contributions solicited

49. Falconer and Gooday address this question most closely. Falconer, "Corpuscles to Electrons" (2001); Gooday, "The Questionable Matter of Electricity" (2001).

indicates the openness of the field of related studies, while the order of papers may well suggest the particular prominence of electrodynamic and spectroscopic traditions. The committee solicited papers from Dutch, German, French, and English physicists directly related to electron theory and cathode rays. (This is in strong contrast to their treatment of the optical fields in which Michelson worked, where they did not see any need to step outside the French community.) The papers concerned came from Lorentz, the German physicist Paul Drude, the Frenchman Paul Villard, and J. J. Thomson. Contributions from Becquerel and the Curies on radioactivity were placed between those of Drude and Villard in the published order, and their studies of the emissions given off by radioactive bodies also bore on the new entity.

The framing of Lorentz's paper, "The Theory of the Magneto-Optic Effect Recently Discovered," indicates quite clearly how important new empirical research could be, even for perceptions of a broadly successful theory. As we learned in chapter 3, Lorentz had been developing a theory of electrodynamics and optics based on the interactions between charged bodies (ions) and the electromagnetic field from 1892—several years before Zeeman proved that a strong magnetic field could affect the spectral lines given off by a body. Zeeman's work was universally described as a "discovery" and precipitated a particularly close interaction between theory and experiment that led to important changes in the electrodynamics of Lorentz and Larmor. Both theorists inferred that the negative ions concerned had a charge to mass ratio much higher than that of hydrogen. Using simple models of the effect of a magnetic field on a rotating ion, they were able to offer reasons to expect not just a broadening but tripling and polarization of the lines, subsequently observed by Zeeman. As we saw in chapter 3, Zeeman's work also significantly increased interest in related experimental research. Michelson, Cornu, Preston, and others all undertook spectroscopic studies that showed still further splitting of spectral lines into as many as nine components. Rather than shaking confidence in the theory as a whole, this was regarded as indicating the need to introduce less simplistic assumptions about the rotation of ions in complex atoms and molecules. By 1900 Lorentz was referring to "ions or electrons," the latter term having been popularized by FitzGerald and Larmor. In Paris, Lorentz gave an overview that focused more on theoretical issues than experiment, and discussed in particular recent suggestions by the German physicist Woldemar Voigt. Reviewing Lorentz's report for the *Physikalische Zeitschrift*, Max Abraham thought the journal's readers would already be familiar with Lorentz's work and Voigt's views through recent publications in that journal. Widespread awareness of this complex of studies is confirmed by the fact that at the Paris congress, Poincaré and Kelvin referred to the electrodynamic theories of Lorentz and Larmor (and both used the term *electron*).[50]

Paul Drude's paper, "The Theory of Dispersion in Metals, Based on the Considerations of Electron Theory," was one report the *Physikalische Zeitschrift* neglected to review, very likely because they judged his work already well known to their readers. Drude was a former student of Voigt's who had carefully weighed up the advantages and disadvantages of mechanical and electromagnetic theories of light from the late 1880s, before becoming an enthusiastic advocate of the electromagnetic view. Lorentz's ion theory helped stimulate Drude to explain the electrical, optical, and thermal characteristics of metals based on the assumption that charge carriers—positive and negative electrons—moved more or less freely within the metal, an approach he began to express from 1898. While electromagnetic theory operated with macrophysical properties, Drude favored the electron theory because it allowed him to probe the microphysical constitution of matter. In addition to his focus on dispersion in Paris, 1900 saw Drude develop his theory in a textbook on optics and articles in both *Physikalische Zeitschrift* and the *Annalen der Physik*.[51]

Both Lorentz and Drude referred to Thomson's work, if at all, as concerning the nature of cathode rays and—still more generally—the properties of gases. After describing how measurement of the distance between the doublets or triplets allowed a determination of the charge to mass ratio, Lorentz simply wrote that it was of the same order of magnitude as that found for the ions constituting cathode rays.[52] In his concluding paragraphs Drude wrote that the concept that seemed essential to the basis of his theory was that of the electron, "which has played such an important role not only in electrolysis, but also more recently in the representation of the properties of gases—mainly thanks to the work of J. J. Thomson."[53] The congress reports on radioactivity by Becquerel and the Curies similarly pointed to the identity between one component of the radiations given off by radioactive elements and cathode rays. The Curies

50. Abraham, "Lorentz, Über die Theorie" (1900).
51. See Goldberg, "Drude" (1970); Renn, "Einstein's Controversy with Drude" (2000).
52. Lorentz, "Théorie des phénomènes magnéto-optiques" (1900), 4–5.
53. Drude, "Dispersion dans les métaux" (1900), 45–46.

went into more detail in a discussion that shows they regarded Thomson as having developed and rounded out (or completed [complétée]) the ballistic theory William Crookes formulated, in which the cathode rays consist of extremely tenuous material particles carrying an electric charge.[54] References of this nature indicate researchers linking diverse fields together through common measures and/or bridging concepts. Sometimes, (as Drude did) subtle differences in those concepts were obscured by describing the work of one context using a label accepted in another (encompassing J. J. Thomson's research within studies of the "electron"). Negotiations over such names reflected different judgments about whether the similarities or differences were more significant, and in Michelson's move from "interferential refractometer" to "interferometer" we have an example that indicates how important it could be to label instrumental as well as natural categories in ways appropriate to the narrowness or breadth of the referent.

Thus those reading the congress volumes met Lorentz and Drude's engagement with electrons in electrodynamic theory, magneto optics and the theory of metals, as well as brief references to the identity between radioactive emissions and cathode rays, before they considered cathode rays in a report from Villard, followed by Thomson's paper, "Indications on the Constitution of Matter Provided by Recent Research on the Passage of Electricity through Gases." The published order reflects the fact that theoretical electrodynamics and spectroscopical research ensured widespread familiarity with the contributions of Lorentz and Larmor, before Wilhelm Röntgen's work brought broader interest to the study of cathode rays. Rather than being shaped primarily by nationality, the most important distinctions here are based on the coherence of different fields of research (electrodynamics on the one hand and cathode rays on the other) that found a diverse range of participants in different nations. However, the decision to ask Paul Villard to write on cathode rays does show the French committee moving between foreign and local contributions to construct an international perspective, for Villard was far less widely known outside France than other contributors. (It is also likely that the division of reports expresses the committee's awareness of the broader ambitions underlying Thomson's engagement with cathode rays.)

54. Becquerel, "Rayonnement de l'uranium" (1900), 69–74; Curie and Curie, "Nouvelles substances radioactives" (1900), 107–8 and 114. The contrast is with ether-based theories that regarded cathode rays as wave phenomena.

Villard was an independently wealthy physical chemist associated with the chemistry laboratory of the École Normale Supérieure, whose most substantial area of research through the 1880s and early 1890s related to gas hydrates, in which he pursued an antiatomist approach common in the French scientific community.[55] After Röntgen's discovery he began publishing first on X-rays and then on cathode rays. Villard presented his studies as continuing the work of physicists like Perrin and Thomson, who had argued for the particulate nature of cathode rays, but did so without accepting or even discussing explicitly Thomson's claim that cathode rays represented a new subdivision of matter. In part this neglect reflects the extent to which Villard's research stemmed from quite different traditions. Villard himself used chemical rather than electromagnetic techniques to explore the manifold phenomena within the cathode tube, and in particular to enumerate the properties of the rays and determine the substance(s) of which they were composed, especially by placing screens coated with chemical reagents such as lead sulfate or oxidized copper in their path. His congress report was organized as a sequential discussion of different properties, drawing liberally on the work of many physicists to describe their phosphorescent effects, mechanical effects, behavior in electric and magnetic fields, their velocity, and so on. Thomson's determination of the invariable nature of the ratio of charge to mass and his demonstration that it was much greater than the ratio obtained from electrolytic phenomena were both discussed, but reinterpreted. For Villard, the key feature was that cathode rays reduced a metal oxide placed in their path. Hydrogen was always present in the tube and was the only reducing gas known. He presented a model in which the glass walls of the tube (which exude water) provided a source of hydrogen that hit the cathode, was electrified and violently repelled. Villard concluded that awaiting the discovery of another simple, reducing gas, it was acceptable to regard the radiant matter at issue as hydrogen (carrying an extraordinarily large quantity of electricity); and he never used Thomson's term *corpuscle*.[56]

Villard's report was immediately followed by Thomson's paper, in which a very different interpretation of cathode rays was used to develop an account of the constitution of matter and a new theory of conduction in

55. On Villard and French work related to cathode rays, see the extremely helpful paper of Lelong, "Paul Villard, J. J. Thomson" (2001).

56. Villard, "Rayons cathodiques" (1900). His conclusion is at 137.

metals. (Since Thomson had declined to make the journey, Paul Langevin gave a resume to the congress. He was excellently qualified for that service, having spent eight months in Cambridge in 1898.) Thomson took an empiricist approach, first laying out the experiments on which his theoretical considerations were based. By referring to his 1897 *Philosophical Magazine* determination of the charge to mass ratio before describing Philipp Lenard and Kaufmann as confirming this result, Thomson's selective citations emphasized his role and avoided potentially awkward priority questions; he cited an 1898 Kaufmann paper rather than Kaufmann's first measurements of 1897.[57] Thomson then outlined his second investigative step, measuring the value of the charge carried (establishing that it was approximately equal to that of the hydrogen ion and independent of the gas in which it was produced). For Thomson these experiments formed an important link in a longer argument, precisely because they prohibited the kind of alternative interpretation Villard offered. Knowing the ratio, and the magnitude of the charge, proved the mass concerned was approximately a thousandth that of hydrogen—the smallest mass so far considered in physics. It is worth quoting Thomson's conclusion: "Thus the study of the passage of electricity through gases leads us to recognize the possibility of the existence of a new state of matter, which resembles the three other states, solid, liquid and gas, in the sense that it consists of a great number of small identical particles, that we will call *corpuscles*, but which differs from ordinary matter . . . " Whereas the masses of ordinary molecules vary with the substance, Thomson went on to write, the mass of corpuscles was invariable, regardless of the substance from which they came or the manner in which they were produced.[58] Thomson's stress on his own role was associated with this particular argument, rightly regarded as highly novel. Responding to Thomson's first formulation of the claim in 1897, FitzGerald had emphasized its significance but also suggested an alternative explanation, assimilating Thomson's particles with Larmor's electrodynamics as "free electrons," whose mass arose

57. Thomson, "Constitution de la matière" (1900), 138. In fact the paper Thomson cited was the third occasion on which Kaufmann gave a figure for the charge to mass ratio (see Kaufmann 1897; 1897 Nachtrag; 1898 Annalen 65, 431). Thomson had followed a similar strategy in his 1899 paper. There he wrote that his 1897 results "are in substantial agreement with those subsequently obtained by Lenard and Kaufmann," without citing specific papers from the German researchers. Thomson, "Masses of the Ions in Gases at Low Pressures" (1899), 547.

58. Thomson, "Constitution de la matière" (1900), 139. The mass of corpuscles was soon shown to vary with velocity.

from the electromagnetic inertia of a moving charge.[59] So was this the discovery of a new state of matter, or of new structures in the ether? Thomson's research over the next three years was designed to justify his claim more fully, but he also worked hard to distinguish his stance from the more abstract and symmetrical treatment of charged particles in electrodynamic theory. Using the term *corpuscle* helped Thomson emphasize the asymmetry between negatively and positively charged matter. Electrodynamic theory had given no reason to suspect such an asymmetry, and Thomson's congress report now used its existence to approach the conductive and thermal properties of metals on the basis of the behavior of corpuscles distributed within the metal and capable of moving freely. His reviewer in the *Physikalische Zeitschrift* faithfully reproduced Thomson's terminology, but also pointed attention to similarities between Thomson's approach and the theories Drude had developed.[60]

Accepting this constellation of reports at their face value, the congress would have left no one in any doubt of the recent confluence of empirical and theoretical research identifying and manipulating negative particles, but could not have settled views on the constitution of a new entity. With both electrodynamic research and cathode ray studies offering a long and complex heritage for current research, they also reveal no settled grounds for a fundamental discovery claim. However, we have seen that authorship of a specific hypothesis—and priority for a set of related experimental studies—was particularly important to Thomson, presenting an argument for a new state of matter. It is significant that Thomson wished to distinguish his hypothesis from the fruits of electrodynamic theory, thereby investing cathode rays—as corpuscles—with a unique importance they did not possess for other researchers, who were in general more interested in asserting connections across different fields. This gave him a greater interest in demonstrating novelty than any others. But of course, none of the scientists closely involved in these fields did accept the research of their contemporaries at face value. We have seen this

59. FitzGerald presented a commentary preceding the publication of Thomson's address at the Royal Institution, discussed in, among others, Smith, "Thomson and the Electron" (2001), 36–38. FitzGerald had concluded: "I may express a hope that Prof. J. J. Thomson is quite right in his by no means impossible hypothesis. It would be the beginning of great advances in science, and the results it would be likely to lead to in the near future might easily eclipse most of the other great discoveries of the nineteenth century, and be a magnificent scientific contribution to this Jubilee year."

60. Guggenheimer, "Thomson, Über Andeutungen" (1901).

especially clearly in the diverse references to Thomson's work in Paris, and it made *naming* the particle significantly easier than settling its constitution. The more widespread use of the term *electron* that we have noted in Paris rapidly eclipsed the niceties of Thomson's language, at least outside the group of physicists closely associated with the Cavendish, even as many physicists accepted the basic empirical dichotomy to which Thomson pointed.

Conclusion

The two chapters of part II of our study have examined two events that are usually only of incidental interest to historians of physics, although they became a stage for the (inter)national commercial and disciplinary concerns of physicists and their instrument makers. Our perspective on the Exposition Universelle and the Congress of Physics in 1900 has highlighted the rise of electricity in broader culture and the electron within the physics community. We have seen that at the point at which their discipline was simultaneously reaching an impressive weight of numbers and unfolding into specializations, physicists worked with books, with instruments, and with concepts to unify both their discipline and its new view of nature. The French Physical Society created a new social organ, too, and the inaugural International Congress of Physics expressed disciplinary ambitions on an unprecedented scale. Leaving Paris, physicists carried a sheaf of papers containing reports of interest (with the complete proceedings to follow soon), and of course a host of impressions. How should we sum up what the event meant to them? Describing the congress to readers of the *Physikalische Zeitschrift*, Hermann Theodor Simon regretted that organizers had not followed the German custom of designating an official pub as a general meeting point, whether for different sections or different nations. Evidently some physicists came to Paris wanting to show, *drink*, and tell! He noted too that in the absence of a printed list of participants and a daily newssheet, its organization had not been easy for non-French participants. More positively, Simon focused on the reports—the essential end result of the congress. He thought the event had brought little new scientifically, but allowed that its three volumes of reports on current fields and questions by the best-known researchers could lay claim to a valuable winnowing and—in the sense that Ernst Mach had described—an economic significance. That would be the basis for the place of the congress in the history of physics.[61] We shall conclude with reflections on what the event and its aftermath reveal about the interplay between nations and sections, organization and economy, and novelty and memory in the physics of 1900.

The congress committee clearly played its greatest hand in the winnowing process with the choices it made about topics and authorship, and even though contributors were solicited from French physicists far more often than others, the distribution of reports across thirteen nations shows them mediating between existing research strengths in different nations. In the fields we have examined most closely, we have seen clear evidence that committee choices promoted more rigorous exchanges across both methods and national boundaries. Although they cut in several different ways, the host community is likely to have benefited most directly from such interchanges. Thus, reliance on French perspectives in optics meant that Cornu voiced (and Lorentz and Michelson attended to) judgments about methods and probable errors in velocity of light that might otherwise have gone unnoticed, while Macé refrained from raising the awkward queries Preston might have put to interferential metrology. Conversely, in the case of the electron the distribution of papers from different research traditions helped to promote a new openness to foreign research within France. Benoit Lelong has emphasized the fact that the congress presented the first treatment of Thomson's ideas in French and juxtaposed Thomson's and Villard's views in the same publication. Lelong sees this to have been an early step in an increasing engagement with Thomson's research in France more generally, that in 1901 led to the category "Gas Ionization" being introduced into the classification of papers in the annual volume of the *Journal de Physique*. By 1906, Villard had abandoned his own view on the constitution of cathode rays, taking up first the language of corpuscles and then, from 1908, that of electrons.[62]

61. Simon, "Physikerkongresse" (1900), 572. In the United States, Webster gave a brief résumé of the work of the Paris congress at the annual meeting of the American Physical Society in December. Hallock, "The American Physical Society" (1901).

62. In a book published at the end of 1900, Villard discussed the ratio of the cathode rays' charge to mass, and for the first time explicitly raised Thomson's argument that this indicated a new subdivision of matter—before going on to restate his own view. At this time Paul Langevin and a group of younger scholars began using techniques associated with Cambridge studies and presenting arguments based on the acceptance of ions rather than avoiding the specter of atomism as many of their teachers and patrons did. Lelong, "Paul Villard, J. J. Thomson" (2001), 150–56.

By encouraging new traverses of the boundaries between diverse methods and different nations, the congress surely spurred many specific instances of this kind of gradual incorporation of new perspectives, instruments, or insights into physicists' existing orientations. But apart from recognizing the extraordinary initiative of the French Physical Society in organizing the event as a whole and noting the benefits that flowed to the host community in particular, we can offer a further observation concerning national participation. This point will echo the general message that we saw many scientists took away from the displays of the World's Fair. German physicists gave ten reports in Paris, more than any other nation apart from France. Still more importantly, the German-speaking community engaged with the congress collectively in a way that no others did. Even before the congress began, the weekly issues of the *Physikalische Zeitschrift* had begun publishing reports on the reports, choosing German experts to comment on the work delivered in Paris. Walter Kaufmann provides an example. Without attending the conference he interpreted its proceedings by writing three short reviews. The first was an account of Henri Pellat's report on national laboratories of science that appeared two days before the congress opened. Kaufmann had contributed to cathode ray research since 1897. Reviewing Villard's report in November, he dismissed it as unsatisfactory for its incompleteness, lack of consistency, and crass contradictions. In contrast he thought the paired reports of Becquerel and the Curies offered the very best way for those concerned with the new radiations to orient themselves in a burgeoning field (and he himself soon began experiments with radioactive materials).[63] With Max Abraham, Heinrich Johannes Boruttau, Gerhard Schmidt, Johannes Stark, and Emile Wiechert, Kaufmann was among more than twenty German physicists to write short reviews condensing the findings of papers given in Paris. Thus, in a mark of disciplinary vigor as striking as the turn of German dynamos and as delicate as the sheen on their scientific instruments, collectively German physicists chewed over the congress proceedings, digested its reports, and in some cases spat them out. By the time the last *Physikalische Zeitschrift* review appeared almost a year later they had worked through more than half the contributions, and moved on.

63. Kaufmann could not allow Villard's argument that cathode rays were composed of hydrogen to stand unchallenged, and conducted a blow-by-blow deconstruction of his treatment of different issues. See Kaufmann, "Pellat, Die Physikalisch Technischen Staatslaboratorien" (1900); Kaufmann, "Villard, Die Kathodenstrahlen" (1900); Kaufmann, "Becquerel, Über die Uranstrahlen" (1901).

The collective organizational vigor that this indicates complements what we have learned about the relative strengths of different national communities on the basis of more quantitative measures. We know that although there were fewer German physicists than American, and less was spent on the average German physicist than on his British colleague, the German community was substantially more productive than any other (producing 3.2 papers per physicist per year).[64] This chapter has demonstrated that German physicists were also better organized across both academic and instrument-making communities. Just as importantly, it indicates that collectively, German physicists were more deeply engaged in understanding the work of their colleagues in different nations. If this interest in traversing boundaries holds, it suggests that circa 1900 innovative work in different nations might have had a better chance of being evaluated in Germany than elsewhere. It also raises the possibility that the success we have come to grant German physics in the early twentieth century might in part be a result of the fact that German physicists were so well organized. I will put the point provocatively. We will later see that German perspectives came to dominate electrodynamics, electron theory, and relativity—despite the widespread attention these fields attracted from physicists throughout the world. Do we celebrate Einstein now in part because he wrote in German and made his way in the strongest physics community of his day? Or, alternatively, would he have been as successful a physicist if he had remained with his family in northern Italy after leaving his Munich secondary school, instead of going on to study mathematics and physics for secondary teachers at the Zurich Polytechnic, as he did from 1896? Understanding all the issues required to investigate such arguments would take us far beyond present purposes, and I do not want to make them in any strong form. Nevertheless, it is worth noting that although Einstein could have read it in its original French, he is likely to have benefited substantially from the German translation of at least one major text in his field of study—Poincaré's *Wissenschaft und Hypothese*. Further, the final chapters of this book will maintain that the particular organization of the elite Solvay conference that gathered together a select

64. To provide an overview, circa 1900 an average of 14,000 marks was spent on each of 215 American academic physicists who wrote 1.1 papers per year; while 114 British physicists received 14,500 marks and wrote 2.2 papers; 105 French physicists received 10,500 marks and wrote 2.5 papers and 145 German physicists each received 10,300 marks and wrote 3.2 papers. See Forman, Heilbron, and Weart, "Physics *circa* 1900" (1975), 6; Kragh, *Quantum Generations* (1999).

group of eighteen physicists in 1911 did in fact play an integral part in the subsequent widespread acceptance of a distinctively German understanding of the rise of modern physics.

Yet as remarkable as what occurred in Paris in 1900 is the rapidity of its passing. Even as we recover the event in its contemporary dimensions, we must ask why it has been forgotten. The World's Fair was over in a matter of months, and the Congress of Physics came and went within a week. Betraying the monumental intentions of its organizers and despite the extraordinary novelty of its disciplinary scale and international scope, the event disappeared from historical memory almost as soon as it was over. Even if reports on the reports continued to appear in the *Physikalische Zeitschrift* until the summer of 1901, and even as the conference volumes made their way into most big libraries around the world, the congress footprint gradually faded. Perhaps it was simply that many of the more valuable contributions were published in full or in part in other venues, and for whatever reason the collective French version remained less visible. That was certainly the case with Kelvin's address. His May lecture to the Royal Institution, "Nineteenth Century Clouds on the Dynamical Theory of Heat and Light," is still well known, while the Paris paper offering the first part of that message has been forgotten. Poincaré's address won many compliments in its day, and in 1900 was serialized in the *Physikalische Zeitschrift*. Then it was split into two chapters in his 1902 book *La science et l'hypothèse*, with German and English editions following in 1904 and 1905. There, many of the themes Poincaré highlighted in Paris received still greater emphasis in chapters on the classical mechanics and on the distinctions between relative and absolute motions. In this new context his 1900 address was read by thousands, who had little reason to suspect the occasion for its original production. While Joseph Larmor offered an interpretative preface to the English edition, Ferdinand Lindemann annotated and updated the German version, and in 1904 Einstein pored over it for several weeks.[65]

Taking the conference volumes down from library shelves wherever I can, I have rarely seen physical evidence that their chapters were read, and while we have noted many instances in which individual reports were taken very seriously, discussions of the conference as a whole are hard to come by in articles published through the next few years. Simon noted

65. Poincaré, "Beziehungen" (1900); Poincaré, *Wissenschaft und Hypothese* (1904); Poincaré, *Science and Hypothesis* (1905).

that the congress saw little new scientifically; but whatever new material individual reports brought to light or however economically they could synthesize specific subjects, it was in their *collective* extent and the attempt to bring physicists of all nations together that the congress was most novel. Perhaps the scale of the event was simply beyond its participants, so that the collective disciplinary vision the congress created was grasped only piece by sectioned piece, dissolving into the very specializations its organizers had sought to surmount. Its proceedings could not unite the discipline they portrayed so completely. In any case, it is evident that in the happenstance of scholarly work, none of the papers given at the congress turned out to fire disciplinary imagination as successfully as the twenty-three current problems Hilbert delivered to the conference in mathematics that same week. The twenty-one-year-old Einstein wasn't in Paris either. Had he celebrated finishing his final exams by attending the congress we would surely know far more about the event now. Betraying the confident hopes of its organizers, the conference simply disappeared from view as an event and record.

PART III

From the Promise of the Electron to Einstein's Theory of Relativity

Beyond Discovery Accounts

Whether we remember the inaugural International Congress of Physics or not, the work represented in its conference halls and published reports was soon continued outside its bounds. In this part we move into the German physics community in order to pursue the further history of the electron. The next two chapters follow that particle several steps beyond the Paris conference as a necessary prelude and foil to our study of relativity, which rapidly became a major concern to research on the electron. Chapter 8 then examines participant histories in order to trace the emergence of a common interpretation of major features of relativity theory, showing how what was initially regarded as an electrodynamic theory offering important insight into the electron—and as the fruit of collaborative labor—came to be viewed as a new theory of universal scope associated with Albert Einstein in particular.

As far as the electron is concerned, histories of science have usually paid relatively little attention to the period between its initial discovery before 1900 and the announcement six years later of experiments on the mass of electrons moving near the velocity of light, which played an important role testing what contemporaries then called the "Lorentz-Einstein" theory of relativity. Thus discovery accounts punctuate our understanding. They leave lines of work just as they have begun to take shape, and (just as brutally) only return if and when the dramatic needs of the next chosen moment permit.[1] Rather than switching gaze in this way, chapters 6 and 7 traverse the entire period from 1900 to 1911, focusing first on the electron and engaging with the principle of relativity insofar as that seemed to bear on the prospects of the electron to a number of ambitious young physicists. Our study of the way Walter Kaufmann, Paul Ehrenfest, and Max Born used experimental and theoretical studies of the electron to test the validity and establish the limits of relativity will function as an investigative probe through a landscape that is usually approached from a rather different perspective. Historians have commonly treated the history of relativity through a broader conceptual matrix oriented toward Einstein's contributions.[2] Focusing on the narrow question of the interrelations between the electron and relativity—but deliberately moving between experiment and theory—will offer several critical advantages.

One fruit of this procedure is the possibility of describing an empirical counterpart to theoretical studies of space and time in measurements of the positions of electrons on photographic plates with exposure times running into days. Recalling a theme explored in earlier chapters, chapter 6 will show that like wavelength studies, experimental accuracy in electron research depended on being able to move from relative to absolute measures, an endeavor that complements theorists' concerns with the principle of relativity and testing absolute space and time. A second argument, pursued across both chapters 6 and 7, will demonstrate that mastering rigidity—materially and conceptually—became unusually significant to experiments on the electron and the elaboration of relativity theory alike. In particular, examining apparatus and measurement protocols will establish that by the early twentieth century interferential techniques were required of high precision experiments in electron physics, thereby highlighting the paradox that as an ether-drift experiment Michelson's work

1. A recent "biography" of the particle treats its earliest discovery with great subtlety but then moves to the role of the electron in Bohr's work, commenting, "the electron ceased to be an inhabitant of the classical world toward the end of 1913." Arabatzis, *Representing Electrons* (2006), 115. The major exception to this neglect is Miller, *Einstein's Special Theory* (1981), see esp. 45–85 and 334–51.

2. The most important studies of the rise of relativity in the German community include Miller, *Einstein's Special Theory* (1981); Goldberg, *Understanding Relativity* (1984); Pyenson, *The Young Einstein* (1985); Pyenson, "The Relativity Revolution" (1987). For a helpful review and overviews, see the papers of Cassidy, Stachel, and Brush: Cassidy, "Understanding the History of Special Relativity" (1986); Cassidy, *Einstein and Our World* (2004); Stachel, "History of Relativity" (1995); Brush, "Why Was Relativity Accepted?" (1999).

demonstrated the *contraction* of all bodies with motion, while as a technique of measurement and standard of length it offered proof of the *invariability* of the dimensions of both that emblematic rigid rod, the international meter, and the apparatus that put relativity to the test by passing electrons through electric and magnetic fields.

As helpful as it is to follow these thematic concerns across experiment and theory, the exercise of approaching relativity in the light of priorities that were widespread in the period is a neglected and perhaps still more valuable task. Historians have long appreciated the importance of the electromagnetic worldview that was inspired by experiments on the variation of the mass of the electron with velocity, and they have often recounted the earliest sparring matches between adherents of that ambitious program and supporters of the principle of relativity or the *relativtheorie*, as German physicists were beginning to identify the emerging theory. But even the most detailed accounts of the experiments and concepts at issue have pursued them with a primary interest in Einstein's work, and with theorists' immediate responses to the challenge that empirical results represented.[3] Focusing on the electron throughout this period will supplement our existing appreciation of the way relativity physics emerged from the complex of electrodynamic theory around 1905 in several valuable respects. Firstly, it will show that to a significant extent experiments on the electron drove theoretical contributions in this field, but also, that having set an agenda, divergent results meant those experiments could not ultimately be brought to any final conclusions. Secondly, it will give new insight into the reception of Einstein's theoretical approach by complementing our understanding of the strengths of relativity with a sustained study of what many contemporaries regarded as the most critical issue the principle faced—its relations to the electromagnetic worldview as these were expressed in treatments of the electron.[4] Thirdly, it will open

3. McCormmach helped highlight the electromagnetic worldview. The overriding interest in Einstein and theory is illustrated by the fact that our best account of the experiments as a whole (Miller's) is developed in a book organized around the different sections of Einstein's 1905 paper, while the most detailed exploration of responses to Kaufmann's 1906 experiment (Hon's) focuses on theoretical contexts in which it is accounted successful or in error. See McCormmach, "Lorentz and the Electromagnetic View of Nature" (1970); Miller, *Einstein's Special Theory* (1981); Hon, "Experimental Error" (1995).

4. While standard studies of the reception of relativity often highlight the electron, they have seldom focused on its significance throughout the period as a whole. The issue is commonly treated as an opportunity to explore evaluative contrasts between Einstein's approach and that of others (often assuming the particular probity of Einstein's perspective without

up the question of authorship in a remarkably open and rapidly changing field of work. The diverse attributions of credit (and blame) we encounter in chapters 6 and 7 will illustrate just why this issue has been critical to historians interested in establishing when Einstein's approach to relativity became widespread, but will also challenge earlier treatments of their significance. Chapter 8 then complements our investigative probe of physicists' treatment of the electron with a broad-ranging study of participant histories of relativity, now exploring the stamp Einstein put on the history of relativity, and establishing when and why others in the German physics community started to regard him as the sole author of the theory. Our study will confirm the basic outlines of the account that has been built up by previous historians, indicating that Einstein's approach had won widespread acceptance in Germany by 1911, and it will contribute to the task of developing a more subtle understanding of the complex process by which this occurred.

Before we do so it will be helpful to investigate the early history of the Nobel Prize in physics. Awarded from 1901, the Nobels clearly fed off and helped cultivate a focus on discovery moments—and soon won a far stronger role shaping the disciplinary memory and public image of physics than the International Congress ever had. Ironically, while the Nobel prizes strongly favored the kind of punctuated vision I have just criticized, the particular awards made in the early twentieth century will show just why it makes sense to put relativity in second place to the electron for a moment. We will also be able to explain when and why Michelson was honored.

Distinguishing Fin de Siècle Discoveries: The Nobel Prizes in Physics

Although national societies had certainly recognized achievement across political borders, the Nobel Prize was distinguished from earlier forms of reward in the scientific community in being both international in conception and fabulously endowed.[5] Nobel's 1896 will set several Scandinavian

clearly establishing if, when, and why his contemporaries were convinced of this). For synthetic accounts, see especially Miller, *Einstein's Special Theory* (1981); Goldberg, *Understanding Relativity* (1984); Pyenson, *The Young Einstein* (1985).

5. At 150,000 crowns or 210,000 francs, Röntgen received approximately fifteen times his annual salary (of about 14,000 marks or 11,500 francs) when he won the first physics prize in 1901. In contrast, Royal Society medals were of symbolic rather than material value and rewarded a whole body of work, while prizes from the French Academy of Sciences were

institutions the task of rewarding those discoveries or inventions that had conferred the greatest benefit on mankind in the preceding year in five different categories—paying "no consideration whatever" to nationality. Prizes in physics and chemistry were to be awarded by the Swedish Academy of Sciences, and governing statutes established in 1900 subtly reworked Nobel's stipulations to recognize the drawn-out and communal nature of scientific work, at least to some extent. Awards could go to the most recent achievements, or works whose significance had only recently been recognized; and they could be divided equally between two works, or shared jointly if two or more people were together responsible for a single discovery. Already-considerable media attention mushroomed once the first prizes began to be distributed. But interest among scientists grew more slowly, and it is worth noting that fame and money alone could not ensure the reputation of the prizes.

For the choice of particular discoveries and individuals to have credibility, the nomination, selection, and decision processes had to mediate between the guiding interests of specific committees that were appointed to assess potential prize winners, the broader disciplinary matrix from which achievements were selected, and the final authority of the awarding institution.[6] This fact renders the prizes a rich and multifaceted source of insight into physics at the turn of the century, giving us an unusual opportunity to explore how members of the Swedish scientific community perceived and portrayed their subject, and how they worked to consolidate and shape the international physics discipline over time. Regarding

---

more fully part of the structure of scientific research (with less funding available in the form of grants). Since his initial discovery, Röntgen had been elected a member of twenty-two German and foreign scientific societies and received seven significant prizes and medals, including the particularly prestigious Rumford medal of the Royal Society of London and the Prix Lacaze of the French Academy of Science (worth 10,000 francs). On rewards, see Crawford, *Beginnings of the Nobel Institution* (1984), 16–18. On academic incomes, see Forman, Heilbron, and Weart, "Physics *circa* 1900" (1975), 41–49.

6. In the case of physics and chemistry, six groups of people could submit nominations: members of the Royal Swedish Academy of Sciences; the three to five members of the Nobel Committee established for that field; prior winners of the prize; Scandinavian professors in the discipline; professors in at least six other universities or similar institutions (to ensure the appropriate representation of different countries); and other scientists on invitation from the Academy of Sciences. Solicited in October, such nominations had to be in by February. The committee then assessed the candidates' merits, advising the Academy of Sciences by the end of September. The academy's decision was to be made by mid-November, and the prize awarded on the anniversary of Nobel's death on the tenth of December. A prize could be deferred one year in cases where no satisfactory candidate was found.

relativity and quantum theory as the defining achievements of modern physics, historians puzzling over brief dedicatory sentences and noting authorial inclinations in confidential committee reports have often discerned an experimentalist bias underlying the physics prizes awarded in the early twentieth century. The award of the 1907 prize to Albert Michelson not for ether drift but "for his optical precision instruments and the spectroscopic and metrological investigations carried out with their aid" has provided grist for this mill, along with the finesse of rewarding Einstein only in 1922 and then for his law of the photoelectric effect rather than relativity.[7] But our historiographical favoritism has left us poorly equipped to understand the broad coherence of disciplinary dynamics in the period, and the public speeches *presenting* specific prizes may well provide even more valuable indications of committee perspectives on the physics of their age than confidential reports, precisely because they consciously and deliberately address both local and disciplinary communities. A very brief overview of the Nobel prizes granted in the early twentieth century will help show what Swedish physicists regarded as the core element of modern physics, while a more detailed examination of the specific context in which Michelson was rewarded will demonstrate the disciplinary grounds for his receipt of the Nobel Prize in 1907—and that year in particular.

The first six years of prizes caught up with events involving the discovery of new radiations or elements from 1895 onward, in two different passes. Between 1901 and 1904 rewards for the discovery of X-rays, the

7. In the light of the argument I make about the significance of the electron below, it is worth noting that Einstein was first mentioned publicly by the Nobel organization in the presentation speech rewarding Kamerlingh Onnes for his low temperature research in 1913. Nordström described the "so-called theory of electrons," as "the guiding principle in physics in explaining all electrical, magnetic, optical, and many heat phenomena," but as challenged by experiments in low temperature regions. These made it more and more clear that a change in the whole theory of electrons was necessary, with Planck and Einstein having already begun theoretical work in this direction. "1913. Heike Kamerlingh Onnes. Presentation by Th. Nordström" (1967), 305. In contrast to the prominence of electron theory, relativity arose behind scenes in a growing number of nominations to Einstein. They began in 1910 with one from Wilhelm Ostwald, and by 1920 had come from the pens of some twenty-three physicists. Sometimes this was in division with Lorentz, as Wien proposed in 1912, 1913, and 1918, and von Laue advocated in 1918 and 1919. Sometimes Einstein was suggested as an alternative after Planck, as Stefan Meyer and Wien proposed in 1918, and von Laue suggested in 1919 (pairing Planck with Sommerfeld). On the complex circumstances surrounding the award to Einstein, see Friedman, *Politics of Excellence* (2001), 129–40.

electron, radioactivity, and argon went to Röntgen; Lorentz and Zeeman; Becquerel and the Curies; and Rayleigh. Revealingly, the committee then consolidated both the international representation and historical and disciplinary grasp of their celebration of what they did not hesitate to call the *electron*. In 1905 they rewarded Philipp Lenard for his research on cathode rays (before Röntgen's discovery of X-rays) and the following year Thomson for his work on the conduction of electricity through gases. Paying attention to the presentation speeches given by the president of the Swedish Academy of Sciences each year would have underlined the wider importance of the electron to the new physics still further. When H. R. Törnebladh presented the 1903 award to Becquerel and the Curies for radioactivity, for example, he mentioned also the "modern theory of the electron," which had been applied with great success to explain cathode rays.[8] A year earlier the principle of conservation of energy had been described as the first basic principle of modern physics, and Faraday as the founder of the modern science of electricity. These references will serve to indicate that the physics of the nineteenth century was modern *before* it came to be regarded as *classical*, a term that surfaces later and less frequently in Nobel addresses.[9]

It is worth underlining that the particular salience of the electron as the core of "modern" physics rested as much on its general purchase as on its currency. Describing Lenard's research two years later, A. Lindstedt wrote of the confluence of theory and experiment in the study of electrons, which "has been gradually developed into one of the foremost theories of modern physics." He described electron theory as being "of the most fundamental importance for the sciences of electricity and of light and for both the physicist and the chemist," with its significance for the constitution of matter.[10]

8. "1903. Antoine Henri Becquerel, Pierre Curie and Marie Sklodowska-Curie. Presentation by H. R. Törnebladh" (1967), 49. For the term *electron*, see note 10 below.

9. For the early reference to "modern," see "1902. Hendrik Antoon Lorentz and Pieter Zeeman. Presentation by Hj. Théel" (1967), 11. Michelson's research, Hertz's experiments, and the studies of Langley (and Lummer, Pringsheim, and Kurlbaum) were described as "classical" in the context of awards to Michelson (1907), Marconi and Braun (1909), and Wien (1911).

10. "1905. Philipp Eduard Anton von Lenard. Presentation by A. Lindstedt" (1967), 103. The speech distinguishing Thomson a year later also used the term *electron*, while Thomson's Nobel lecture pointedly described his detection of "corpuscles." "1906. Joseph John Thomson. Presentation by J. P. Klason" (1967); Thomson, "Carriers of Negative Electricity" (1967), 145.

Comments of this nature suggest also that the Swedish committee would hardly have recognized themselves as favoring experiment over theory. Yet studying the backstage process of nominations and committee machinations that led to the eventual choice of candidates each year, Robert Marc Friedman has described Michelson's award as illustrating the narrow experimentalist bias that dominated the Uppsala school of physics in Sweden and found a strong voice on the Nobel Prize committee for physics in the person of Bernhard Hasselberg. Hasselberg was an expert in spectroscopy and Sweden's representative on the International Bureau of Weights and Measures, and he later described Michelson's as the best award to date. Should we regard this as a case of individual sympathies trumping broader perspectives on the value of Michelson's contributions? Michelson was first considered seriously in 1904, when he was nominated by Edward Pickering alone.[11] The committee report stated that the rigor and originality of Michelson's work put him at the forefront of contemporary research, but selected Rayleigh instead.[12] Three years later Michelson was nominated by just three physicists, the German expert in refractometry Hermann Ebert, the Dane Krystian Prytz, and the American astronomer George Ellery Hale. This fact has led Friedman to suggest that many in the wider physics community did not share Hasselberg's view that the rigor of Michelson's work placed it "on the highest level of contemporary scientific research."[13] In contrast, a brief study of the fortunes of Michelson's research after 1900 will show that there were strong reasons why Michelson won the award when he did, that these had as much to do with the contributions of others as they did with his original research, and that they cut across the local politics of feuding segments of the Swedish physics community.

11. Pickering stated he proposed Michelson, "In view of your great work in determining the Velocity of Light and your varied applications of the interference of light," and noted that earlier research could also be considered despite Nobel's time stipulation. Pickering to Michelson, 12 November 1903, ECP-HUA, as cited in Livingston, *Master of Light* (1973), 233–34. See also the correspondence in which Hasselberg declined Pickering's initial self-nomination. Hasselberg to Pickering, 30 October 1903, ECP-HUA UA V 630.17.10; and Michelson to Pickering, 14 and 16 November 1903 and 10 January 1908, UA V 630.17.7

12. The committee explicitly noted that the choice to be made depended on everyone's "individual" notions of the "greater or lesser scientific or practical significance" of different discoveries. Friedman, *Politics of Excellence* (2001), 42.

13. Ibid., 41–47, quote on 47. Rutherford, Lippman, and van der Waals had received seven or six nominations, and Planck, Wien, and Poincaré also received notable support. See 43 and 300 in ibid.

I have shown already that one of the major characteristics of Michelson's career was the singularity of his studies. By 1907, however, his isolated position had changed fundamentally in one important field: the spectroscopical community had finally put the relations between gratings and interferometers to test. Despite his vigorous advocacy of interferential methods, Michelson himself never stressed the discrepancy between his absolute determinations of wavelength of several spectral lines in 1893 and the best measures delivered by diffraction gratings. But by 1902 the battle had been brought to the center stage of physical and astronomical spectroscopy. In that year Charles Fabry and Alfred Pérot used the results of their etalon interferometer to argue that Henry Rowland's table of spectroscopic wavelengths was subject to periodic errors. Pitting interferometric methods against gratings, the French researchers also set the increasingly stable production of laboratory spectra against differences noted between terrestrial and solar spectral lines (and attributed to the effect of pressure). They advocated replacing Rowland's use of sodium as an absolute standard and his map of the solar spectrum as a reference, with Michelson's cadmium wavelength as a primary, absolute standard, to be supplemented with measures of lines in the iron spectrum. Just as Preston had challenged Michelson's observations of the Zeeman effect, Louis Bell responded by arguing that Fabry and Pérot had relied on a new instrument and untested claims. In contrast, the errors of grating spectroscopy were well recognized and could be managed by experienced spectroscopists. Interferometry might well offer a superior method for absolute measures, but Michelson's determination had never been verified. Bell thought the discrepancies probably arose from the fact that the possible sources of error in interferometric measurements had not received adequate consideration.[14]

With Johannes Hartmann and Lewis Jewell soon coming out in favor of a revised version of Rowland's standards, the debate was serious enough for Hale to make the discussion of standards of wavelength a

14. Fabry and Pérot, "Measures of Absolute Wave-Lengths" (1902); Bell, "On the Discrepancy" (1902); Fabry and Pérot, "A Reply" (1903); Bell, "The Perot-Fabry Corrections" (1903). In 1902 Otto Lummer described Fizeau as the founder of interference spectroscopy and Michelson as having built it up into a discipline. However, he thought Fabry and Pérot had given Michelson's methods certain foundations and compelling force because their use of two homogenous waves achieved a doubling of rings at a point where with Michelson a complete disappearance of the interference fringes entered. Lummer, "Die planparallelen Platten als Interferenzspektroskop" (1902), 172–73. For evidence of wider interest in interferometric measures of length, see also Gumlich, *Präcisionsmessungen* (1902).

central platform of his plan to create a new International Union for Co-operation in Solar Research. Hale set his project in play by organizing a conference of astrophysics in conjunction with the World's Fair in St. Louis in 1904. Inviting the Stockholm physicist Svante Arrhenius gave fuel to a long-standing feud that pitted Arrhenius's appreciation for theory against Hasselberg, whom Arrhenius deprecated for collecting and publishing diverse astrophysical and geophysical data without adequate theoretical analysis. Hearing of the invitation, a furious Hasselberg wrote Hale a lengthy letter condemning the popular and hypothetical nature of Arrhenius's recent textbook on cosmic physics, and announcing that neither he nor Ångström would be willing to serve as Swedish representatives in any international union if Arrhenius were included. Hale managed that conflict by asking Arrhenius to speak on meteorological issues, while the terms of the public debate on wavelength measures were set out in contrasting papers delivered to the conference.

It is revealing that Hale urged Michelson to attend, writing "you, above all men, ought to be present to take part in the discussion." It is still more revealing that Michelson was reluctant to advocate his own standards, and in fact did not make the journey south.[15] His methods would be represented by others. Fabry and Pérot's arguments were presented in the name of the French Physical Society by Henri Poincaré (in St. Louis to deliver an address on mathematical physics to the second International Congress of Physics), and Jewell spoke in favor of correcting Rowland's tables. Importantly, the new International Union offered a means of resolving the issue within a formal and international structure by setting up a committee to consider whether to produce new standards or correct Rowland's existing tables.[16] While different parties argued their case in the *Astrophysical Journal*, a growing consensus for reform was clearly signaled a year later when the union voted to establish a three-tiered system

15. Hale was ready to change the date or place of the meeting to suit the physicist. Hale to Michelson, 8 July 1904 (quoted) and 24 June 1904, HOA.

16. For the Swedish conflict, see Friedman, *Politics of Excellence* (2001), 45–47. For the International Union for Cooperation in Solar Research and the debate over spectroscopic standards, see the excellent studies of Charlotte Bigg, and contributions to the *Astrophysical Journal*: Bigg, "Behind the Lines" (2002), 29–45; Bigg, "Spectroscopic Metrologies" (2003); "International Co-operation in Solar Research. Minutes of the Meeting of Delegates to the Conference on Solar Research, Held in the Hall of Congresses, St. Louis, September 23, 1904" (1904); Hale, "Co-Operation in Solar Research" (1904); Crew, "Standard Wave-Lengths" (1904); Fabry and Pérot, "Nouveau système de longueurs dónde étalons" (1904); Jewell, "Revision of Rowland's System" (1905).

of standards based on interferometric methods.[17] The practical organization of the system was deferred to its next meeting in Meudon in 1907, but a committee was appointed to select the standards and organize the determination of their wavelengths in terms of the primary standard in at least two laboratories. Fabry and Pérot were already on their way. They had first floated the possibility of redetermining the meter in cadmium wavelengths as early as October 1901, and they began organizing the endeavor in 1904.[18] Now they had a deadline to meet. Working with Benoît at the International Bureau of Weights and Measures, they were able to publish the result of a replication of Michelson's 1893 measurements before the 1907 meeting—where the International Union for Cooperation in Solar Research formally adopted the wavelength of cadmium as the absolute primary standard for spectroscopic measurements.

Both Hasselberg and Arrhenius undoubtedly followed these events closely. Hasselberg had earlier based a highly accurate series of metallic spark spectra on Rowland's standards; and Arrhenius's assent to the award shows that Michelson's work had won appreciation across even the bitter division between theoretical and experimental inclinations in Swedish physics. We can see, too, that recent events made a significant difference to the evaluation of Michelson's research. In the 1880s Michelson had described his goal to redetermine the length of the meter as feasible and practical; once achieved it had been described alternatively as rendering the standard of length indestructible, or as an unverified measurement from an untested instrument.[19] Now it was of current significance

17. A primary standard would be based on the wavelength of a suitable spectral line, with the number defining the wavelength of this line being fixed permanently and defining a new unit to be called the Ångström ([angstrom] as close as possible to 10^{-10} meters). Secondary standards would be established at distances of less than 50 angstroms, referred to the primary standard by an interferometer method; and tertiary standards would be selected at distances between 5 and 10 angstroms, with their wavelengths to be obtained by interpolation with the help of gratings.

18. Comité international des poids et mesures, *Procès-verbaux* (1902 [1901]), 121–22.

19. A representative view from the early twentieth century may be gathered from Wood's textbook on optics. He thought Michelson's determinations "will doubtless stand for a long time, if not for ever, as the standards from which all other lines will be measured," but noted that difficulties estimating "visibilities" and interpreting the results attended Michelson's interferometrical spectroscopy. Wood, *Physical Optics* (1905), 217 and 219. Increased interest in the method is reflected in Gehrcke, *Die Anwendung Interferenzen in der Spektroskopie und Metrologie* (1906). Responding to the events of 1904–07, in 1909 Schuster expanded his treatment of interference metrology in general and Fabry and Pérot's research in particular: Schuster, *Theory of Optics*, 2nd ed. (1909), iii, 74–76, and 131–41. It is worth noting that

and had been confirmed. As Hasselberg noted delivering the address to celebrate Michelson's prize, Fabry and Pérot's work had given a result that differed by only 0.1 wavelength from Michelson's (0.00006 mm). This demonstrated both the accuracy of Michelson's earlier research and the constancy of the material standard in Sèvres. The care taken in the execution and preservation of the prototype gave it the appearance of constancy, but only an absolute measurement of length that was itself independent of any physical element could guarantee it. Hasselberg wrote: "It is to Michelson's eternal honor that by his classical research he has been the first to provide such proof."[20]

Hasselberg's address foregrounded improvements in methods and the means of making observations and experiments as "the very root, the essential condition, of our penetration deeper into the laws of physics—our only way to new discoveries." He introduced the interferometer as a group of measuring instruments before writing of their applications in general and then detailing Michelson's metrological work and its significance for spectroscopy. Looking to the future, Hasselberg wrote that the link between spectral structure and the molecular structure of luminous bodies meant Michelson's research and instruments had brought physics to "the threshold of entirely new fields of research."[21] As well known as Michelson's ether-drift experiment was, Hasselberg felt no need to refer to the experiment or outline the significance of its null result for electrodynamic theory. Later he told Hale the award to Michelson was:

> the best of all which have been made up to date. Our earlier laureates Röntgen, Lorentz, Zeeman, Becquerel, Curie, Rayleigh, Lenard and J. J. Thomson are indeed men of eminent scientific merits, but for my part I must consider the work of Michelson as more fundamental and also by far more delicate. Perhaps this is in some way an opinion of sympathetic for a work closely connected with my own specialty; but I cannot but prefer works of high precision and in this respect those of Rayleigh are the only ones which could be compared with Michelson."[22]

Rolt's comprehensive discussion of gauges and fine measurements indicates that interferometry had made significant inroads into both scientific and industrial metrology by the 1920s. Its introduction into broader industrial practice was materially enhanced by the events of World War II and demands of precision weaponry in the 1950s. See Rolt, *Gauges and Fine Measurements*, vol. I (1929).

20. "1907. Albert Abraham Michelson. Presentation by K. B. Hasselberg" (1967), 162.

21. Ibid., on 160 and 163.

22. Hasselberg to Hale, 29 December 1907, Hale Observatories Archives, Pasadena.

Michelson's acceptance address framed the significance of his research in similar terms to Hasselberg, and he too left ether drift unmentioned.[23] While the omission is certainly revealing, studying the spectroscopic rationale for Michelson's award has shown that our ignorance of major developments in physical and astronomical spectroscopy has made Michelson's award look far less widely motivated and less currently significant than contemporaries knew it to be.

Our survey of early Nobel prizes has shown that the electron surfaced again and again, linking a wide range of fields and becoming emblematic of "modern" physics. Pursuing that particle in the following chapters will bring us back to Michelson's work in two different ways. It will show that the apparatus used in experiments on the electron highlight the increasing purchase of interferometric techniques in precision experimentation; while the results those experiments yielded gave renewed significance to the ether-drift experiment for those who supported the principle of relativity.

23. Michelson, "Recent Advances in Spectroscopy" (1909), also printed in Michelson, "Recent Advances in Spectroscopy" (1967 [1909]).

CHAPTER SIX

The Empirical Electron: Space and Time on a Photographic Plate

Electron Physics: Framing the Future

The many Nobel Prize presentation speeches that referred to the electron in the early twentieth century underline the great breadth and centrality of research on that particle, but they do not convey the specific promise that drove many researchers into the field: the hope that the electron could form the core to a new worldview. Soon after the Paris Congress of Physics shut its doors, this hope coalesced around the possibility that the mass of the electron might be electromagnetic in origin. Guillaume had expected the open-ended phenomena discussed in section 5 of the congress to be readily absorbed within earlier sections. Instead, the electron became the focus of proposals to invert the foundations and rewrite the hierarchies of the discipline. The central idea on which this hope was founded is disarmingly simple. An electrostatic field surrounds a charge at rest; when it moves, it generates in addition a magnetic field. As early as 1881, J. J. Thomson had drawn attention to the fact that a charge in motion must accordingly pass through its own electromagnetic field, with a consequent decrease in its velocity—just as if it had gained mass. At the annual meeting of German scientists (the Naturforscherversammlung) in September 1900, Lorentz addressed what he described as the "extraordinary importance" of the question whether besides the "apparent mass" resulting from their motion as charged bodies, ions also possessed an "actual mass" in the ordinary sense of the word. Experiments on ions moving near the velocity of light might decide the issue. One of the physicists present had already proposed taking several steps still further. Wilhelm Wien rose to outline his endeavor to provide electromagnetic foundations for the whole of physics by neglecting the ordinary mass entirely, emphasizing that particular assumptions about the form of the electrical charge were critical to the kind of test Lorentz envisaged.[1] The electromagnetic program was extremely general in its projected scope. Its proponents aimed, for example, to replace mechanics as the key discipline in physics and to provide a field theoretic approach to gravitation on the model of electromagnetic theory. It also drew strength from the increasing utility of the electron in theoretical treatments of a wide variety of phenomena such as the electron theory of metals, thermionic emission, the photoelectric effect, and explanations of the place of elements in the periodic table. Indeed, as the Nobel citations will have indicated, to fully appreciate the unifying role of electron theory in the early twentieth century would require historical research across an extremely broad base.[2] Here I will focus solely on the sharp end of the reductionist program for a new worldview, and its interactions with relativity.

The most direct and powerful sustenance for the foundational importance of the electron came from experimental studies of the way the charge to mass ratio of the electron varied with velocity. They were spearheaded by the papers Walter Kaufmann began publishing on high velocity electrons from 1901. We met Kaufmann briefly in the previous chapter. Kaufmann had studied electrotechnology and physics in the Technische Hochschulen and universities of Berlin and Munich before earning his doctorate in Munich in 1894.[3] In 1896 he became an assistant in the Physical Institute at the University of Berlin. Like many following the announcement of Röntgen's discovery of X-rays, he took up research on

1. Lorentz, "Scheinbare Masse der Ionen" (1900). Wien had earlier outlined his views in a contribution to Lorentz's festschrift. On the electromagnetic program, see especially McCormmach's influential paper and recent discussion initiated by Seth's demonstration of its salience for Sommerfeld's approach to quantum theory. McCormmach, "Lorentz and the Electromagnetic View of Nature" (1970); Seth, "Quantum Theory and the Electromagnetic World View" (2004); Katzir, "On 'The Electromagnetic World-View'" (2005); Seth, "Response to Shaul Katzir" (2005).

2. John Heilbron, for example, has suggested that the development of atomic theory from the 1890s through to 1925 "can be construed as the gradual invention of the physics of electrons." Heilbron, *Historical Studies* (1981), 11.

3. von Kossel, "Walter Kaufmann" (1947); Campbell, "Kaufmann, Walter" (1973).

FIGURE 6.1 Walter Kaufmann. Courtesy Niedersächsische Staats- und Universitätsbibliothek, Göttingen.

the cathode rays that were responsible for their production. As we have seen, in 1897 he determined their charge to mass ratio at about the same time as J. J. Thomson, completing experiments on the magnetic deflection of cathode rays in April.[4] Like Thomson, Kaufmann moved fast to

4. Kaufmann was clearly puzzled by results indicating a constant charge to mass ratio far greater than that of hydrogen, even when using diverse cathode materials, gases, and potential differences. For just the same reasons Thomson argued the rays disclosed a new, corpuscular state of matter, Kaufmann argued they were inconsistent with the present ionic theory and the assumption that they were particles that gained their charge on contact with the cathode and were then accelerated away. Kaufmann, "Die magnetische Ablenkbarkeit der Kathodenstrahlen" (1897), 552. Kaufmann referred to Thomson's preliminary publication on cathode rays in *Nature* (on 544), having learned of Thomson's publication shortly before concluding his experiments. J. J. Thomson's son commented on the irony of the different conclusions Thomson and Kaufmann drew in George Thomson, *J. J. Thomson* (1966), 65.

consolidate understanding of cathode rays. First he investigated a further property that bore on the emission hypothesis by showing they could be deflected by the electric field of a second cathode. Then he refined his determination of the charge to mass ratio. Kaufmann had completed papers on both these tasks by October 1897, the point at which J. J. Thomson began publishing his major studies of cathode rays.[5] For the next decade Kaufmann remained in the forefront of research on what soon became identified as the "electron." Indeed, as we shall see, his studies set the direction of the field.

In 1899 Kaufmann moved to an assistant position in Göttingen, where he took the next step on the German academic ladder by undertaking his habilitation.[6] He also joined the advisory board for the journal *Physikalische Zeitschrift* on its founding in October 1899, and as we saw in the previous chapter, contributed to the interpretation of the Paris congress from afar. His early papers show Kaufmann mastering a set of experimental resources and a field of scholarship. He scrupulously mentioned the respects in which Thomson's papers bore on his independent research, expressing admiration for the bold and fruitful hypotheses that Thomson and his school pursued but noting criticisms also. Thomson sometimes treated as recognized fact what were hypothetical conclusions; and the school of Cambridge physicists often published their research with incomplete numerical material, almost always giving only final results so that it was hardly possible to form an independent judgment on the accuracy achieved—or achievable.[7] Kaufmann's remarks indicate a possibility

5. Kaufmann and Aschkinass, "Über die Deflexion der Kathodenstrahlen" (1897); Kaufmann, "Nachtrag zu . . . 'Die magnetische Ablenkbarkeit'" (1897). These papers were framed as investigations of the properties of particles that could not be identified with electrolytic ions or the molecules of the kinetic theory of gases and as supporting a modified form of the emission hypothesis first proposed by Crookes. Thomson's *Philosophical Magazine* paper had appeared after the first set of experiments had been completed. See 589 and 597.

6. With the agreement of the faculty of a specific university, the publication of a major research project earned young academics the right to lecture as a Privatdozent. Something of a waiting station, the post signaled their suitability for a professorship but carried no salary. Candidates had to survive on their own resources and what student fees they obtained from their lectures until a call came to take up a professorship.

7. On hypotheses, see Kaufmann, "J. J. Thomson, Die Entladung der Elektrizität durch Gase (Besprechung)" (1901). The occasional confession that some phenomena did not fit into a hypothesis was preferable to over-eager explanatory attempts. This was a review of the updated German edition of Thomson's 1897 book (translated in 1900). Kaufmann commented on the publications of Thomson, Rutherford, McClelland, and Zeleny in his habilitation lecture. Kaufmann, "Ionenwanderung in Gasen" (1899), 24.

that will have been evident in our discussions of Michelson, Rowland, and Cornu, of understanding differences in different fields and schools of research in part by examining the way they manage accuracy and error.[8] Kaufmann's papers consistently provided far more than results alone, and we will see that in the case of his definitive publication this very openness made it possible for discussion of his results to focus on an examination of potential sources of error.

High-Velocity Electrons and the Electromagnetic Mass

Kaufmann's early research on cathode rays and discharge in gases was rather broad in scope and showed a predilection for phenomenological caution, rather than exploring more speculative approaches by postulating specific theories of the internal mechanism of events like discharge.[9] By the spring and summer of 1901 he had turned to the radiation emitted from radioactive sources (building on his studies of Becquerel's and the Curies' congress reports). Recent research had established that to a rough order of magnitude Becquerel rays had the same charge to mass ratio as cathode rays. The quantitative differences between the two—in the degree to which they were deflected by a given electric or magnetic field, or their penetrability—could be explained by the suggestion that both were particles of the same kind, with Becquerel rays moving at significantly higher velocities.[10] Conceptually, identifying Becquerel rays with cathode rays as *electrons* gave Kaufmann the confidence to base his research project on the properties of the particles themselves, rather than investigating more generally the different contexts in which they were produced.

Although Kaufmann focused on the behavior of electrons at high velocities, this closely defined aim facilitated the pursuit of a new, reductive generality. The program found clear expression in an address, "The

8. This issue deserves more serious consideration, along the lines Olesko, Wise, and Gooday have established in their valuable studies. Olesko, *Physics as a Calling* (1991); Olesko, "Precision, Tolerance, and Consensus" (1996); Wise, ed., *The Values of Precision* (1995); Gooday, *Morals of Measurement* (2004).

9. In addition to the papers cited above, see Kaufmann, "Gasentladungen, I" (1899); Kaufmann, "Gasentladungen, II" (1899); Kaufmann, "Ionenwanderung in Gasen" (1899); Kaufmann, "Geisslersche Röhren" (1899); Kaufmann, "'Widerstand' leitender Gase" (1899); Kaufmann, "Elektrodynamische Eigentümlichkeiten leitender Gase" (1900).

10. Note that Kaufmann also published a translation of Marie Curie's dissertation. Curie, *Untersuchungen über die radioaktiven Substanzen* (1904).

Development of the Concept of the Electron," delivered to the 1901 Naturforscherversammlung in Hamburg.[11] There Kaufmann gave an account of the history of the electron that focused on the role that charged ions had found in theories of light and electrodynamics, before the Zeeman effect and studies of cathode rays led to the identification of a material particle much lighter than that of hydrogen. Interestingly, he added that Max Planck's work on the theory of the absolutely black body had recently given a value for the charge of the electron from a third and quite independent line of investigation.[12] In fact, one of the principal contexts in which physicists learned of Planck's work was to note its relevance to electron physics—a field of much wider concern than blackbody radiation. The diversity of circumstances in which the concept of the electron had been successfully deployed was important. "Everywhere and in all states of aggregation," Kaufmann said, "the smallest part of the visible world hitherto known" had been shown to play an important role in electrical and optical phenomena. But the spontaneous, natural emission of the electron in Becquerel rays formed the "keystone" to the argument—because the presence of the particle in such radiation, "even in the absence of external optical or electrical influences" offered "a direct proof of their permanent existence."[13]

Further, while Kaufmann noted that the specific mechanism responsible for their emission remained a "complete riddle," the behavior of the particles at these prodigious velocities appeared likely to offer insight into their true nature. If the entire inertia of a moving charge could be explained as an electromagnetic effect due to that motion, and if material atoms consist of conglomerates of electrons, the sterile effort to reduce electrical to mechanical phenomena could be replaced by research in the other direction, endeavoring to explain mechanics on the basis of electromagnetic theory. Proof that gravitation propagated over time or depended on the position and velocity of gravitating bodies would offer

11. Kaufmann, "Entwicklung des Elektronenbegriffs" (1902). An English translation was published as Kaufmann, "Development of the Electron Idea" (1901). Falconer uses the paper to explore differences between British and Continental understandings of the history of the "electron" in Falconer, "Corpuscles to Electrons" (2001).

12. Kaufmann, "Development of the Electron Idea" (1901), 97. When Kaufmann discussed cathode rays he listed seven people as responsible for the conclusion that nothing more than a modification of Crookes's hypothesis was required and stated that Wiechert had probably first proposed that the particles concerned were the same as those determined through the Zeeman effect.

13. Ibid.

crucial experiments for the new view; and perhaps the periodic table could be explained on the basis of stable groupings of electrons. Kaufmann concluded: "Although much may appear hypothetical, it is clear from what has been said that the electrons—those minute particles whose size is to that of a bacillus as the size of a bacillus to, say, the whole earth, and whose properties we may yet measure with the greatest precision—that these electrons are one of the most important foundations of our whole world structure."[14] Kaufmann referred to the electromagnetic program as one that had first been stated by Johann Zöllner thirty years earlier, and recently improved by Lorentz, J. J. Thomson, and Wilhelm Wien. He would attend to its empirical basis.

Earlier studies had explored the deviation of Becquerel rays in electric and magnetic fields.[15] Kaufmann introduced a new arrangement that would make an advantage out of what had previously been regarded as a liability: the inhomogeneity of the velocity spectrum of the rays (that is, the fact that the particles emitted showed a great variation in velocity). His central technique was analogous to the method of "crossed spectra" that August Kundt deployed in observations of anomalous dispersion. Kaufmann used as small a radioactive source as possible, and allowed only a tight bundle of the radiation emitted to pass through a small aperture in a lead diaphragm and subsequently fall on a photographic plate perpendicular to their path (fig. 6.2). In the absence of any fields the rays form a point on the photographic plate. When a magnetic field is applied (oriented perpendicular to the direction of the rays), the image left is a straight line, showing that fast-moving electrons are barely deviated from the original point while those moving more slowly suffer considerable deviation. If an electric field is applied at the same time, so as to deviate the electrons in a direction perpendicular to the magnetic deviation, then a curved line results. (In this arrangement the magnetic and electric

14. Ibid.

15. Kaufmann referred to papers from Giesel, Dorn, and Becquerel and described the charge to mass measurements of the latter two as "somewhat crude." Kaufmann, "Magnetische und electrische Ablenkbarkeit der Becquerelstrahlen" (1901), 143. This paper was translated in an important anthology (and there described as "a classic investigation that after sixty years is still cited in the textbooks of modern physics"), which may help explain why Kaufmann is often remembered by physicists for this (qualitative) demonstration of the variation of mass with velocity, rather than his later, more accurate and more controversial studies. Boorse and Motz, eds., *The World of the Atom* (1966), 502–12. Kaufmann's early experiments have been described most fully by Miller, *Einstein's Special Theory* (1981), 47–54 and 61–67.

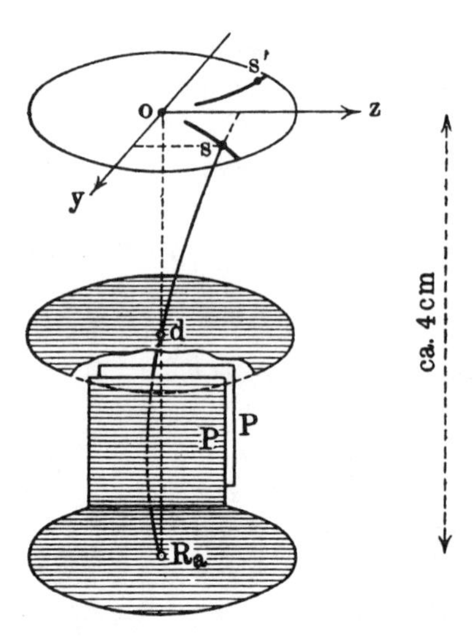

FIGURE 6.2 A schematic depiction of Kaufmann's arrangement. Radiation is emitted from R_a. Undeflected, it passes through the diaphragm d and hits the photographic plate at O (traveling in the direction of the x-axis). A magnetic field oriented in the direction of the y-axis deflects the radiation in the direction of the z-axis, while an electric field created by the condenser plates *PP* (oriented in the same direction as the magnetic field) deflects the radiation in the direction of the y-axis. S is a point on the resultant curve; s' an equivalent point on the curve created when the direction of the electric field is reversed. From Kaufmann, Coehn, and Nippoldt, *Magnetismus und Elektrizität* (1914), fig. 1134. See figs. 6.4–5 for the final version of the apparatus.

fields are parallel to each other, rather than crossed as in J. J. Thomson's experiments.) If both fields remain of constant strength throughout the course of the experiment, each point on the resulting curve corresponds to electrons with a completely definite velocity v and a completely definite charge to mass ratio. A single photographic plate thereby yields a whole series of observations. From them, Kaufmann wrote, "one can read the relation between the charge to mass ratio and velocity directly."[16] Despite the confidence of this statement, both the precision of the measurements involved and what it *meant* to read the relations between the velocity, mass, and charge of the electron were to change dramatically over the next decade.

Kaufmann was after precision, and the dimensions of his apparatus reflected this aim. Having as small a radioactive source as possible was a

16. Kaufmann, "Magnetische und electrische Ablenkbarkeit der Becquerelstrahlen" (1901), 143–45 on 144–45.

start (he first used a piece of radium bromide approximately 1 mm long and 0.3 mm thick). Given the difficulty of maintaining a constant electric or magnetic field over any volume, Kaufmann made the apparatus as small as possible and placed it within an evacuated flask, with a total path trajectory of just over 4 cm.[17] The difficulty of achieving constant conditions and taking precision measurements meant the spatial environment and distance measurements involved had to be controlled very tightly. But considerable time was required to develop enough data points to generate a visible curve. Kaufmann ran his experiments for three to four days, reversing the electric field halfway through that period in order to provide a check on the homogeneity of the field. Ideal conditions would leave symmetrical images on either side of the central point. Finally, developing the photographic plates was a delicate procedure; a weak developing agent needed sufficient time to diffuse to the deeper parts of the photographic plate without fogging the surface (for the fastest Becquerel rays penetrated to deeper levels of the plate without being absorbed). The images obtained were fairly weak but still clear enough to measure distances to 1/200 cm.[18]

These were the material dimensions of Kaufmann's apparatus, but as a Privatdozent in a busy laboratory he had to share resources and time. The batteries he used were constantly sought after, so Kaufmann conducted his experiments on holidays and from Saturday to Monday, and his early papers could present data from only a few runs.[19] Still, he got considerable support. At different times through three years' research in Göttingen, Kaufmann thanked Dr. Giesel of the Braunschweig Chininfabrik (Braunschweig Quinine Factory) for providing him with radium bromide as a radioactive source, and the Curies for an exceptionally active sample of radium chloride (which decomposed rather quickly).[20] Professor des

17. The distance between the radioactive source and the diaphragm was 2.07 cm and from there to the photographic plate was another 2 cm. The electric field was applied between plates 1.5 mm apart and 1.775 cm long, and acted only for the first part of the electrons' trajectory. The magnetic field was created by an electromagnet outside the apparatus because of the difficulty of running it within a vacuum. It acted over the entire path of the electrons.

18. Kaufmann, "Magnetische und electrische Ablenkbarkeit der Becquerelstrahlen" (1901), 148.

19. Kaufmann, "'Elektromagnetische Masse' der Elektronen" (1903), 92.

20. See Kaufmann, "Magnetische und electrische Ablenkbarkeit der Becquerelstrahlen" (1901), 144; Kaufmann, "'Elektromagnetische Masse' der Elektronen" (1903), 90–91. Beate Ceranski has studied the network of radium research in Germany. See Ceranski, "Radioaktivitätsforschung 1896–1914" (2004).

Coudres gave access to the high voltage batteries of the electrotechnical division of the Physics Institute. Both professors Emil Wiechert in Göttingen and Wilhelm Hallwachs from the Technische Hochschule in Dresden gave advice on potential multipliers to increase the voltage supplied to produce the electric field.[21] Two students (Hartmann and Kuntze) helped take measurements of the positions of points on the curve.[22] For later trials with a stronger electric field, the Göttingen Academy of Sciences loaned instruments and the materials required to build a vacuum pump.[23] At this time Kaufmann even received aid from the Prussian Ministry for Culture, which made it possible to supplement the Physics Institute batteries with a high voltage battery of 1,600 accumulators (Kaufmann thanked the Ministerial Director Friedrich Althoff by name).[24] Kaufmann's instruments and procedures clearly involved him in a considerable network of material and social resources. To take up Poincaré's metaphor, this is what it took to strain every nerve and spend the funds required to bring a new book to the library of science (as limited as those funds were). The theoretical implications of Kaufmann's work attracted even wider attention. From the beginning he conducted his research with an eye to this broader context, knowing that he was bringing precise experiment to bear on an open question that raised great hopes.

Kaufmann published preliminary reports in 1901, describing his aim to determine the velocity and charge to mass ratio of Becquerel rays as precisely as possible, and draw conclusions as to the proportion between "actual" and "apparent" mass.[25] Both the concepts at issue and the apparatus involved were to be continually refined over the next five years. His measurements showed that the charge to mass ratio decreased markedly with increasing velocity. A strict formula for the field energy of rapidly moving electrons had been derived by the Cambridge physicist George Searle, assuming that the electron was equivalent to a sphere with its charge distributed over an infinitely thin shell. Using Searle's model, Kaufmann's results suggested that at low velocities the proportion of apparent to

21. Kaufmann, "Magnetische und electrische Ablenkbarkeit der Becquerelstrahlen" (1901), 146.

22. Ibid., 150.

23. Kaufmann, "'Elektromagnetische Masse' der Elektronen" (1903), 91.

24. Ibid.

25. Kaufmann first described the technique and by November offered initial results. Kaufmann, "Methode" (1901); Kaufmann, "Magnetische und electrische Ablenkbarkeit der Becquerelstrahlen" (1901), on 144.

actual mass was 1 to 3, or as he put it, "the apparent mass is of the same order of magnitude as the actual mass, and in the case of the fastest Becquerel rays significantly larger."[26] But the theoretical assumptions concerning the distribution of the charge were critical, "Since we know nothing about the constitution of the electron and a priori are not justified in applying to the electron itself the electrostatic laws which in the end should be explained with its help, it is naturally possible that the energy relations of the electron can be represented by other assumptions." Accordingly, Kaufmann called for calculations of the field energy for other distributions of charge.[27]

The plea was heeded extremely rapidly. In 1902 a fellow Privatdozent in Göttingen published a groundbreaking paper on the dynamics of the electron. Max Abraham had completed his PhD under Planck in Berlin and served as his assistant for three years before moving to Göttingen in 1900. Abraham argued that the earlier theoretical work on which Kaufmann relied was unsatisfactory, and the dynamics of the electron required an essential completion of these studies. Most importantly, Searle's discussion of the field and field energy of an electrically charged conductor of ellipsoidal form was only sufficient to calculate what Abraham termed the *longitudinal mass* of the electron, that is, the inertia opposing its acceleration in the direction of its motion. The *transverse mass* or inertia opposing acceleration perpendicular to its path remained undetermined (because it did not change the energy).[28] But it was the transverse mass that was at issue in Kaufmann's experiments. Abraham introduced a vector for the "electromagnetic momentum" in addition to the electromagnetic energy, which enabled a simplification of the calculations of the energy and the longitudinal mass. Finally, among all the possible distributions of surface and volume charge on an ellipsoid with three axes of symmetry he derived formulas for a spherical electron and compared them with Kaufmann's experimental results. His results showed that it was not necessary to assume that such a spherical electron possessed any "material" mass:

26. Kaufmann, "Magnetische und electrische Ablenkbarkeit der Becquerelstrahlen" (1901), 155.

27. Ibid. Kaufmann also raised an issue that was to be important through the rest of his work, contrasting the considerable relative accuracy of the values obtained (comparing the position of different points on the curve) with the more difficult to achieve absolute accuracy of the values he obtained for the velocity and charge to mass ratio (which he thought were certain to about 5 percent). See 152.

28. Abraham, "Dynamik des Elektrons" (1902), 20–21. See Goldberg, "The Abraham Theory of the Electron" (1970); Miller, *Einstein's Special Theory* (1981), 55–61.

its entire mass could be explained as the dynamical effect of its electromagnetic field, that is, as "electromagnetic" mass.[29]

Kaufmann came back into print to echo Abraham's view. The close relationship between theoretical and experimental work—and reciprocal spur each offered to the development of the other—was illustrated particularly clearly at the Naturforscherversammlung later that year.[30] Kaufmann outlined the new situation following Abraham's work, but admitted that an early comparison had shown no good agreement. They had first evaluated Abraham's theory using absolute values derived from the experiment, in which an error of 2 percent in the determination of the ratio between the measured velocity of the electrons and the velocity of light would lead to a corresponding error of 19 percent in the determination of the mass of the electron. Now Kaufmann outlined a new method of data analysis based on the relative positions of points on the curve, using the method of least squares to determine the most probable values of the field strengths and apparatus dimensions.[31]

Speaking next, Abraham's paper showed that Kaufmann's experimental results were a direct motivation for new theory, but also that mechanics as it was currently understood provided important guidance in constructing a new mechanics based on the new particle, the electron. His earlier publications had used both a "material" and "electromagnetic" mass in the equations of motion, but it was now necessary to found the dynamics of the electron on a purely electromagnetic basis. The concept of electromagnetic mass established that mass—a quantity hitherto regarded as constant in ordinary mechanics—both varied and was vectorial in nature rather than scalar. Despite this distinction, Abraham's studies involved the exploitation of what he described as: "remarkable analogies

29. Abraham, "Dynamik des Elektrons" (1902), 40–41. In a pointed footnote playing on the use of the German word *wahr*, meaning true or real, and often applied to what Kaufmann had described as the "actual" mass, Abraham indicated his reasons for introducing a new terminology: "The widely used designations 'apparent' and 'true' mass can lead to confusion. The 'apparent' mass is in the mechanical sense real, while the 'true' mass is probably unreal," on 24.

30. Kaufmann, "Elektromagnetische Masse des Elektrons" (1902), 295. Kaufmann also corrected a calculational error and gave more attention to the empirical accuracy of his earlier experiments based on the measure of accuracy provided by the theoretical expectation that the velocity of the fastest electrons should converge to the velocity of light without exceeding it, 296.

31. He also let the audience see recent photographic plates. Kaufmann, "Elektromagnetische Masse des Elektrons" (1903).

between the principles of the dynamics of the electron on the one hand and the principles of the customary dynamics of material bodies on the other hand, analogies that will be important for the future electromagnetic founding of mechanics."[32] Note both the search for analogies, which would soon be important for a dynamics based on the principle of relativity, and Abraham's terminology. The foil for his development of a new mechanics is the "ordinary" or "customary" mechanics. Rather than departing from an "old" or "classical" mechanics, he is exploring ways to transform *present* theory. Abraham's reasons for favoring a rigid sphere electron were enhanced by his demonstration that a rigid ellipsoid with a uniform charge distribution could move in stable force-free translation only if its velocity were parallel to its major axis. This was a highly significant point that he and others would consistently raise against Lorentz's deformable electron: according to Abraham, such an electron would require the addition of extra, nonelectromagnetic forces to retain stability.[33]

Kaufmann's search for empirical improvement led to a further paper in the Göttingen *Nachrichten* in 1903, which became a platform for significant theoretical contributions. The two main requirements for greater accuracy were to achieve as clearly defined a curve as possible, and as great a deviation as possible. The first requirement depended on using the smallest and most active radioactive source available. Using a tiny sample of pure radium bromide he could now obtain delicate but satisfactorily clear curves even reducing the exposure to just twenty hours. Kaufmann had shown plates to an audience before. Now he published an image that became emblematic of his work, a reproduction of "Plate 19" (see fig. 6.3). Achieving a greater deviation of the electrons meant working with higher field strengths, and the need to both increase the electric field strength and keep it completely constant demanded more resources than Kaufmann

32. Abraham, "Prinzipien der Dynamik des Elektrons" (1903), 57. Note that this report was published in the *Physikalische Zeitschrift*; in 1903 Abraham also published an extended version with the same title in *Annalen der Physik*. In discussion Max Planck raised the question of how Cohn's alternative approach to electrodynamics would approach the questions at issue, and concerning the nature of the theoretical limitation of Abraham's results to "quasi-stationary motion" in which only slow changes in velocity were considered, 63. Abraham's work has been discussed in Miller, *Einstein's Special Theory* (1981), 55–61; Goldberg, "The Abraham Theory of the Electron" (1970); Janssen and Mecklenburg, "Electromagnetic Models of the Electron and the Transition from Classical to Relativistic Mechanics" (2004).

33. Abraham, "Prinzipien der Dynamik des Elektrons" (1903), 108–9. (In the *Annalen der Physik*.) See Miller, *Einstein's Special Theory* (1981), 55–61.

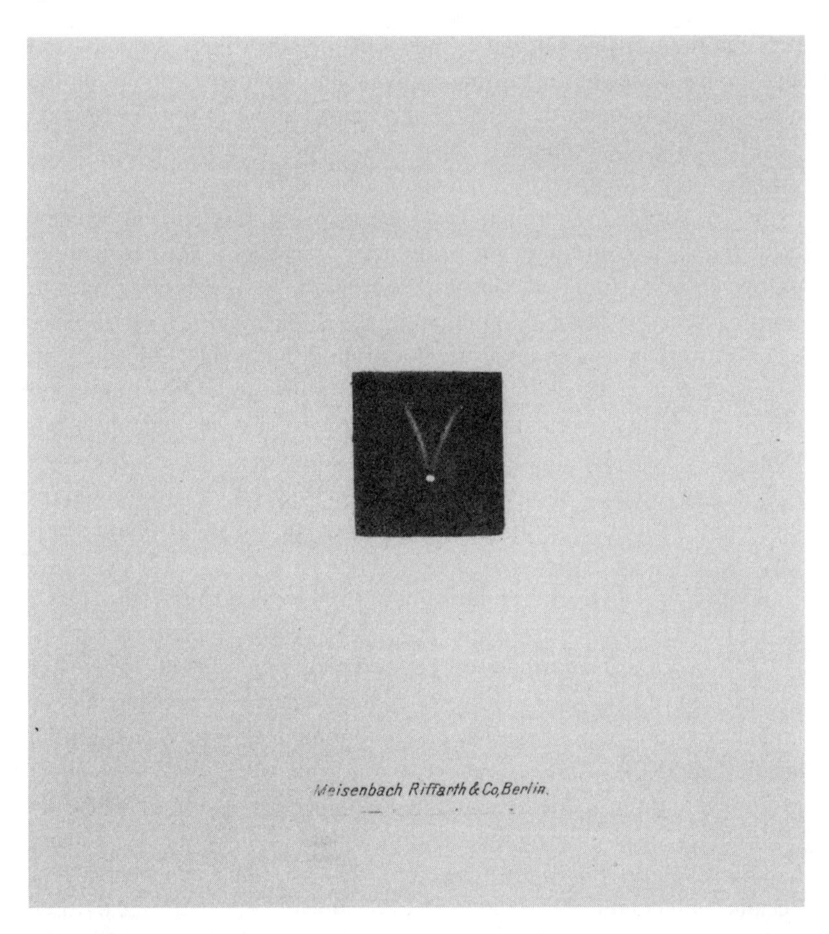

FIGURE 6.3 Kaufmann's Plate 19 from 1903. Kaufmann described this plate as offering by far the most clear and problem-free image (it also yielded a value for the charge to mass ratio of slow moving electrons that deviated from that of cathode rays by only 2.5 percent). He did state, however, that the reproduction did not come close to the clarity of the plate itself. From Kaufmann, "'Elektromagnetische Masse' der Elektronen," (1903), 102–3.

had at first been able to command.[34] In addition, Kaufmann paid careful critical attention to the difficulty of measuring the coordinates of points on the curve, and to his application of a least squares method to his data analysis—which Carl Runge criticized and reworked in a paper that

34. It was this that led to the involvement of the Ministry of Culture and the Göttingen Academy of Sciences described earlier. Kaufmann, "'Elektromagnetische Masse' der Elektronen" (1903), 91.

exemplifies his view of the role of mathematical methods in clarifying the import of empirical research. Runge was soon to become Germany's first and only professor of applied mathematics in Göttingen.[35]

"On the Constitution of the Electron": Testing Rival Electron Theories

On leaving Göttingen to take up a professorship at the University of Bonn in 1903, Kaufmann could look back on an extraordinary achievement. He had demonstrated empirically that "not only Becquerel rays but also cathode rays consist of electrons, whose mass is of a purely electromagnetic nature."[36] In Bonn he joined the experimental physicist and spectroscopist Heinrich Kayser, who wanted an extraordinary professor "who could treat all parts of theoretical physics with their connections."[37] That

35. Measurement of the area near the origin was both difficult and gave values for the velocity over that of light. Kaufmann had excluded this part of the curve from the calculations of the result. Runge praised Kaufmann's "excellent" work but maintained that this rendered his application of the method of least squares unorthodox and problematic (together with a question of weighting in Kaufmann's evaluation of a constant). In view of the fundamental importance of the measurements, with Kaufmann's consent Runge reworked the data treatment of four plates. See ibid., 93–94 and 97; Runge, "Elektromagnetische Masse der Elektronen" (1903). See also their correspondence: Kaufmann to Runge, 10 and 19 October 1903, in DMA, Archiv HS 1948/78/1; and Runge to Kaufmann, 18 October 1903, Archiv HS 1948/79.

36. Kaufmann, "'Elektromagnetische Masse' der Elektronen" (1903), 103.

37. Jungnickel and McCormmach, *Intellectual Mastery*, vol. 2 (1986), 144. The fact that Lorberg continued as an extraordinary professor of theoretical physics until his death in 1906 made this somewhat difficult. Kaufmann's teaching was indeed comprehensive. He began a cycle with lectures on "Wave Theory and Optics" in the summer semester of 1903 before moving on to "Theoretical Mechanics," then "The Theory of Heat," and then courses on both "The Theory of Magnetism and Electricity" and "The New Radiations (Cathode-, Röntgen- and Radium-rays)" in the winter of 1904/05, before beginning again with wave theory and optics. His lectures as a Privatdozent in Göttingen had shown considerable range, from "Exercises in the Production and Handling of Simple Demonstration Devices" to treatments of his field of research, "Electrical Phenomena in Gases," and courses on other interests such as "Physical Units and Constants" and the "Physical Foundations of Music." Kaufmann's pedagogical concerns also surfaced in occasional notes to the *Physikalische Zeitschrift* in which he discussed demonstration aids for the laboratory and lecture hall, and in 1909 and 1914 he jointly authored two comprehensive volumes devoted to electricity and magnetism in the tenth edition of the famous Müller-Pouillet's *Textbook on Physics and Meteorology*: Kaufmann, "Neue Hilfsmittel für Laboratorium und Hörsaal (mit Demonstrationen)" (1907); Kaufmann, "Eine einfacher Vorlesungsapparat" (1911); Kaufmann and Coehn, *Magnetismus und Elektrizität* (1909); Kaufmann, Coehn, and Nippoldt, *Magnetismus und Elektrizität* (1914).

desire reflects the rising disciplinary importance of theory in Germany (but also its often subordinate institutional position, given the lower status of extraordinary as against ordinary professorships), and Kaufmann seems to have focused on theoretical issues in teaching courses on wave theory and optics, theoretical mechanics, and the theory of heat. Even more importantly, dealing with Kaufmann's research had become nearly indispensable for theorists of electrodynamics. In 1904 H. A. Lorentz published a major paper, "Electromagnetic Phenomena in a System Moving with Any Velocity Smaller Than That of Light." Lorentz's work was clearly motivated in the first place by the optical and electric experiments that continued to show no evidence of the effect of an ether wind to the second order. As he put it, the first of these was "Michelson's well-known interference experiment"; more recent examples included the experiments of Lord Rayleigh and DeWitt Brace on double refraction and Frederick Trouton and H. R. Noble on the couple expected on a turning condenser.[38] A further motivation was the criticism that Poincaré had expressed at the 1900 Congress of Physics, commenting on the artificial nature of Lorentz's introduction of a new hypothesis to explain the negative result of the Michelson experiment.[39] Lorentz now offered what he regarded as a more satisfactory approach along lines he had first framed in 1899. Starting from the fundamental equations of the theory of electrons, he developed a set of transformation equations that incorporated a variable for "local time" and were exact to the second order. They described the relations between a system at rest in the ether (for which Lorentz used the term *true time*) and a second moving with a constant velocity relative to the first; and as Lorentz wrote, they showed that such a translation of a system would "*of itself... produce*" the deformation at issue.[40] In comparison with his earlier treatments, Lorentz's physical argument was now extremely general as a result of supplementing the translation equations with two general assumptions. The first was "*that the electrons, which I take to be spheres of radius R in the state of rest, have their dimensions changed by the effect of a translation.*" The second was the assumption that the forces between uncharged particles and those between electrons

38. Lorentz, "Electromagnetic Phenomena" (1937 [1904]), 172. On Lorentz, see Janssen, "Lorentz's Ether Theory and Special Relativity" (1995); Janssen, "Reconsidering a Scientific Revolution" (2002).

39. Lorentz, "Electromagnetic Phenomena" (1937 [1904]), 173.

40. Ibid., 183. Lorentz's emphasis.

and uncharged particles were influenced in just the same way by a translation.[41]

In the second part of the paper Lorentz applied his new methods to the calculation of the electromagnetic momentum of a single electron, reaching the third major motivation to his studies: Kaufmann's experiments. Referring to Abraham's earlier work, Lorentz derived expressions for the longitudinal and transverse electromagnetic mass of the electron. Like Abraham and Kaufmann, Lorentz was now also ready to suppose "*that there is no other, no 'true' or 'material' mass.*"[42] However, his formulas for the electromagnetic mass of an electron that deformed in motion differed from Abraham's rigid sphere electron. Lorentz stated the challenge: "Now, as regards the transverse mass, the results of Abraham have been confirmed in the most remarkable way by Kaufmann's measurements of the deflexion of radium-rays in electric and magnetic fields. Therefore, if there is not to be a most serious objection to the theory I have now proposed, it must be possible to show that those measurements agree with my values nearly as well as with those of Abraham."[43] Lorentz addressed this question by reworking the data reduction for curves III and IV from Kaufmann's 1902 experiments, and plates 15 and 19 from Kaufmann's 1903 paper (as calculated by Runge). In each case Lorentz was able to find appropriate constants that established a fit between the results and his own theory that was, as he put it, "satisfactory."[44]

41. Ibid., 182–83. Lorentz's emphasis. This implied Lorentz's earlier deformation hypothesis but was more general because the only limitation imposed on such motion was that it be with a velocity less than that of light. See 183–84.

42. Ibid., 185. Lorentz's emphasis.

43. Ibid., 192–93. As Kaufmann stated the contrast in 1906, the expressions for the transverse mass are: for Abraham

$$\mu_t = \frac{3}{4}\mu_0 \frac{1}{\beta^2}\left[\frac{1+\beta^2}{2\beta}\ln\left(\frac{1+\beta}{1-\beta}\right) - 1\right],$$

for Lorentz

$$\mu_t = \mu_0\left(1-\beta^2\right)^{-1/2},$$

and for the theory Bucherer was soon to outline,

$$\mu_t = \mu_0\left(1-\beta^2\right)^{-1/3},$$

where μ_0 is the mass for electrons moving infinitely slowly and β is v/c. See Kaufmann, "Konstitution des Elektrons" (1906), 529–30.

44. See Lorentz, "Electromagnetic Phenomena" (1937 [1904]), 192–95. He was referring to Kaufmann, "Elektromagnetische Masse des Elektrons" (1903), 55, Tabelle III and IV; Runge, "Elektromagnetische Masse der Elektronen" (1903), 328–29.

Writing to Lorentz in July 1904, Kaufmann conceded the point and promised to repeat his measurements with increased accuracy.[45] Kaufmann was surprised his experiments could support two theories postulating such different fundamental structures for the electron, but minor differences offered some hope for distinguishing between them.[46] Lorentz achieved a satisfactory fit to the shape of the curve by assuming values for the velocity of the electrons that were between 5 and 7 percent smaller than those that Abraham's theory had yielded. This opened the way to a comparison, but it would have to overcome what Kaufmann had identified as the prime weakness of his earlier experimental results. Hitherto his *relative* measures of the positions of different points on a single curve had been considerably more accurate than Kaufmann's *absolute* determinations of the velocity of the electrons, since the latter relied on an extensive chain of inferences and measurements. Now Kaufmann would have to establish the accuracy of the experiment in its "absolute" dimensions. His experiment would have to be carried out "with far greater precision than previously."[47] We will see later that Kaufmann assimilated Lorentz's theory into the same categories he used in his own experimental work.

Achieving far greater precision required new apparatus, and it was more than a year before Kaufmann could offer a preliminary report in November 1905 and send his definitive paper to the *Annalen der Physik* on New Year's Day in 1906. The new design and protocols differed from his earlier experiments solely in that "all particularities are arranged with the goal of attaining the greatest possible precision."[48] In fact, he sought conceptual as well as material precision. Kaufmann opened with an extensive and subtle treatment of the broader theoretical context, as rapidly changing as that was with recent contributions from Alfred Bucherer and Albert Einstein (and we consider the nature and significance of Kaufmann's conceptual work in chapter 8, when we examine the interpretative role of physicists' "research histories"). Then followed an exhaustive

45. Kaufmann to Lorentz, 16 July 1904, on deposit at the Algemeen Rijsarchief, The Hague, The Netherlands. As cited in Miller, *Einstein's Special Theory* (1981), 75.

46. Kaufmann, "Konstitution des Elektrons" (1905), 949; Kaufmann, "Konstitution des Elektrons" (1906), 493.

47. Kaufmann, "Konstitution des Elektrons" (1905), 950. Kaufmann's earlier papers had exhibited a ceaseless search for accuracy in all facets of his work. In 1902 this was driven by the possibility of using the expectation that the fastest electrons would converge at the speed of light without exceeding that velocity, and we discussed the measures he took in 1903 above. See Kaufmann, "Elektromagnetische Masse des Elektrons" (1902), 296.

48. Kaufmann, "Konstitution des Elektrons" (1906), 495.

discussion of experimental procedures. Finally, Kaufmann set out his results with admirable clarity, and left no room for ambiguity in stating his conclusions. Historians' perceptions of Kaufmann's work have been dominated by those conclusions and the varied responses they drew from contemporaries, but the extensive attention paid to his results was only possible because of the care Kaufmann devoted to all facets of his research. Here we will consider Kaufmann's experimental procedures and the material culture of his apparatus as an example of what precision experimentation meant in the early twentieth century, before analyzing the theoretical and experimental environment in which his results were discussed.

Kaufmann's mechanic was the highly respected Max Wolz, who had set up his workshop in Bonn after training at the Zeiss works in Jena and working with Carl Bamberg in Berlin. (Wolz had displayed just six instruments in the collective German display in Paris, including geodetic instruments and a spectrogram measuring device. Made according to the directions of Kayser, it simplified the precision measurements of spectrographic plates.[49]) Kaufmann described Wolz as having constructed his apparatus with exceptional diligence. In 1897 Thomson had used a paper scale pasted on his glass cathode ray tube. In 1901 Kaufmann had paid no specific attention to the construction of his apparatus and measured the distances between the source, diaphragm, and photographic plate with a compass and ruler. Now, both the "absolute invariance" of all dimensions and the need to measure them precisely were essential requirements in the design and fabrication of his apparatus.[50] That meant bringing the optical techniques Michelson had helped advance into play. For example, the lower surface of the diaphragm, now constructed out of platinum rather than lead, was optically plane and rested on three columns. The common tangential plane to the top of these columns was rendered parallel to the inside base of the instrument by optical means also. (A plane parallel glass plate was rested on them and observed from above illuminated by sodium light.) The photographic plate was carried on three further columns that rose through holes bored in the diaphragm. These two

49. *Weltausstellung in Paris 1900: Sonderkatalog der deutschen Kollektivausstellung für Mechanik und Optik,* (1900), 63 (for geodetic instruments including levels and a demonstration theodolite) and 107–8 (for the spectrogram measuring instrument).

50. Kaufmann, "Konstitution des Elektrons" (1906), 496 (for the references to diligence and absolute invariance); Kaufmann, "Magnetische und electrische Ablenkbarkeit der Becquerelstrahlen" (1901), 148 (for the compass and ruler).

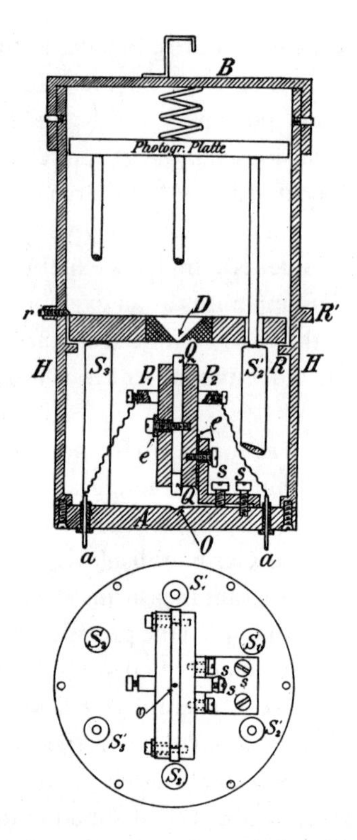

FIGURE 6.4 Kaufmann's 1906 apparatus in section and viewed from above. The radioactive source is situated at *O*; electrons pass through the diaphragm *D* to the photographic plate. *D* rests on three columns *S*; and three further columns *S′* pass through the diaphragm to carry the photographic plate. The condenser plates are P_1 and P_2, separated by four pieces of quartz *Q* cut out of a single plane-parallel plate. From Kaufmann, "Konstitution des Elektrons," (1906), 496, fig. 1. The illustration was printed at natural size, as it is here. It has often been reproduced; Einstein may have been the first to do so in Einstein, "Relativitätsprinzip," (1907), 438, fig. 1.

distances, from the source to the diaphragm and from the diaphragm to the photographic plate, were measured with a Zeiss gauge made available by the firm itself.[51] Kaufmann described the measurement methods and results in the body of the paper and detailed the measurement and observation protocols in his first appendix. The final value for such dimensions

51. The plane parallel glass plate was placed on the first and second set of columns in turn, with measurements being taken of the distance from the top surface of the glass plate to the base of the apparatus at several different points, compensating for the impossibility of measuring the distances directly in the center of the apparatus.

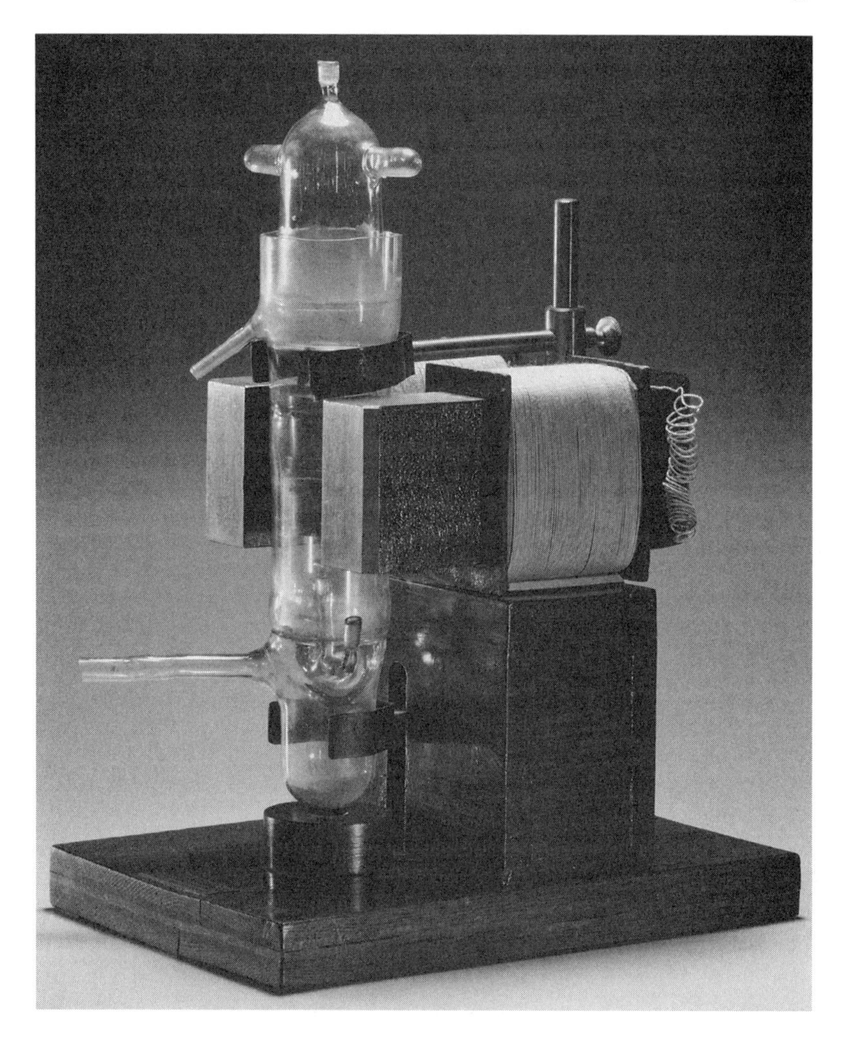

FIGURE 6.5 Kaufmann's apparatus is now held at the Deutsches Museum in Munich. This illustrates the total arrangement. The cylinder containing the radioactive sample, condenser, and photographic plate is held within an evacuated glass tube and positioned between the poles of the magnet. The armature of the magnet was removed for his 1906 experiments. Courtesy Deutsches Museum, Munich.

was given to the micron (thousandths of a millimeter): the distance from source to diaphragm was 2.0048 cm, and from diaphragm to photographic plate was 1.9693 cm. Other important dimensions were the length of the condenser plates, their position, and the distance between them (0.12434 cm ±0.00011).

One major difference with his earlier work concerned replacing the electromagnet with a powerful, old permanent magnet, which obviated the difficulty of maintaining a constant current over several days in an electromagnet. Kaufmann measured its absolute field strength over the radiation path, using an empirical equation derived from these measurements to work out the field integrals.[52] Direct measurements of the electric field over its path were not possible, so Kaufmann made a scale model of the top half of the condenser and the diaphragm above it, magnified 29 times. Timing the oscillations of a metal mirror 5 mm in diameter enabled him to plot the distribution of the field within the condenser plates and beyond their edges.[53]

Once a single experimental run had been completed and the photographic plate developed, a small Abbe comparator was used to measure the positions of points on each photographic curve, with individual values representing the mean of ten measurements. Plates from five runs were measured, and all measurements were then reduced to the same standard field strength, corresponding to a potential difference of 2,500 volts over the condenser plates. A single "balanced" curve (*ausgeglichene*) was then produced by combining the individual points from different plates, taking account of different weights that Kaufmann assigned (proportional to the average field across the condenser for the plate concerned). This balanced curve was as free as possible from the individual discrepancies of specific curves and formed the basis for the comparison of the different theories.[54] To facilitate the calculations required, Kaufmann considered only deflections that were infinitely small relative to the dimensions of his apparatus and reduced his actual observations appropriately. To put his comparison on as secure a footing as possible, Kaufmann then used three different ways of comparing theory with observation. One involved comparing his apparatus constants with the curve constants that established the best fit between a theory and the reduced observations. A second involved comparing the value that his observations indicated each theory would predict for the charge to mass ratio for infinitely slow electrons (ε/μ_0) with a value for that ratio extrapolated from the earlier experiments of Hermann Simon (noting that there was as yet no complete

52. It gave a magnetic field of approximately 140 gauss, was homogenous within 2 percent over the path of the radiation, and remained constant over time. Kaufmann, "Konstitution des Elektrons" (1906), 508–13.

53. Ibid., 513–17.

54. Ibid., 517–24.

agreement concerning the ratio for slow electrons). The third used apparatus constants in conjunction with the figure for ε/μ_0 extrapolated from Simon and convenient values of the measured deflection in the *z* direction, to compare the value each theory calculated for the deflection in the *y* direction with the reduced value actually observed.[55] So much for the material culture of precision. We will see later how it was tested in the discussion that Kaufmann's work occasioned.

By the time Kaufmann finished his experimental research the theoretical situation had developed considerably. As we have seen, the empirical tests of the structure of the electron had already called forth two theories from Abraham and Lorentz. A third was proposed in 1904 by Alfred Bucherer—since 1899 a Privatdozent at Kaufmann's university in Bonn—in order to mediate between the perceived limitations of both theories (and French researcher Paul Langevin proposed a similar electron theory independently). With its rigid electron, Abraham's theory on the one hand was in conflict with the ether-drift experiments. On the other hand, following Abraham's argument (now stated most fully in his 1905 textbook, *The Electromagnetic Theory of Radiation*), Lorentz's theory was thought to require an extra, nonelectromagnetic force in order to retain equilibrium, for without it the deformable electron could explode owing to the repulsive forces between the constituents of its charge. (Poincaré had addressed this criticism by introducing a supplemental, nonelectromagnetic potential, which has become known as the Poincaré stress or pressure, that ensured the electron's stability for arbitrary motion.)[56] Responding to these limitations, Bucherer and Langevin proposed an electron that was deformed, with a length contraction in the direction of its motion equivalent to that of Lorentz's. But their electron retained its volume, expanding in the lateral dimension and resulting in a different expression for the transverse mass of the electron.

A further contribution came with the publication of Albert Einstein's paper, "On the Electrodynamics of Moving Bodies," on 26 September 1905. Like Lorentz, Einstein motivated his approach from a general consideration of the mechanical, electrodynamic, and optical consequences

55. For detailed discussions of the different comparisons, see Miller, *Einstein's Special Theory* (1981), 228–32; Hon, "Experimental Error" (1995), 190–94.

56. Abraham, "Grundhypothesen der Elektronentheorie" (1904), 578; Abraham, *Theorie der Elektrizität*, vol. 2 (1905), 201–8, esp. 204–8. On the Poincaré stress, see Miller, *Einstein's Special Theory* (1981), 80–84.

of motion. However, he focused attention far more closely on the operations of measurement and the significance of the velocity of light in evaluating the relations between events in systems at rest and in uniform motion in relation to one another. Rather than starting as Lorentz did with the equations of electrons, the primitive concepts in Einstein's theory concerned the basic tools of measurement. He wrote: "The theory to be developed is based—like any other electrodynamics—on the kinematics of the rigid body, since the statements of each theory concern relations between rigid bodies (coordinate systems), clocks and electromagnetic processes."[57] Famously, Einstein referred to ether-drift experiments generally, without naming any in particular (even the most famous of them) and concluded that the light ether would prove superfluous. Like Lorentz he considered the dynamics of the electron in the final section of his paper. Unlike Lorentz, however, Einstein did not discuss specific experiments in this field either. While he derived expressions for the longitudinal and transverse mass and considered the general relations that allowed experimental tests of the properties of electrons in motion, Einstein did not compare his theory with Kaufmann's earlier results. In this case the lack of direct citations functioned as a protective strategy. Einstein surely knew Kaufmann's research, but as Miller has pointed out likely saw no advantage in drawing attention to what would have been a rather large difference of 13 percent between his predictions for the transverse mass of the electron and Kaufmann's 1903 values.[58] Rather, Einstein stressed one feature of his work that distanced it, at least in some respects, from the electromagnetic program so often directly associated with research on electron theory. Einstein opened his section on the dynamics of the electron by defining the electron as a *point formed*, charged particle. While he would go on to describe the variation of the mass of the particle with velocity, he noted that those changes were independent of its charge, being valid for a ponderable material body in general.[59]

Kaufmann, meanwhile, *was* ready to comment on Einstein's approach—and its relations to Lorentz's work. In November he wrote that Einstein's conclusions were formally identical to those of Lorentz's theory, and noted that his own results tested the validity of the fundamental assumptions

57. Einstein, "Elektrodynamik bewegter Körper" (1905), 892.

58. This is the kind of result Einstein would have obtained if he made such a comparison directly rather than incorporating Runge's least squares analysis. See Miller, *Einstein's Special Theory* (1981), 333–34.

59. Einstein, "Elektrodynamik bewegter Körper" (1905), 919.

of Lorentz and "consequently also" of Einstein, describing these as "the attempt to ground the whole of physics, including electrodynamics and optics, on the principle of relative motion."[60] In his definitive paper Kaufmann went into still more detail. While the significance of the links and distinctions he drew will be discussed more fully in the following chapter, it is important to note one major feature at this point. We have become used to thinking of the question of absolute and relative measures conceptually, in terms of space and time and the contrast between Newton and Einstein. For many contemporaries a discussion of this kind had been initiated in the dialogue between Lorentz and Poincaré, with Poincaré focusing on the status of the principle of relativity (and Lorentz providing a system of equations that related the frame of reference of the ether at rest to frames of reference in uniform motion). Exploring the material culture of Kaufmann's experimental work has shown that a contrast between relative and absolute measures was central to his understanding of his research. Relative meant precise but isolated measures; and absolute measures were significantly more demanding because they required linking a more extended network of measurements. In his overview of theories of the electron, Kaufmann interpreted Lorentz's 1904 discussion of corresponding states and his contrast between true time and local time in terms quite different from Lorentz's own. Kaufmann described Lorentz using "absolute velocities" (relative to a rest ether and required in Lorentz's calculations) and "relative velocities" (which Kaufmann noted were alone observable).[61] Kaufmann thus assimilated Lorentz's work to the kind of categories that he himself used in his experimental work and that both Poincaré and Einstein had employed in their treatments of electrodynamics. While noting specific differences between Lorentz's and Einstein's approaches, Kaufmann thereby helped render Lorentz's electron theory part of an ongoing development in which the principle of relativity played a critical role. A new relativity physics was being constructed across both experiment and theory.

Kaufmann stated his conclusions directly after summarizing current theory: "*The results of the measurements are not compatible with the Lorentz-Einstein fundamental assumptions. The equations of Abraham and Bucherer represent the results of the observations equally well. A decision between both through measurement of the transverse mass of β-rays appears*

60. Kaufmann, "Konstitution des Elektrons" (1905), 954 and 956.

61. Kaufmann, "Konstitution des Elektrons" (1906), 490–93.

impossible at present."[62] The Lorentz-Einstein approach had come off worst in all three of Kaufmann's methods of comparing theory and observation.

Responding to Kaufmann: "A Matter of Life or Death for Different Electrodynamic Theories"

The varied responses that Kaufmann's 1906 experiment received from a large number of physicists have quite rightly been of vital interest to historians of relativity, for they bring out key issues in the attitude of the physics community to Einstein's work, especially in Germany. They have also provided fodder for innumerable discussions of the relations between theory and experiment in philosophical accounts of scientific research (although the range of responses precludes any simplistic treatment of those relations). Both Poincaré and Lorentz took Kaufmann's conclusions seriously, expressing a readiness to give up the approach they had worked so hard to refine (while on other occasions they offered reasons for suspending judgment). Lorentz wrote to Poincaré that "unfortunately my hypothesis of the flattening of electrons is in contradiction with Kaufmann's new results, and I must abandon it."[63] In 1908 Poincaré wrote that Kaufmann's experiments "*have shown Abraham's theory to be right.* Accordingly, it would seem that the Principle of Relativity has not the exact value we have been tempted to give it."[64] In contrast, as a convinced adherent of the electromagnetic worldview, the applied mathematician Arnold Sommerfeld eagerly passed on news of the latest results to Wilhelm Wien and showed original photos from Kaufmann in a talk to the local scientific society.[65]

62. Ibid., 495. Kaufmann's emphasis.

63. Lorentz to Poincaré, 8 March 1906, as cited in Miller, *Einstein's Special Theory* (1981), 334. See also Hon, "Experimental Error" (1995), 210–22.

64. Poincaré, *Science and Method* (1914), 228. Poincaré's emphasis. As cited in Hon, "Experimental Error" (1995), 202. Hon considers the tensions between different facets of Poincaré's varied responses to Kaufmann's experimental work in detail on 198–204.

65. Sommerfeld had heard from Kaufmann himself: Kaufmann to Sommerfeld, 4 November 1905, in Deutsches Museum, Munich, Archiv HS 1977–28/A,161. For his reaction, see Sommerfeld to Wien, 5 November 1905, 14 December 1905, and also 23 November 1906 (where Sommerfeld states that Röntgen regarded the experiment as insufficiently accurate to support the decision). In Sommerfeld, *Wissenschaftlicher Briefwechsel*, vol. 1 (2000), Docs. 96, 97, and 102.

Intellectual discussions on the relations between theory and experiment have an institutional complement, and Sommerfeld's career reinforces a point we have already seen illustrated in the extensive network within which Kaufmann pursued his experimental research. Work on the electron interested a broad spectrum of disciplinary communities in Germany. It proved of particular importance, however, in forming an intellectual agenda for those who specialized in theory. After training in mathematics and physics in Königsberg and Göttingen, Sommerfeld had habilitated in Göttingen with Felix Klein, becoming probably the most successful missionary of Klein's disciplinary aims to bridge mathematics and technology through an intellectual focus on "physical mathematics" and the institutional promotion of applied mathematics. Sommerfeld taught mathematics at the Mining Academy in Clausthal (1897–1900) and mechanics at the Technische Hochschule in Aachen (1900–1906), where he enthusiastically developed research programs in technical mathematics (working on the resonance of unbalanced bridges, for example). From 1898, Sommerfeld edited the volumes on physics for Klein's encyclopedia of mathematical sciences. The intellectual engagement in electron theory that this stimulated resulted in a series of studies of the Abraham electron from Sommerfeld's own pen, beginning in 1904—and these helped make him a theoretical physicist. In 1906 Sommerfeld took a coveted position as professor of theoretical physics in the University of Munich, which had previously been held by Ludwig Boltzmann (known for the statistical microphysics of his kinetic theory of gases) but had remained unfilled for a dozen years.[66]

Undoubtedly the most important immediate response to Kaufmann's work, however, was that of Max Planck, and it came in the form of a defense of the principle of relativity, which Planck described as having been recently introduced by Lorentz, and in a still more general form by Einstein.[67] Max Planck was one of the first people to identify themselves as

66. For want of suitable candidates apart from Boltzmann, indicating the paucity of the field of theoretical physics. Eckert and Märker offer a sensitive portrayal of the different stages of Sommerfeld's transition from applied mathematics to theoretical physics, achieved in large measure through his work on electron theory. Ibid. On Sommerfeld's papers on electron theory, see also Pyenson, "Physics in the Shadow of Mathematics" (1985), 120–28. For a rich study of Sommerfeld's work articulating a model of theoretical physics different from that of Planck and Einstein, see Seth, "Principles and Problems" (2003).

67. Planck, "Prinzip der Relativität" (1906), esp. 136, 137. For Planck's reasons for accepting relativity, and his role in the development of the theory more generally, see Goldberg, "Max Planck's Philosophy of Nature" (1976); Pyenson, "Physical Sense in Relativity"

a theoretical physicist and concentrate on theory to the exclusion of experimental work, and was particularly impressed by the universal and absolute features of physical law embodied in thermodynamics (the subject of his 1879 PhD dissertation). An extraordinary professor at Kiel from 1885, he moved to Berlin in 1889 and had been made ordinary professor by 1992. He initially took up blackbody radiation in 1894 as an opponent of Boltzmann's statistical approach to the kinetic theory of gases. Because Planck's attitude toward the electron was important in his approach to relativity—while the fault line between Planck's and Boltzmann's approaches later proved critical to the formation of both quantum theory and concepts of classical theory—it will be helpful to outline broad features of his approach here. Planck hoped to find a counter to Boltzmann's merely probabilistic interpretation of the second law of thermodynamics, intending to show that like the first law, the second law of thermodynamics was absolute. Believing entropy always increases, Planck explored the possibility of proving this by considering the relations between thermodynamics and radiation in the context of the blackbody. Matter and gases might often be understood atomistically, but the laws of radiation were continuous. There, Planck thought, one might find enough purchase to develop an absolute thermodynamics drawing on a continuum mechanics. Wary of models of microphysical phenomena and suspicious of electron theory in the 1890s, Planck represented the walls of the blackbody as a set of charged resonators—a deliberate idealization that reflected the search for universal laws constraining physical phenomena rather than microphysically plausible mechanisms. We have seen that Planck's early support for a law proposed by Wien had faced critical scrutiny in papers presented to the Congress of Physics in Paris, and by late 1900 it had become clear that Wien's law was empirically inadequate. In October and December 1900, Planck formulated and derived a successful law that involved drawing selectively on Boltzmann's work. Doing so, Planck celebrated the two natural constants his law involved, and the quantitative links it established between electromagnetic theory and the properties of electrons and atoms. He became a convinced atomist, but did not abandon his commitment to an absolute interpretation of the second law.[68]

(1985); Pyenson, "The Relativity Revolution" (1987), esp. 96–101. Planck's philosophy and disciplinary role in Germany are discussed in Heilbron, *Dilemmas of an Upright Man* (1986).

68. For a deft overview of Planck's path, see Heilbron, *Dilemmas of an Upright Man* (1986). The more detailed studies of Kuhn, Needell, and Darrigol are addressed in relation to

As an editor of the *Annalen der Physik*, the journal in which Einstein published, Planck was well aware of the range of the Bern physicists' work. From 1901 on, Einstein had published a series of papers on capillarity, kinetic theory, and statistical mechanics. In 1905 a paper on the nature of light and a discussion of Brownian motion had passed over Planck's desk in quick succession. The former drew highly speculative implications out of blackbody theory, which Planck's later comments show he clearly distrusted; but Planck was extremely impressed by the paper on the electrodynamics of moving bodies that arrived on 30 June (and was published in late September). He presented a colloquium on it in the winter semester of 1905/06 (with his new assistant Max Laue in the audience) and then published a paper showing how the Newtonian equations of motion could be generalized in accord with the principle of relativity.[69] In chapters 8 and 9 we will consider the significance of the terms in which Planck linked Lorentz and Einstein's work on the one hand, and related new and old forms of mechanics on the other, as we consider participant histories of relativity and trace the development of concepts of classical mechanics. Planck thought the simplicity and generality of the principle of relativity warranted investigation from many points of view, and he believed it should be tested thoroughly enough to see whether it could stand up to severe scrutiny. Kaufmann's empirical disproof threatened that possibility, so Planck combed through Kaufmann's data reduction and went to the 1906 Naturforscherversammlung in Stuttgart to advocate the continued relevance of the Lorentz-Einstein approach.

Planck applauded the comprehensive clarity and numerical openness of Kaufmann's publication, which put anyone in the position of testing his conclusions. This was particularly important since, Planck stated, "the questions to which Kaufmann's experiments are dedicated are a matter of life or death for different electrodynamic theories."[70] Using a different method of calculating the results, Planck obtained values that differed from Kaufmann's only inessentially.[71] More critically, however, he noted that the observed values closest to the origin led to a value for the velocity of the electrons above that of light, if Kaufmann's apparatus constants

---

the development of concepts of classical physics in chapter 9 (see references cited in chapter 9 note 2).

69. Planck, "Prinzip der Relativität" (1906), 137 and 140.

70. Planck, "Kaufmannschen Messungen" (1906), 754.

71. Ibid., 754–57. Planck did not reduce the measured deviations to infinitely small deviations.

were used. That helped Planck argue that before the experiment could be regarded as definitive some essential gaps needed to be filled in concerning the theoretical meaning of the measured values, though he was on too uncertain ground to suggest any physically realistic possibilities himself.[72]

The discussion that followed showed how fully the electromagnetic worldview dominated perceptions of the electron. Circulating the five original plates and a curve graphing his conclusions, Kaufmann expressed his pleasure that Planck's analysis obtained equivalent results through different calculations. While both major theories fell outside his error bounds of 1–2 percent, Kaufmann thought the extent to which Lorentz's theory deviated from observations ruled it out of consideration (10–12 percent in contrast to 3–5 percent for Abraham's theory).[73] The comments of Bucherer, Abraham, and Sommerfeld suggested that the main objection to the Lorentz-Einstein theory of the electron was not that it represented experiment less satisfactorily than rivals, but Abraham's argument that it required a nonelectromagnetic force to maintain equilibrium.[74] Planck, the sole supporter of relativity at the Naturforscherversammlung, was not able to meet this objection directly. In response to his questioners he maintained rather that the electromagnetic view of nature, like the principle of relativity, was simply another postulate of physics—and he inclined to the latter.[75] Thus, rather weakly, he treated the question as one involving a basic contrast in values: it was a matter of preference.

By now we can see that experiments on fast-moving electrons and the prospects of the electromagnetic worldview had played a significant role shaping Lorentz's theoretical work and occasioned and helped frame early reactions to Einstein's work on relativity. It has been customary for historians telling the story from this point on to focus particularly on the further conceptual development of relativity and to describe landmarks of its gradual acceptance within Germany and the rest of the academic world, largely from the perspective of adherents of relativity. Here we will follow a somewhat different tack, first testing the principle of relativity from the point of view of those experimentalists who followed Kaufmann in measuring the variation of the electron's mass with velocity in the years between 1906 and 1911. In chapter 7, we then turn to theorists—

72. See ibid., 758.

73. Ibid., 759–60.

74. Ibid., 760–61. The episode is discussed in Miller, *Einstein's Special Theory* (1981), 232–35.

75. Planck, "Kaufmannschen Messungen" (1906), 761.

both supporters and opponents of relativity—whose primary interest was to understand the implications of Einstein's work for electron theory and the electromagnetic view of nature. Rigidity emerges as a key issue in both cases: the practical rigidity of metal plates in experimental studies, and the conceptual consistency of a relativistic treatment of rigidity in theoretical contributions. Understanding the perspective of physicists who worked to extend electron theory will give us a strong awareness of the problems facing relativity and will enhance our appreciation of why its wider acceptance took the time it did.

The first empirical challenge to Kaufmann's results was indirect, resulting from Adolf Bestelmeyer's studies of slow-moving electrons in Göttingen. Using cathode rays produced by X-rays falling on a platinum plate, Bestelmeyer obtained a charge to mass ratio of slow-moving electrons 8 to 9 percent less than the value Kaufmann had used from Simon's earlier determination. Bestelmeyer had not paid particular attention to precision, but this difference was far more than the 1–2 percent error to which he thought his experiment was subject, so Bestelmeyer concluded that new and more precise experiments were required before a definitive value could be assumed. Then he investigated the consequences of his result for Kaufmann's test. In fact, the lower charge to mass ratio increased the deviation between measured and theoretical values for all three theories by about the same amount, but Bestelmeyer focused on the prospects for Lorentz's formula. The whole question, he said, concentrated on working out which mistake one had to assume Kaufmann had made in determining the electric and magnetic fields in order to bring the calculated points within the error limits of the observed curve.[76]

In Stuttgart, Planck had declined to speculate about a plausible physical basis for any gap in the interpretation of Kaufmann's experiment. By June 1907 he was ready to follow Bestelmeyer's lead, suggesting that the radiation passing through the condenser plates may have ionized any residual gas, thereby reducing the effective electric field. Together with

76. Bestelmeyer, "Spezifische Ladung und Geschwindigkeit" (1907), 443 (for the "whole question"). It is worth noting that many did not discern a significant challenge to belief in the electromagnetic origins of the mass in such considerations. For example, writing from Cambridge the experimental physicist Norman Campbell noted that the difference between the observed and calculated curves might be greater than the "error of experiment," but argued: "All will agree that it is better to assume that the error arises from an ignorance of the right assumption [concerning fundamental points in electrodynamics] than that some small fraction of the whole mass is of a nature different from the rest." Campbell, *Modern Electrical Theory* (1907), 201–2.

Bestelmeyer's new figure for the charge to mass ratio, that might "increase the chances for the *relativtheorie* a little."[77] But in order to harmonize with the deformable electron, the effective electric field had to be about 11 percent lower than the value Kaufmann had used (while Abraham's theory required a decrease of 8 percent). That was a lot—and Kaufmann had already considered whether errors in his determination of the fields might have affected his results. He was quick to reply with calculations showing that far fewer ions could be ionized by any alpha particles given off than would be required to produce such a reduction, and that the diffusion and reabsorption of ions in the electrodes was vanishingly small and could be neglected.[78] Just how closely his work was being followed is illustrated by the fact that Johannes Stark immediately entered the debate to question the assumptions underlying Kaufmann's reply, although he was rapidly rebuffed. The different means proposed might decrease the electric field by one hundred thousandth, rather than the tenth that Planck's assumption required. Kaufmann concluded that the objection against the Relativtheorie still stood, as long as no new sources of error had been found, and that further enlightenment could be expected from the new experimental studies of β-rays being carried out by several observers independently of one another.[79] The most important of these came from Alfred Bucherer, a Privatdozent at the University of Bonn in which Kaufmann had carried out his experimental work.

Alfred Bucherer's Results and the Challenges of Exact Experiment

Bucherer had devised an experiment in which the effect of the magnetic field on the behavior of the electron could be explored at all angles to

77. Planck, "Nachtrag zu der Besprechung der Kaufmannschen Ablenkungsmessungen" (1907), on 213–14.

78. Kaufmann, "Bemerkungen zu Herrn Plancks: 'Nachtrag zu der Besprechung der Kaufmannschen Ablenkungsmessungen'" (1907).

79. Kaufmann, "Erwiderung an Herrn Stark" (1908). He was replying to Stark, "Bemerkung zu Herrn Kaufmanns Antwort auf einen Einwand von Herrn Planck" (1908). The alliance formed here between Stark, Planck, and Einstein was important in the early history of relativity, but fractured remarkably in the 1920s, when anti-Semitism played into a number of other divisions (between theory and experiment, and Berlin as a center and other German universities) that became increasingly critical to Stark. See Hentschel, *Physics and National Socialism: An Anthology of Primary Sources* (1996).

the motion of the electron. That would enable a direct test of Bucherer's formulation of a new principle of relativity (which distinguished between "active" and "passive" frames of reference), as well as providing grounds for a comparison between the Lorentz-Einstein principle and Abraham's theory.[80] In Bucherer's experiment, a radioactive source was placed at the center of the two plates of a circular condenser, which was itself positioned exactly in the center of a cylinder. Photographic film was pressed against the interior of that cylinder, and the entire apparatus rested inside a solenoid that created a uniform magnetic field parallel to the condenser plates. Electrons emitted from the source could propagate in all possible directions in relation to the magnetic field, so the magnetic force acting on them took all possible values. But in each direction the compensating effects of the crossed electric and magnetic fields allowed only electrons of a specific velocity to reach the end of the condenser plates (the rest were halted by hitting one of the plates). These rays then passed through the magnetic field alone, being deflected in a roughly circular path, before reaching the cylinder wall and darkening the photographic film.

Bucherer could therefore measure the mass of rays with very different velocities (and determine directly the charge to mass ratio of slow-moving electrons, without relying on other physicists' measurements of cathode rays). Observing the angle relative to the magnetic field at which electrons were undeflected quickly ruled out Bucherer's theory, so his experiments became a test between the Abraham and Lorentz-Einstein theories. His final measurements were of electrons moving at five different velocities between 30 and 70 percent of the velocity of light. The accuracy with which he could measure the apparatus constants was critical, and Bucherer believed he had gone as far as present technology permitted (his apparatus, too, was built by the Wolz firm in Bonn). Bucherer presented his results at the 1908 Naturforscherversammlung in Cologne, an occasion on which the mathematician Hermann Minkowski gave a new cast to his own treatment of relativity—or the "world postulate," as he preferred to call it—in a lecture titled "Space and Time." The combination of both events provided significant impetus to the wider acceptance of relativity (and we will consider Minkowski's work further below).

80. In several respects Bucherer sought to overcome perceived limitations of Kaufmann's apparatus as well: Bucherer, "Messungen an Becquerelstrahlen" (1908); Bucherer, "Bestätigung des Relativitätsprinzips" (1909). For a discussion of Bucherer's experiment and responses to it, see Miller, *Einstein's Special Theory* (1981), 345–52.

Bucherer's decision fell in favor of the Lorentz-Einstein theory, and for some physicists that alone provided enough ammunition to counter Kaufmann's earlier experiment. In the audience, Minkowski expressed his joy at the result (but also affirmed that the Abraham rigid electron constituted such a theoretical monstrosity in the harmony of Maxwell's equations, it would have to fail). Elsewhere, Poincaré and Lorentz both noted Bucherer's experiment, with Lorentz writing "in all probability, the only objection that could be raised to the hypothesis of the deformable electron and the principle of relativity has now been removed."[81] Nevertheless, the proceedings in Cologne illustrated just how much the discussion of Bucherer's research was dominated by Kaufmann's earlier work, and raised what proved to be persistent queries about the probity of Bucherer's contribution. Wilhelm Wien asked whether Bucherer could offer reasons for the different results to which Kaufmann's experiments had led. In response, Bucherer emphasized how difficult it was to measure such a small curve and noted that inspecting one of Kaufmann's photographic plates himself he had discovered an asymmetry of 5 percent in one position (which Kaufmann confirmed). At least in that instance, Kaufmann's electric field had not been homogeneous. Bucherer identified two potential sources of error in determining the electric field: Kaufmann had assumed his condenser had an infinite resistance rather than measuring it (which might introduce an error of 1 percent), and Bucherer questioned the rigidity of the condenser plates. These had to be kept apart from each other by small pieces of quartz and were held in position by two screws (fig. 6.6). Interference studies of Bucherer's apparatus indicated that if the pressure was placed in the middle of the plates rather than directly over the quartz crystals, they bent slightly. Bucherer did not show that this potential difficulty had plagued Kaufmann's experiments, but it is worth emphasizing firstly that the *rigidity* of the materials concerned was an important practical issue in precision measurement, and secondly, that testing rigidity required interference techniques. A final potential source of error concerned Kaufmann's magnetic field, which might have varied over time (Kaufmann had removed the armature from the permanent magnet he used).[82]

Following a different line of questioning to Wien, Adolf Bestelmeyer drew attention to a possible source of error in Bucherer's experiment

81. Bucherer, "Messungen an Becquerelstrahlen" (1908), 762; Poincaré, "Science and Method" (2001 [1908]), 525–26; Lorentz, *The Theory of Electrons* (1909), 329.

82. Bucherer, "Messungen an Becquerelstrahlen" (1908), 759–61.

FIGURE 6.6 Detail of Kaufmann's apparatus from above. Removing the photographic plate and platinum plate from their supporting columns has revealed the parallel plates of the condenser. Bucherer suggested they might have been slightly deformed, being held in place by two screws that were not positioned directly over the quartz crystals that separated them from each other. The upper two of four quartz crystals are visible; the screws in question are directly below them. Courtesy Deutsches Museum, Munich.

itself: particles following a curved path—with velocities other than the value for which the electric and magnetic fields compensated each other—might escape the condenser and reach the photographic plate, displacing the observed curve.[83] Bucherer's formal publication in the *Annalen der Physik* failed to settle Bestelmeyer's doubts. In fact, it prompted a far more damaging slew of objections. Firstly, Bestelmeyer doubted whether the experiments of a single researcher using one arrangement could ever justify the claim to decisively settle so important a question as the foundations of the theory of electricity and perhaps of mechanics. Even more critically, Bestelmeyer contended that Bucherer's conclusion rested almost solely on only one of his thirteen experiments, and that his treatment of the measurements, protocols, and sources of error was far from adequate. Bucherer's procedures had obviated the advantages of the methods he had used; and Bestelmeyer showed that uncompensated electrons could play a far greater role in the photographic results than Bucherer allowed. According to Bestelmeyer one *could* speak of a proof of the theory of relativity in the results of the optics of moving bodies. But to do so

83. Ibid., 760–62.

on the basis of the mass of the electron required a greater number of experiments, a narrower choice of velocities registered in one experiment, and a more expansive treatment of protocol and sources of error. He was tutoring Bucherer in the rigors of precision experimentation.[84]

The thrust and counterthrust we have seen surrounding the experiments of Kaufmann, Bestelmeyer, and Bucherer shows that debate over results proved far more difficult to resolve conclusively than it was to raise and dismiss objections. Younger experimentalists in Berlin found this also, as they cut their teeth on experiment and criticism at the turn of the decade. But it is also clear that the ground had shifted, subtly but surely, since Kaufmann's original experiments. Following Kaufmann's lead, from 1906 experiments on electron mass were most often described as a test between the Abraham rigid body model of the electron and the "relativitätstheorie" or "relativtheorie" of Lorentz and Einstein. Typically they assumed the identity of the theories of Lorentz and Einstein as far as experimental consequences were concerned; and the Lorentz theory of the electron often formed the focus of any discussion of the *relativtheorie*. Nevertheless, such experiments were accorded high importance as much for their test of the principle of relativity as for the decision they might enable between theories of the electron.[85]

No Clear Result: Research Reviews, 1910–14

By the end of the decade the debate over mass-velocity experiments on the electron had assumed something of a schizophrenic demeanor. On the one hand, the experiments were extraordinarily demanding of time and resources. Yet despite the care lavished on their performance, their

84. Bucherer had numbered experiments through to thirteen, but given the results of only four together with the mean of two others, without any discussion of why these were chosen. In the first four values of

$$\frac{\varepsilon}{m_0}$$

listed, the differences between the two theories were not great enough to choose one above the other, so Bucherer's conclusion rested almost solely on experiment number three. Bestelmeyer, "Bemerkungen" (1909).

85. Hupka, "Trägen Masse bewegter Elektronen" (1910), 169; Heil, "Diskussion der Versuche über die träge Masse bewegter Elektronen" (1910); Hupka, "Frage der trägen Masse bewegter Elektronen" (1910); Heil, "Diskussion der Hupkaschen Versuche" (1910). For a discussion of Hupka and Heil's work and the role Planck played in their public exchange as editor of the *Annalen*, see Pyenson, "Physical Sense in Relativity" (1985), 202–3.

divergence revealed a gap between the error estimates experimentalists based on the accuracy with which different components of their apparatus and observations could be measured, and empirical achievement—that no one could bridge. On the other hand, the very vigor of the debate following each experiment made it possible to escape the life-or-death verdict Planck had described; and in the event many physicists made their decisions about the theories at issue on quite other grounds. Examining three overviews will illustrate this situation and conclude our study of experiment.

Consider first two accounts by proponents of relativity, one in a review article on the experimental foundations of the subject, the second incorporated in the first textbook on relativity as a whole. In 1910 Jakob Laub's exhaustive research review described individual experiments in detail and went into a blow-by-blow discussion of the give and take on many issues before highlighting the importance of repeating Kaufmann's measurements with exactly the same protocols but improved means (a higher vacuum). Such an exact repetition was required because only then could the reader form a judgment about the accuracy and weight of the measurements, and Laub thought experimentalists in this region were working at the limits of perception, or perhaps relationships were in play that were not yet fully appreciated.[86] In contrast to Laub's attention to detail, it is revealing that in the first textbook on relativity published in 1911, Max Laue discussed the experiments on the dynamics of the electron in just three sentences: "Kaufmann, who was the first to approach these experiments, proved that the transverse mass of the electron depended on velocity. The accuracy was not sufficient, however, for a decision between both theories. And even if very significant experiments from Bucherer, Ratnowski and Hupka later appeared to speak in favor of relativity theory, opinions on their conclusiveness are so divided, that relativity has probably still not received unconditionally reliable support from this side."[87] Laue then briefly reviewed the empirical support relativity had received from other sides. The law of induction, aberration, and the Doppler effect all spoke for the possibility of a principle of relativity,

86. Laub, "Experimentellen Grundlagen des Relativitätsprinzip" (1910), 440–59 and (for his conclusions) 462–63.

87. See Laue, *Relativitätsprinzip* (1911), 17–18; Laue, *Relativitätsprinzip*, 2nd ed. (1913), 18 (for the quotation, where he has incorporated Ratnowski between Bucherer and Hupka.) Ratnowski had published preliminary results from his dissertation research on the variation in mass of cathode rays as a function of velocity in *Comptes rendus* in 1910. Laue thought

as did all electromagnetic and optical experiments on the earth's surface, with the exception of Fizeau's experiment (which seemed to provide a palpable demonstration of the existence of a resting ether). But to Laue, more important than each individual experiment was the extraordinary scientific deed that enabled such apparently contradictory results to be explained from one perspective—the theory of relativity, which entered deeply into our whole physical worldview and touched on the epistemological basis of physics.[88]

Kaufmann surveyed the lessons of the decade too. His 1914 textbook on electricity and magnetism pointed out that according to Einstein's so-called Relativtheorie any mass depended on velocity, whether mechanical or electromagnetic in nature, and noted that the program to provide an electromagnetic basis for physics had lost a great deal of impetus as a result. Describing the rise and fall of the rigid sphere electron, Kaufmann stated that his own experiments had led to no clear result. Their concordance with Abraham's theory and then-current values for the charge to mass ratio of electrons at rest had subsequently given way to a more exact value for that ratio. Kaufmann noted that the means Planck suggested to bring his results into accord with the predictions of relativity were insufficient; and despite many unpublished efforts theoretically and experimentally, he himself had been unable to find any other cause. According to Kaufmann, Bucherer had provided a final confirmation of the relativistic theory of the variation of mass, with experiments recently repeated and expanded by Schaefer and Neumann (Hupka's experiments on cathode rays also confirmed relativity theory).[89] Soon after the end of the First World War, Kaufmann told Sommerfeld he was among those who required the stimulus of colleagues and discussion to be encouraged and

something had come from the dynamics of the electron: it had led to the conclusion that classical mechanics and its principle of relativity is not generally valid, because of the variability of mass.

88. Laue, *Relativitätsprinzip*, 2nd ed. (1913), 18–19.

89. Kaufmann, Coehn, and Nippoldt, *Magnetismus und Elektrizität* (1914), 1223–28. The reference to his unpublished efforts is on 1226. The experiments of Schaefer and Neumann between 1914 and 1916 indeed provided broader grounds for a consensus view on the outcome of the experiments. However, subsequent reviews have upheld Bestelmeyer's criticisms and concluded that while certainly consistent with relativity, none of the previous electron experiments were in fact precise enough to distinguish between the different theories. See Zahn and Spees, "A Critical Analysis of the Classical Experiments on the Relativistic Variation of Electron Mass" (1938); Faragó and Jánossy, "Review of the Experimental Evidence for the Law of Variation of the Electron Mass with Velocity" (1957).

stimulated to their own creative work. The years in Göttingen (from 1899 to 1903) had been the best of his career.[90]

Conclusion

In the introduction to Part III a brief study of the Nobel prizes awarded in physics underlined two important points about physics in the early twentieth century. The recently discovered electron helped define its modernity, and that particle was widely perceived to provide the most important element unifying physical research across a broad range of fields. Our historical understanding of the physics community has often been too strongly driven by later developments to fully appreciate the implications of the broad canvas physicists saw in the electron soon after 1900, but by focusing on the sharp end of experimental research on the mass and dynamics of the electron, this chapter has delivered a new perspective on the importance of experiment to the emergence of relativity in Germany. Historians have long recognized that the results of Walter Kaufmann's experiments on fast-moving electrons played a critical role in shaping early views of the prospects of the principle of relativity and Einstein's work. Yet the significance of experimental work rarely rests on its results alone, and our examination of Kaufmann's apparatus and methods in this chapter has helped recover facets of his practice that have rarely attracted attention, although they were integral to the achievement and articulation of those results. In particular it has disclosed several revealing continuities

90. Already before the war Kaufmann's work on the Müller-Pouillet textbook and the isolation of Königsberg had pushed him slowly toward a merely contemplative participation in scientific progress, and since 1915 foreign literature had been impossible to obtain. Kaufmann struggled with Paul Volkmann for space in the institute. When the Nazi party dismissal of Jewish civil servants opened up new positions in physics, he advanced the cause of Eduard Steinke, noting that Steinke was not politically engaged. (Kaufmann was himself Jewish but also near retiring age.) He also commented that in 1934, students had begun to buzz and clatter when he mentioned E...n's name in lectures. Kaufmann had calmly told them that he was speaking of the physicist E...n and not the political dilettante. Then they were quiet, and nothing similar occurred again. Kaufmann described his scientific personality and increasing disengagement from active research in Kaufmann to Sommerfeld, 20 January 1920, Deutsches Museum, Munich, Archiv HS 1977–28/A,161. For his battle with Volkmann, see Volkmann to Sommerfeld, 16 January 1925 and Richard Gans to Sommerfeld, 25 October 1927, Archiv NL 89, 014 and 008, respectively. The support for Steinke is in Kaufmann to Sommerfeld, 10 March and 9 April 1935, Archiv HS 1977–28/A,161. For the comments on E...n see Kaufmann to Sommerfeld, 10 March 1935, Archiv HS 1977–28/A,161.

in physical research, both over time and across material and conceptual dimensions of relativity physics. They concern the dynamics between theory and experiment, the increasing importance of interferometry over time (exemplified in the need to demonstrate material rigidity), and the articulation of regimes of relative and absolute knowledge.

Perhaps the most fundamental point is that the development of electrodynamic theory and theories of the electron in the decade after 1900 was in general strongly shaped by the emergence of more precise studies of the variation of mass of the electron, in what can best be described as an intimate dialogue between experiment and theory, and between experimentalists and theorists. In view of the eventual importance of these events in shaping concepts of theoretical physics in the hands of Planck and Einstein (explored more fully in later chapters), it is important to note at this point both that the multifaceted contributions of Kaufmann and Bucherer should trouble any attempt to draw sharp distinctions between those roles, and that interpreting experiment largely through its results leaves us ill equipped to understand the full dimensions of experimental work.

My second point is that paying due attention to Kaufmann's apparatus and procedures has demonstrated the increasing importance of optically precise techniques of machining and measurement. In his search for the greatest accuracy possible, Kaufmann substituted Zeiss gauges and optically plane components for dimensions initially measured with a compass and ruler. We saw too that critiquing the relations between different results, diverse apparatus, and competing theories, the rigidity of condenser plates came into question with an argument based on the position of two screws and interferometric studies of the distortion of metal plates. Albert Michelson would have applauded this demonstration that accurate mechanical work should now rely on precision optics; and we have noted that the 1907 replication of his interferential determination of the length of the prototype meter in Sèvres was widely regarded as offering the first convincing demonstration that the dimensions of that rigid rod had remained invariable over more than a decade.

Indeed, it is ironic that in 1907 Albert Michelson was honored with a Nobel Prize for his work in interferometry, yet found no need either to draw attention to the fact that his instrument had first been developed in an experiment designed to detect ether drift, or to outline the way its null result had been incorporated into electrodynamic theory. Chapter 8 will show that in the very same year Albert Einstein described the Michelson-Morley experiment as central to the development of relativity, in a research review that helped shape perspectives on relativity physics as an emerging field. Quite apart from acknowledging the significance of the experiment in the evolution of Lorentz's earlier work, we can already see strong reasons for Einstein's tactic: the ether-drift experiment played an important role in the justification of relativity because relativistic predictions then stood in conflict with high-profile experiments on the variation of the mass of the electron with velocity. Michelson's and Einstein's rather different concerns would lead them to cultivate quite different memories.

Recognizing the sense in which interferometry could both provide convincing demonstrations of the invariability of material dimensions *and* fuel the argument that all bodies contract with motion is important because it offers a more complete understanding of the range of physical work than we have gained from studies focused on theory and crucial experiments. Similarly, the work of Kaufmann in particular has illustrated a third important point. Like Michelson, Kaufmann distinguished relative from absolute measures. Thus experimentalists articulated a conceptual pairing similar in nature to the contrast between absolute and relative space and time that Poincaré and Einstein described in physical theory. However, in contrast to the lesson Einstein would draw about the ether, to a large degree empirical success depended on being able to move successfully from relative to absolute measures, rather than denying the validity of the absolute. Noting experimental and theoretical cultures of the relative in this way goes beyond drawing attention to the elaboration of a common theme in markedly different contexts. Our study of Kaufmann's conceptual work has also demonstrated a still more intimate relation. Although then an opponent of Lorentz's theory, by reinterpreting Lorentz's 1904 work in the terms he used in his experimental research, Kaufmann had helped render the work of Poincaré, Lorentz, and Einstein part of a common enterprise oriented toward the distinction between relative and absolute motions in space and time. To appreciate the grounds from which relativity physics emerged, our understanding of the relations between theory and experiment must draw on far more than results alone.

CHAPTER SEVEN

Relativity and Electron Theory, 1905–1911

Relativity and the Electromagnetic Worldview, 1905–11

If, despite their potential to spell life or death, the results of experiments on the electron ultimately offered no decisive guidance, it is time to investigate more fully the role theoretical perspectives on that entity played in the fortunes of relativity physics. We have seen that within Germany at least, Einstein helped ensure vigorous discussion of his electrodynamic theory by providing predictions of the variation of the mass of the electron with velocity (noting, as he did, that the effect was a general one, independent of whether the body carried any charge). Over time, that contribution stimulated considerable debate on the conceptual relations between the principle of relativity, the electromagnetic worldview, and electron theory. A great deal of our knowledge of the period from 1905 has been gained by examining the work of the major champions of relativity: Einstein, Planck, and Minkowski; and to a large extent our historical accounts have been driven by the logic of the constructive advances these theorists were able to make with relativity. It has clearly been very important to understand how the trail leads to a new relation between energy, mass, and the velocity of light, to a generalized relativistic mechanics, and to a powerful new formalism. Yet the electron has often drifted out of sight of such histories of relativity, or been approached largely from Einstein's somewhat unusual perspective. Once again, stepping to one side of the established historiographical trail and keeping our eyes on the electron will deliver new insight into the complex process through which relativity came to prominence in Germany. We will take Paul Ehrenfest and Max Born as our guides, two young theorists with a nose for critical issues and the ambition to push both electron theory and relativity into new regions. They will lead us to events that one of the leading historians of relativity has described as crucial to the demarcation between the views of Lorentz and Einstein, and to the eventual acceptance of relativity within Germany.[1]

Most analysts have regarded the central issue in the propagation of relativity to rest in recognizing Einstein's interpretation of relativity as a kinematic theory based on relating concepts of simultaneity, space, and time to measurement processes in given frames of reference. On this view, a proper understanding of relativity required distinguishing Einstein's work from the earlier electrodynamic theory of Lorentz, based as that was on the electron as a fundamental constituent of matter. It is surprising then, that earlier historians have not paid more critical attention to the longer-term dialogue between relativity and the electron. Doing so through Ehrenfest and Born will show that we need to introduce more subtlety into our understanding of what it meant to understand and agree with Einstein, and into our appreciation of the conceptual relations between relativity theory and the electromagnetic worldview. In particular, Born will provide an example of one physicist who sought to combine the two approaches that many physicists and most historical accounts have regarded as essentially conflicting—and did so to important effect. Our narrow focus on one critical issue here will be supplemented in the following chapter by a more broadly based study of changing perceptions of relativity that brings us to an understanding of how that field came to be identified principally with Einstein, rather than with the many other physicists who contributed to its development.

1. Summarizing his understanding of the development of relativity, Miller writes: "the consensus of the physics community was that Einstein's first paper on electrodynamics was a solid contribution to Lorentz's theory of the electron. It was primarily Ehrenfest's foundational analyses of 1909–1910 (after Einstein [in 1907] had set Ehrenfest straight) of a rigid body that by about 1911 demarcated between the views of Lorentz and Einstein." Miller, *Frontiers of Physics* (1986), 238. See also Miller, *Einstein's Special Theory* (1981). Goldberg and Pyenson also draw attention to the events I will discuss, but similarly do so without appreciating the full range of issues exhibited in the work of Ehrenfest and Born. Their overviews of the reception of relativity are: Goldberg, *Understanding Relativity* (1984), chap. 6; Pyenson, "The Relativity Revolution" (1987).

FIGURE 7.1 Paul Ehrenfest as a student, October 1901. Literally, the transcription reads: "If you are scared–Grandma!" Courtesy AIP Emilio Segrè Visual Archives.

Paul Ehrenfest began his studies at the Technical University in Vienna in the winter semester of 1899–1900, where he attended Ludwig Boltzmann's lectures on the mechanical theory of heat and became one of the last students of the man who had done so much to develop the kinetic theory of gases. In the peripatetic life of a student in the German-speaking universities, Ehrenfest studied in Göttingen in the following semester (Boltzmann, also restless, had gone on to Leipzig for what turned out to be a short stay). Ehrenfest traveled briefly to Leiden in order to meet H. A. Lorentz in 1903, and by 1904 had followed Boltzmann back to Vienna in order to finish a dissertation on the motion of rigid bodies in fluids and the mechanics of Hertz. There he married a fellow student and later collaborator, the Russian-born Tatyana Alexeyevna Afanassjewa. His studies of Boltzmann prepared Ehrenfest to see real differences between Boltzmann's combinatorial techniques, and what Planck had made of them in his blackbody theory. Entering correspondence with Planck, in 1905 and 1906 Ehrenfest published critical studies that made him—with Albert Einstein and James Jeans—one of the first physicists to argue that Planck's theory required energy quantization. In chapter 9 we will explore what the contributions of all these physicists reveal about the concept of "classical" theory. Ehrenfest returned to Göttingen briefly in 1906 before spending five years in St. Petersburg, where the Ehrenfests concluded a study of statistical mechanics for Felix Klein's *Encyklopädie der mathematischen Wissenschaften*. These early contributions have helped give Ehrenfest the reputation of being an extraordinarily sensitive physicist—and he is also celebrated for his role in the acceptance of relativity on the basis of short papers he published in 1909. Here we will show a new side to his work by counterposing Ehrenfest's published papers with the letters he wrote to friends and colleagues.[2]

Ehrenfest's studies of blackbody radiation meant he was more aware than most of the full scope of Einstein's research, and it is revealing that his first paper on relativity took the form of a direct challenge. In 1907 Ehrenfest tackled Einstein on the question Planck had dodged at the Naturforscherversammlung. Abraham had argued that a nonspherical electron moving in uniform translation at an angle to its main axis suffered a torque due to its own field, which had to be compensated for by an external force. Writing that in the formulation in which Einstein had published it, Lorentzian relativity dynamics was generally regarded as an enclosed system, Ehrenfest wanted to know if a deformable electron of arbitrary shape could undergo force-free uniform translation in any direction (note also that referring to customary understandings of Einstein, he cited

2. On Ehrenfest see especially Klein, *Paul Ehrenfest* (1970); Kuhn, *Black-Body Theory* (1987); and a recent study of the difficult professional and emotional situation that Ehrenfest faced between 1904 and 1912 searching for a position: Huijnen and Kox, "Paul Ehrenfest's Rough Road" (2007). Correspondence discussed below will complicate substantially the picture of Ehrenfest conveyed by Klein, Kuhn, and Miller's study of relativity. It is appropriate to note that Klein utilized some archival material and commented that this was made available gradually, but based his discussion of Ehrenfest's scientific work largely on published articles (in distinction to his treatment of friendships, career prospects, and other more personal matters). Kuhn uses Ehrenfest's notebooks extensively. Some but not all of the correspondence I discuss comes from AHQP/EHR Reel 32, material found after the main Archive for the History of Quantum Physics was filmed.

Kaufmann's publications). If it could not, the principle of relativity had to be supplemented by a new hypothesis to exclude such an electron, for it offered a way of detecting absolute rest. If such force-free motion was possible, Ehrenfest wanted to see how this could be done without drawing on a wholly new axiom.[3] Ehrenfest's question shows him taking the epistemological demands of relativity seriously, holding the theory to its own standards. But just as importantly it came from the heart of the electromagnetic view of nature, raising once again the cost of giving up purely electromagnetic conservation of energy and angular momentum. Ehrenfest wanted Einstein to clarify the relations between his work and the central concerns of many in the German physics community.

In reply Einstein summarized the way he thought relativity addressed physical research in general and the questions facing electron theory in particular. The two principles he had invoked were not to be conceived as a closed system. Rather, in themselves they only contained statements about rigid bodies, clocks, and light signals. The theory of relativity went further only by requiring relations between otherwise seemingly unrelated laws (by stipulating that satisfactory laws had to retain the same form when subjected to the Lorentz transformations, for example). Einstein provided an example of how one should proceed: to derive the laws for the acceleration of a rapidly moving electron one applied the transformation equations to the laws of the slowly accelerated electron. But those laws themselves had to be assumed or obtained from experience. Looking to the future, if the framework of electricity was regarded as a rigid body (that is, as a body not deformed by external forces in the way Ehrenfest described), Einstein wrote the problem of the motion of the electron in relativity could be solved without arbitrariness only if the dynamics of the rigid body were known with sufficient accuracy—and that goal lay far ahead.[4]

Einstein did at least go into some detail on the relations between relativity and any theory of the electron, but like Planck in Stuttgart he failed to give a direct answer to Ehrenfest's query. Their engagement with research in other fields, in particular blackbody radiation and quantum theory, had led both Planck and especially Einstein to doubt the sufficiency of the electromagnetic view of nature. As a result both were ready to accept relativity without meeting directly the objection that a relativistic theory of the electron required a nonelectromagnetic energy. It is important at

3. Ehrenfest, "Translation deformierbarer Elektronen" (1907).
4. Einstein, "Bemerkungen zu der Notiz von Hrn. Paul Ehrenfest" (1907), 206–7.

this point to take a few moments to explore the implications of this stance and examine the grounds on which it was based. In 1907 Einstein's distance from the electromagnetic worldview even went as far as conjoining the two major approaches that most of his contemporaries pitted in contrast to each other. Coining a label few would have recognized, he wrote that the present "electromechanical" worldview was insufficient to explain a very particular list of phenomena: the entropic properties of radiation, the emission and absorption of radiation, and specific heats.[5] Many physicists would have joined Einstein's skepticism about the sufficiency of mechanics as the basis for a worldview. However, most of those would have strongly distinguished a potential electromagnetic worldview from mechanical foundations. Further, in 1907, even among those who were aware of the common thread in quantum theory that linked the list of phenomena Einstein offered, very few shared the critical attitude toward Maxwell's theory that underlay Einstein's views.

In 1905 Einstein's first paper on quantum theory had opened with his observation that the treatment of energy in the continuous, field theory of electromagnetism that Maxwell had established differed profoundly from the approach taken in kinetic theory and related fields. He had gone on to mount an argument (based on an analogy between the statistical theory of gases and the form of Wien's law of blackbody radiation for short wavelengths), that in some circumstances at least the energy of light had to be treated as having an atomic constitution. Accordingly, Einstein believed that Maxwell's theory was insufficient—in the limited regime in which Wien's law was known to be accurate—although he also expressed his opinion that the wave theory of light would probably never be replaced by another theory.[6] Indeed, some months later he relied on Maxwell's theory when conjoining that theory with the principle of relativity to develop his new electrodynamics. Einstein's critical attitude toward the limitations of Maxwell's theory set him apart from his colleagues, as he acknowledged describing his light quantum paper as "very revolutionary" to his friend Conrad Habicht.[7] Max Planck, for example, was convinced by Ehrenfest, Jeans, and Einstein that the energy of the oscillators invoked in his blackbody theory must be quantized. But light was another matter (as was

5. Einstein, "Trägheit der Energie" (1907), 372.

6. Einstein, "Erzeugung und Verwandlung des Lichtes" (1905), 132.

7. Einstein to Conrad Habicht, 18 or 25 May 1905, Doc. 27 in Einstein, *Collected Papers*, vol. 5 (1993).

the specific heats of solids, another area Einstein addressed). Given the extraordinary success of Maxwell's theory, it would require more than an argument based on formal analogies to convince Planck that a quantum theory of light was required.[8] Ehrenfest's 1907 query suggests that despite his participation in the discussion of quantum theory with Einstein and Planck, he at least had not abandoned faith in the electromagnetic perspective. Given this, it is worth asking whether Ehrenfest is likely to have been persuaded by Einstein's brief response to his query.

The foremost historian of the early reception of relativity has assumed Ehrenfest found Einstein's reply convincing. Arthur I. Miller accords Ehrenfest's later work a major role in clarifying the relations between Lorentz and Einstein's views, writing, "For the most part during 1907–1911, Einstein left the defense of his view to Paul Ehrenfest, who became a good friend. But first Einstein had to set Ehrenfest straight regarding the thrust of Ehrenfest's own note."[9] The assumption indicates how easy it has been to take Einstein's work as normative, and hence almost self-evidently persuasive. Ehrenfest's next paper on relativity was published only in 1909, two years after the exchange with Einstein. As we shall soon see, in that paper he might indeed be said to defend relativity (if not Einstein); but his correspondence in the interim will give a new understanding of the development of his thinking.

Ehrenfest gave his private views on Einstein free reign in letters with Walter Ritz, whom he first met as a student in Göttingen in 1901. Ritz had begun his studies in the Zurich Polytechnic (where he was a colleague of Einstein's), but traveled to Göttingen in 1901. There he worked under Voigt and completed his dissertation on the theory of spectral lines in late 1902.[10] Rather than accepting an electron-theoretical approach, Ritz treated the atom as an elastic continua whose vibrations gave rise to spectra. Although circumstances soon took Ehrenfest and Ritz in different directions, they made contact whenever possible and in 1903 traveled together to spend six weeks attending H. A. Lorentz's lectures and seminars in Leiden.[11] After working in Bonn and Paris, Ritz spent several

8. This became clear in his discussion of Einstein, "Das Wesen und die Konstitution der Strahlung" (1909), 825–26.

9. Miller, *Einstein's Special Theory* (1981), 235.

10. On Ritz, see Pyenson, "Einstein's Early Scientific Collaboration" (1985), 225–28; Forman, "Ritz, Walter" (1975); Martínez, "Ritz, Einstein, and the Emission Hypothesis" (2004).

11. In 1912 Lorentz proposed that Ehrenfest succeed him on his retirement. Ritz was one of four people Ehrenfest identified as having a particularly important role in his development

years trying to restore his failing health before returning to Göttingen in the spring of 1908. In addition to his widely known work in theoretical spectroscopy, Ritz was deeply concerned with constructing an emission theory of light in this period as a result of his view that an alternative electrodynamics and optics was required to satisfy the mechanical principle of relativity. His correspondence with Ehrenfest began in 1903 and broached electron theory on several occasions, especially in response to a new contribution to the emerging relativity physics from the mathematician Hermann Minkowski.

The number theorist Hermann Minkowski had been brought to Göttingen from Paris in 1902 at the joint insistence of Felix Klein and David Hilbert (who was a close friend). Klein's work to enhance the disciplinary importance of mathematics by stressing its applications certainly did not end with books and editorial projects, but extended both to faculty appointments in Göttingen and the seminars they offered. With Emile Wiechert and the Privatdozent Gustav Herglotz, in 1905 Hilbert and Minkowski gave a seminar on electron theory in which they discussed the work of Poincaré, Abraham, and Sommerfeld in particular.[12] Minkowski made the field his own, and by 1908 had drawn on Poincaré's geometrical, group theoretic approach and Einstein's emphasis on the relativity of time to present a new formal approach to relativistic electrodynamics in which the time variable appeared symmetrically with the space variables. Minkowski discussed the transformation group G, introducing the parameter c and showing that in the limit when $c = \infty$ the group is that appropriate to Newtonian mechanics, while giving c the finite value of the velocity of light delivered the new relativistic mechanics defined by the Lorentz transformations. Among the novel features of the major paper Minkowski published in 1908, "The Fundamental Equations for Electromagnetic Processes in Moving Bodies," was the confident statement that

in a letter he wrote describing his background to Lorentz. The others were Gustav Herglotz, Abraham Joffe, and his wife Tatyana. Klein, *Paul Ehrenfest* (1970), 188.

12. For a close study of the seminar, see Pyenson, "Physics in the Shadow of Mathematics" (1985). I set Minkowski's work in a broader framework and discuss the historiographic treatment of his contributions in the section on "Minkowski's Distinctions" in chapter 8. Note especially that many authors have strongly distinguished both Minkowski's "formalist" understanding from Einstein's "physical" approach, and Minkowski's interest in a theory of matter from Einstein's kinematic approach. Miller, for example, has maintained that Minkowski's exhibition of the contraction phenomena using electrons rather than measuring rods "displayed his incomplete understanding of the differences between [Lorentz and Einstein's] theories." Miller, *Einstein's Special Theory* (1981), 241.

his work offered the first completely clear treatment of the fundamental equations of the electrodynamics of moving bodies on the basis of relativity, and his stress on distinguishing his approach from Lorentz's earlier work. In chapter 8, I will set Minkowski's views in the context of a broader discussion of the authorship of relativity theory, and explore his use of a historical framework to convey important conceptual distinctions. For the moment it is important to note that (together with his later lecture on space and time) Minkowski's paper substantially increased interest in relativity. After it appeared, for example, papers on relativity theory in the *Annalen der Physik* responded to or further developed Minkowski's work—almost without exception. In particular a number of authors focused on the distinction Minkowski drew between the electrodynamics derived from the principle of relativity and Lorentz's treatment of magnetized bodies.[13] But despite the immediate attention it received, proponents of the electromagnetic view were likely to have difficulties with Minkowski's work—along familiar lines.

Interrogating Relativity: The Ehrenfest-Ritz Correspondence

Without having explicitly addressed concerns surrounding the electromagnetic nature of the inertial mass of the electron, Minkowski's work was vulnerable to the same kind of criticisms that Abraham and Ehrenfest had leveled at Lorentz, Planck, and Einstein. Paul Ehrenfest's first,

13. Those papers include Einstein and Laub, "Elektromagnetischen Grundgleichungen" (1908); Einstein and Laub, "Ponderomotorischen Kräfte" (1908); Mirimanoff, "Grundgleichungen der Elektrodynamik bewegter Körper" (1908); Einstein, "Bemerkung zu der Arbeit von D. Mirimanoff" (1908); Mirimanoff, "Bemerkung zur Notiz von A. Einstein" (1908). Philipp Frank offers an example of someone who investigated Einstein's thought only after an initial introduction to relativity through Minkowski. A 1908 paper is based entirely on Minkowski's work, referring to the principle of relativity as the most important of Minkowski's three axioms; this paper also introduces the term *Galilean transformations* to describe the transformation equations appropriate to Newtonian mechanics. In a second paper, Frank referred to the "Lorentz-Einstein principle of relativity." In a more extensive 1909 study he referred to the work of both Einstein and Minkowski, stating that Minkowski had given the most elegant treatment of the application of the Lorentz transformations to the electromagnetic field equations. Two years later he considered different derivations of the transformation equations and the assumptions underlying them (with Hermann Rothe) and for the first time examined closely Einstein's derivation. Frank, "Relativitätsprinzip" (1908); Frank, "Relativitätstheorie und Elektronentheorie" (1908); Frank, "Relativitätsprinzips im System der Mechanik und der Elektrodynamik" (1909); Frank and Rothe, "Transformation der Raum-Zeitkoordinaten" (1911).

unpublished response to "The Fundamental Equations" makes this point very clearly. "In Minkowski's work," Ehrenfest wrote, "I find just the same incomprehensibility as in Einstein: when I advanced my question, Einstein answered: I have never maintained that my postulate of relativity leads to a full determination . . . but Minkowski now maintains that *his* theorems lead to a full determination—to me his concept of 'force' is *completely* ununderstandable (just as with Einstein)."[14] In this first draft of a letter to an unnamed friend (whom we can identify as Ritz from internal evidence), Ehrenfest could therefore only repeat for Minkowski the question he had already put to Albert Einstein, and he jotted it down with emphases and question marks.[15]

Ehrenfest's first two drafts of this letter did not satisfy him and he came back to it a third time in order now to register a substantial change of views:

> Minkowski's electrodynamics impresses me enormously—not as a physical theory but as a mathematical-metaphysical construction—I don't believe it has much to do with empirically given facts—however, if it doesn't contain any contradiction—it is a world of wonderful elegance—that is a great, great difference to the ununderstandable mollusk-stuff of Einstein. Now I am curious to see how
>
> 1. Minkowski without a *kinematics* of rigidity or deformation gives his electrons stability—tentatively it appears to me that it must necessarily dissolve (for example a rest spherical electron, because of the static electrical repulsion).[16]

14. "In der Minkowski Arbeit finde ich genau dieselbe Unverständlichkeit wie bei Einstein: Als ich bei Einstein meine Frage vorbrachte, antwortete Einstein: Ich habe niemals behauptet, dass mein Relativ-postulat zu einer vollen Determination führt—(So wörtlich in Einsteins 'Erwiderung' in den Annalen)—Minkowsky behauptet aber jetzt, dass *sein* Ansatz zu einer vollen Determination führt—Da mir sein Begriff der 'Kraft' *vollkommen* unverständlich ist (ebenso bei Einstein) sehe ich gar nichts . . . " Ehrenfest, undated draft (1st) of letter to "Dear Friend," AHQP/EHR, Reel 32 m57. This and other correspondence from Paul Ehrenfest has been completely overlooked by previous historians, perhaps in reliance of the completeness of Martin Klein's coverage of manuscript sources in his important study, *Paul Ehrenfest.*

15. " . . . ich kann nur einfach dem Herrn Minkowsky meine Frage wiederholen: gegeben ein *in Ruhe nicht* kugelförmiges Elektron—kann sich dieses nicht kugelförmige Elektron nach allen Richtungen hin gleichförmig, *rein translatorisch* bewegen (ohne zu kollern) wenn 'von außen' keine Kräfte angreifen?—Wenn *nein*—wie steht es dann mit dem Relativitätsprinzip?—dann haben wir ja in einem solchen Elektron ein Instrument um absolute Ruhe zu constatieren.—*Wenn ja*—Wie lautet dann das Princip von der Erhaltung des elektromagnetischen Drehimpulses?" Ehrenfest, undated draft (1st) of letter to "Dear Friend."

16. "Minkowskis Elektrodynamik imponiert mir enorm—nicht als physikalische Theorie sondern als mathematisch-metaphysische Construction—ich glaube nicht etwa, dass sie etwas

Ehrenfest did not reach a second question, but it is clear that the elegance of Minkowski's work has inspired a newly positive engagement with relativistic electrodynamics. Without yet accepting the theory, Ehrenfest wishes to probe its consistency and turns immediately to the stability of the electron. Revealingly, like Einstein, Ehrenfest focuses on a kinematics of rigidity or deformation.

Their letters show that Ritz and Ehrenfest shared similar concerns. Both stated objections that focused on questions in electron theory left unanswered by the principle of relativity. How could one decide the nature of force in relativistic electrodynamics? What did the principle have to say about the structure of the electron? The deformation of a particle with changes in velocity? The issue of electromagnetic mass? We can see subtle changes in their conception of the theory, however. To some degree at least, both initially linked together the work of Einstein and Lorentz, with Ritz referring to the "Lorentz-Einstein theory" in late 1907, for example.[17] After the publication of Minkowski's paper in 1908, both wrote of the principle of relativity—and clearly associated this primarily with Einstein and Minkowski. Equally clearly, that they now distinguished Einstein's work from Lorentz's did not imply increased appreciation for Einstein's contributions. Rather, both continued to express a strong dislike for the theory, even as their responses to Minkowski diverged. In contrast to Ehrenfest's appreciation for the elegance of Minkowski's formulation, Ritz saw nothing positive in his work: "Minkowski extends with power the principle of relativity a la Einstein. Hopefully this nonsense will disappear in 10 years. A wholly miserable means of information. One could salvage any theory, no matter how false, in such a way. The question whether generally the mass in this theory has an e.m. [electromagnetic] origin, which you raised, is still open, also

mit den empirisch gegeben Thatsachen zu thun hat—aber—wenn sie nun nicht einen Widerspruch enthält—ist sie eine Welt von wunderbarer Eleganz—das ist ein großer großer Unterschied gegen das unverständliche Molluskenzeug von Einstein. Jetzt bin ich neugierig wie / 1.) Minkowski ohne Starr oder Deformations-*kinematik* seinen Elektronen Stabilität gibt—sie müssen wie mir scheint vorläufig erbarmungslos zerfließen (z.B. ein ruhendes kugelförmiges Elektron wegen der statisch-elektrischen Abstoßung)." Ehrenfest, undated draft (3rd) of letter to "Dear Friend," AHQP/EHR, Reel 32 m57.

17. At this time Ritz thought the theory so undetermined that while it delivered information about deformation at given velocities, it had nothing to say about changes in deformation in the case of acceleration. He concluded: "Mir ist diese ganze Theorie so zuwieder und unklar dass ich diese Frage nicht entscheiden kann." Ritz to Ehrenfest, 17 September 1907, AHQP/EHR, Reel 24.

in Minkowski. I believe now that it is probably to be answered negatively."[18]

These letters underline the value of a general methodological insight, also illustrated in our study of the way different physicists used Michelson's experiments and instruments. Ehrenfest's and Ritz's responses show an active, if partial, interrogation of the papers of Einstein and Minkowski, attributing meaning—or lack of meaning—on the basis of the research goals and theoretical concerns that Ehrenfest and Ritz themselves held to be significant. Such responses indicate clearly that in the communal endeavor of science, meaning rests in the hands of later researchers. Just as Einstein's exposition of relativity failed to convince Ehrenfest, Minkowski's papers were clearly instrumental in fashioning these physicists' understanding of relativity. The following sections of this chapter will offer the first stage of my argument that it is only in the period from 1908 that a standardized reading of the physical consequences of relativity was developed in Germany. While ultimately this was to be based quite closely on Einstein's research, it was developed largely in response firstly to the work of Minkowski and secondly to explicit discussions of the relations between relativity and electron theory. The argument will be set in a broader context and brought to a conclusion in the following chapter.

Max Born's Endeavor: The Principle of Relativity and Inertial Mass

We have noted the debate surrounding experimentalists' publications evaluating different electron theories and seen that Ehrenfest focused on the relations between relativity and electromagnetic conservation of energy. Apart from these papers, early studies of relativity in the German physics journals were confined to a number of specific areas, and it is highly significant that they were largely peripheral to electron theory. Einstein explored the extremely general relation he recognized between energy and mass shortly after submitting his first paper on the relativity principle.

18. "Minkowski breitet mit Macht am Relativitätsprinzip á la Einstein. Hoffe dieser Unsinn ist in 10 Jahren verschwunden. Ganz miserables Auskunfts-mittel. So könnte man jede noch so falsche Theorie retten. Die Frage ob überhaupt nun die Masse in dieser Theorie e.m. Ursprungs noch sei, von dir aufgeworfen, ist noch immer offen, auch bei Minkowski. Ich glaube jetzt eher dass sie negativ zu beantworten ist." Ritz to Ehrenfest, 17 December 1908, AHQP/EHR, Reel 24.

Max Planck developed relativistic mechanics and (with his student Kurt von Mosengeil) investigated the application of relativity to blackbody theory. The young physicists Max Laue and Jakob Laub published papers primarily concerned with applications of relativistic computational techniques to already known fields of optics (and it is worth noting that Laue had written his PhD dissertation on interference phenomena in plane-parallel plates, stimulated particularly by the work of Otto Lummer).[19] As I noted earlier, following the publication of Minkowski's papers, a number of physicists including Einstein and Laub and Philipp Frank wrote articles in which the relations between Einsteinian or Minkowskian electrodynamics and that of Lorentz were clarified. Thus Einstein's reply to Ehrenfest in 1907 was the furthest any adherent of relativity went (at least in print) to investigating the relations among relativity, electron theory, and the electromagnetic worldview—until the twenty-seven-year-old Max Born began addressing that complex of issues early in 1909.

Max Born trained in mathematics and physics in Breslau and Göttingen in particular, and like Laue and Laub participated in the 1905 seminar Hilbert and Hermann Minkowski ran on electrodynamics.[20] After a doctoral dissertation applying variational calculus to problems in elasticity (a topic he pursued somewhat unwillingly at the suggestion of Felix Klein), Born traveled to Cambridge and then Breslau to develop his interests in physics. According to the testimony of Born and Stanislaw Loria, knowing that Max Planck supported Einstein's work helped inspire a group of young physicists in Breslau to turn to relativity. When Born experienced difficulties with his research in electrodynamics, he wrote to Minkowski. The latter replied that he too was currently working on topics in this field and suggested Born come to Göttingen to pursue his academic career further. Born heard Minkowski lecture on Space and Time to the Naturforscherversammlung in September 1908 and then had the benefit of a few weeks of daily contact, but any substantial collaboration was curtailed by Minkowski's tragic death from a ruptured appendix in January 1909.[21]

19. Einstein offers an overview of early studies in his important 1907 review. Einstein, "Relativitätsprinzip" (1907).

20. On Born, see Staley, "Max Born and the German Physics Community" (1992); Im, "Max Born und die Quantentheorie" (1991); Greenspan, *The End of the Certain World* (2005).

21. Minkowski's interest in Born is discussed in Walter, "Minkowski, Mathematicians, and the Mathematical Theory of Relativity" (1999), 47–48.

FIGURE 7.2 Max Born as a student in Göttingen. Courtesy AIP Emilio Segrè Visual Archives. Gift of Jost Lemmerich.

The title of Born's first publication in the field announced his primary concern: "The Inertial Mass and the Principle of Relativity." The first admission to table was that the electrodynamics recently built up from the principle of relativity had not provided a satisfactory explanation of inertial mass. The equations of motion that Einstein, Planck, and Minkowski had given constituted natural generalizations of the Newtonian equations of motion, adapted to the principle of relativity. They modified the concept of mass without giving it an essentially electrodynamic explanation.[22]

22. For example, Minkowski's derivation of the equations of motion of electricity introduced the inertial mass into the integral for the kinetic energy from the start. Born, "Träge

Born was clearly aware that according to relativity a dependence of mass on velocity followed as a direct consequence of relativistic kinematics.[23] His present concern was to indicate the possibility of providing a purely electromagnetic explanation of mass—within a relativistic framework—and he would soon go further to provide an atomistic foundation for such an explanation.

Recognizing a shared aim and common endorsement of the electromagnetic view of nature very likely led Ehrenfest to initiate correspondence. Born replied to his request for a reprint in March 1909, remarking that their space-time threads had once previously come close to speaking distance. In the summer semester of 1904 they had both attended a seminar Hilbert and Minkowski gave on mechanics, but as a young student Born had not dared to speak to Ehrenfest then.[24] Now a rich exchange ensued, with important consequences for the development of relativity. The second stage of this dialogue is well known to historians, having been played out in the journals. In August 1909, Born published an ambitious model of the electron in the *Annalen der Physik*, which quickly drew a

Masse" (1909). Born's paper showed that a basic principle for the dynamics of moving charges could be formulated that contained only electromagnetic quantities, not the inertial mass, and that yielded the known equations of motion of electricity. This involved a very clear covariant formulation of Schwarzschild's action principle in Minkowski's four-dimensional notation, illustrating the formal utility of this development of relativity and extending Born's use of the calculus of variations to the variational principles of dynamics. For the appropriate secondary and boundary conditions the mass action integral vanishes and the inertial mass only appears subsequently in the form of a Lagrangian multiplier. In the correspondence discussed below, Ehrenfest challenged the attempt. In reply, Born conceded the point. It was not possible to "parry" with the final formula he had obtained, for, "as you correctly remark, *E*, *M*, are considered the total field strength, and which one does not know." Born's Lagrangian multiplier was only a "highly pitiful surrogate for an upstanding, honorable mechanical or electromagnetic mass." "Endlich ist mein Lagrangesche Multiplikator μ ein höchst erbärmliches Surrogat für eine anständige, ehrbare mechanische oder elektromagnetische Masse, das ich nur aus dem Grunde dem Publikum nicht vorenthalten wollte, weil wenigstens kein mathematischer Fehler in den Uberlegungen ist (was sie von vielen anderen derartigen Publikationen unterscheidet). Dass man allerdings mit den Endformeln auch gar wenig auffangen kann, gebe ich ebenfalls zu; denn, wie Sie richtig bemerken, bedenken *E, M* die totalen Feldstärken, und die kennt man eben nicht." Born to Ehrenfest, 5 July 1909, AHQP/EHR, Reel 17.

23. This is made explicit in a following paper in which Born refers back to this work. Born, "Theorie des starren Elektrons" (1909), 5.

24. Born to Ehrenfest, 17 March 1909, AHQP/EHR, Reel 17. ("Inertial Mass" had appeared in the *Annalen* on 2 March 1909.) "Übrigens sind sich unsere Raum-Zeitfaden schon einmal auf Sprechdistanz nahe gekommen; das war im Sommersemester 1904 wo wir gleichzeitig an einem Seminar von Hilbert-Minkowski über Mechanik teilnahmen; damals aber wagte ich als junger Student nicht, Sie anzureden." Ibid.

critical response from Ehrenfest in the *Physikalische Zeitschrift.* Together with two subsequent articles on interpretational issues in relativity, it is Ehrenfest's reaction to Born's model of the electron that prompted Miller to accord Ehrenfest an important role defending Einstein's relativity and distinguishing Einstein's work from Lorentz's theory. We have already noted that at least for Ehrenfest and Ritz the seeds of this process of demarcation lay in the work of Minkowski, and that in their case drawing such a distinction did not involve a favorable appreciation of Einstein's work. Further, it will be clear that whatever views Ehrenfest expressed in 1909, Einstein had not set him "straight" in 1907 as Miller suggests.[25] A study of the first stage of the exchange between Born and Ehrenfest will enable us to chart with some precision the route through which Ehrenfest was, in fact, persuaded by the principle of relativity.

After receiving "Inertial Mass," Ehrenfest evidently wrote to Born raising a number of queries about relativity. His letter gained Born's confidence. In a fifteen-page reply Born considered Ehrenfest's questions, discussed in detail his understanding of relativity and the electromagnetic worldview, and outlined his recent work. Relativity was both controversial and subject to numerous different interpretations; Born's letter helps us investigate some of the different kinds of pedagogical and persuasive work required to propagate and stabilize the meaning of innovative theories. Ehrenfest had clearly minced no words—Born wrote he did not agree with his "hatred" of the principle of relativity. Still, he hoped to have gained a "valiant ally" in the struggle against the "confusion of concepts and impudence of hypotheses that now make their way in physics."[26] Indeed, Born thought they shared a great deal in their outlook, venturing to acknowledge quite openly: "Einstein himself is to a great extent guilty of this mischief, that the one brilliant idea is nearly offset by an abundance of mathematical, logical, physical infamies and tactlessnesses. You see I am not embarrassed [or bothered, *geniere*]."[27] Einstein's work is distinct

25. For Ehrenfest as the defender of the relativistic viewpoint, see Miller, *Einstein's Special Theory* (1981), xxvi, 235–36, and 245–50.

26. "Wenn ich mit Ihrem Hass gegen das Relativitätsprincip auch nicht einverstanden bin, so hoffe ich doch in Ihnen einen tapferen Bundesgenossen gewonnen zu haben im Kampfe gegen die Verworrenheit der Begriffe und die Unverfrorenheit der Hypothesen, die sich jetzt in der Physik breit machen." Born to Ehrenfest, 5 July 1909, AHQP/EHR, Reel 17.

27. "... so wage ich doch ganz offen zu behaupten, daß gerade Einstein an einem großen Teil dieses Unheils selbst schuld ist, daß die eine geniale Idee von der Fülle der mathematischen, logischen, physikalischen Schandthaten und Takt-losigkeiten beinahe aufgewogen wird. Sie sehen, ich geniere mich nicht." Ibid.

and highly visible for both Ehrenfest and Born, but could be greeted with ambivalence even by a firm supporter of relativity.[28] Having laid out common ground, Born assured Ehrenfest he had already thought through his objections—till now he considered himself to be on very solid and highly practical pathways. Ehrenfest was quite right to think that for Einstein, Planck, and Minkowski, mass is not electromagnetic, electricity is structureless, and consequently "one does not understand what an 'electron' should be essentially, why, if it exists, it does not explode with an audible crack, etc. etc."[29]

Born endorsed the aims of both the electromagnetic view of nature and relativity. Without regarding relativity to constitute either an electromagnetic theory or an explanation of the electron, he sought an electromagnetic foundation to mechanics, impelled by his sense for the unity of conceptions of nature and by the belief that tracing electrodynamics back to the hidden motions of mechanics caricatured rational explanation. (We will see below that he had an answer to Abraham's argument also.[30]) It is revealing that Born opposed electromagnetic and mechanical explanations without conceiving a further possibility. In contrast both Einstein and Planck indicated a readiness to take relativity as a guide in pursuing an open alternative, as yet only partially articulated.

28. If the "physical infamies" Born referred to were related to the broader context in which Einstein set his research on electrodynamics, Born may have had a way of indicating his caution by referring to the principle or electrodynamic principle of relativity in his publications, rather than to the *theory* of relativity, with the broader implications of that designation. It should also be noted that Born did not take up the terminology Minkowski had offered in "The Fundamental Equations" with its distinction between the theorem, postulate, and principle of relativity, or use the term "World-Postulate" Minkowski employed in referring to relativity in the conclusion to his "Space and Time" lecture.

29. "Also, Sie haben völlig recht: Bei Einstein, Planck, Minkowski ist / 1) die Masse *nicht* elektromagnetisch, / 2) die Elektrizität völlig strukturlos / Infolge dessen versteht man nicht, was 'Elektronen' eigentlich sein sollen; warum sie, wenn sie existieren, nicht mit hörbarem Krach explodieren etc. etc." Born to Ehrenfest, 5 July 1909. Sommerfeld had assured Wien in 1906 that Wien was mistaken if he thought Einstein's theory excluded the electromagnetic mass. In Einstein's approach the variation in mass did not speak against ponderable mass, and the foundations of Einstein's theory had to be considerably extended before one could handle arbitrary motions of the electron. Sommerfeld to Wien, 23 November 1906, DMA, Archiv NL 56, 010, Doc. 102 in Sommerfeld, *Wissenschaftlicher Briefwechsel*, vol. 1 (2000), 255–57.

30. "Ich meinesteil halte sie für geradezu notwendig, mein Sinn für Einheitlichkeit der Naturauffassung ist außerordentlich entwickelt. Eine Zurückführung der Elektrodynamik auf Mechanik aber mit all ihrem Wust von 'verbogenen Bewegungen' etc. scheint mir ein Zerrbild einer vernunftmäßigen Naturbetrachtung." Born to Ehrenfest, 5 July 1909.

If relativity had not yet provided an explanation of the electron, what was one to do? The question had lain behind Born's research for over a year, and his answer was soon to be published. Writing to Ehrenfest, Born stated that long study had convinced him the principle of relativity was, "the expression for a generalized kinematics, which contains the ordinary in all its departments as a boundary case, and stands to it in somewhat the same relation as a hyperbolic Cayley metric (non Euclidean geometry with negative curvature) to the ordinary Euclidean geometry."[31] Born's mathematical analogy surely reflects his training, but might also alert us to the epistemological resource that widespread discussion of non-Euclidean geometries offered for the later rise of relativity, both in the scientific community and more broadly through popular books like Poincaré's *Science and Hypothesis*. Born thought the clear articulation of this kind of relationship could help Ehrenfest accept relativity. Late in the letter he spoke of the relations in terms Minkowski had made possible: "If you have recognized how everywhere the considerations of the group G_c (Minkowski) clarify the theorems concerning the group G_∞, i.e. ordinary mechanics, then I think your hatred of the principle of relativity will soon disappear."[32] Following the example of Minkowski's discussion in his "Space and Time" lecture, Born inverts the historical order of the relationship between ordinary mechanics and relativity—and ignores the unusual nature of the implications of relativity—to assert that the general group G_c enables the more familiar but limited special case of G_∞ to be understood more clearly.

Nor was the history of mathematics the only resource on which Born drew to convey the character and dimensions of the relationship between relativity and ordinary kinematics. He went on to represent relativity as "an extraordinary, brilliant abstraction, almost comparable with the idea

31. "In mehr als einjähriger Arbeit habe ich mich nämlich davon überzeugt, daß das sog. 'Relativitätsprincip' der Ausdruck für eine verallgemeinerte Kinematik ist, die die gewöhnliche in *allen* ihren Zweigen als Grenzfall in sich enthält und zu ihr etwa in demselben Verhältnis steht, wie eine hyperbolische Caylei'sche Maßbestimmung (nichteuklidische Geometrie mit negativem Krümmungsmaß) zu der gewöhnlichen euklidischen Geometrie." Born to Ehrenfest, 5 July 1909. Clearly Born does not show the basic confusion between Lorentz's and Einstein's work that historians have attributed to both Minkowski and Born.

32. "Wenn Sie erkannt haben werden, wie überall die Betrachtung der Gruppe G_c (Minkowski) die Sätze über die Gruppe G_∞, d.h. die gewöhnliche Mechanik, aufhellt, so wird, denke ich, Ihr Hass gegen das Relativitätsprincip bald vergehen." Ibid.

of Copernicus."[33] Born's analogies might enhance appreciation of relativity in at least two ways. Firstly, they give instances of prior abstractions of a (roughly) similar conceptual kind. Both non-Euclidean geometries and the Copernican system had allowed phenomena understood by existing formulations to be reinterpreted. In the former case (as in relativity) the prior theory could now be regarded as having validity only within a strictly limited point of view. Secondly, as well as this conceptual level, such analogies suggest an evaluative scale against which to judge the broader significance of the present innovation—and the comparison between relativity and the Copernican system was soon to become a common theme in expositions of relativity. We will explore the way Max Planck used it in our study of the histories of relativity.

Relativity, Rigidity, and Acceleration: Max Born's Rigid Body Electron

Recognizing specific relations between the principle of relativity and *previous* theory might help one understand the new theory. Still more significant was the research strategy this suggested for the future. Regarding the principle of relativity as the expression for a generalization of ordinary kinematics gave Born great confidence in the consequences of the theory. Indeed, he thought this relationship made it completely impossible for the new kinematics to lead to any logical contradictions, or contradictions to experience, just so long as it was applied consistently.[34] With this deep confidence, there were two sides to Born's work with relativity. One was critical, pointing out respects in which others had applied the principle inconsistently. More positively, Born sought to extend electron theory by finding relativistic analogues for concepts central to both ordinary kinematics and previous formulations of the dynamics of the electron. There were, after all, major limitations in the theory as it stood.

Born's search for an electrodynamic understanding of inertial mass required addressing the problem Einstein had identified for Ehrenfest, and that meant finding a relativistic analogue for the rigid body—for arbitrary

33. "Ich sehe darin eine ganz außerordentlich geniale Abstraktion, fast vergleichbar mit der Idee des Kopernikus." Ibid.

34. "Es ist danach ganz unmöglich, daß die neue Kinematik, wenn man sie konsequent anwendet, zu irgend welchen logischen Widersprüchen führen kann, ja nicht einmal zu Widersprüchen mit der Erfahrung." Born to Ehrenfest, 5 July 1909.

accelerated motion. We will first consider treatments of rigidity before turning to acceleration.[35] In the system of Newtonian mechanics, the rigid body had proved to be highly valuable for its simplicity both kinematically, as a continuous mass system of only six degrees of freedom, and dynamically, in that it allowed the forces exerted on it to be resolved into a number of "resultant" forces and moments, knowledge of which sufficed for the description of its motion. Yet, as Born commented, it was now possible to see that the great importance invested in the concepts of the rigid body and rigid connections in Newtonian mechanics was closely united "to the fundamental conceptions of space and time on which this discipline is built."[36] Accepting the "electrodynamic principle of relativity" that Lorentz, Einstein, Minkowski, and others had established, the basis for Newtonian kinematics now had to be given up. This did not mean the new kinematics could dispense with an appropriate concept of rigidity.

There was no difficulty in considering systems moving uniformly in relation to one another, and previous authors had employed a concept of rigidity without offering a specific definition—although in 1907 Einstein had begun to express a cautious distrust of the concept. Soon after stressing the centrality of rigidity in his reply to Ehrenfest, Einstein emphasized the difficulties facing treatments of the dynamics of a rigid body by discussing a simple thought experiment. Consider a rigid rod, observed from a reference system in which it is at rest. Applying equal but opposite forces to each end over a very short time does not produce any motion of the body. Observed from a reference system moving in a direction parallel to the main axis of the body, however, the impulses at either end do not act simultaneously; rather, the impulse at one end is delayed, compensating the impulse that first acted at the other end only after some time. There appears to be a violation of the energy principle. Einstein decided that the impulse spreads with a finite velocity, associated with a change of state of unknown quality in the body, producing an acceleration on the body in a short period of time unless this effect is compensated by some other force acting on the body within that time. He used the occasion

35. Born, "Träge Masse" (1909), 572. Born's work on the rigid electron and the ensuing debates in the German community have been discussed by Klein, *Paul Ehrenfest* (1970), 150–56; Miller, *Einstein's Special Theory* (1981), 243–53; Goldberg, *Understanding Relativity* (1984), 191–96. Other authors have focused on the episode for its implications for Einstein's path to general relativity: Stachel, "Rigidly Rotating Disk" (2002); Maltese and Orlando, "Definition of Rigidity" (1996).

36. Born, "Theorie des starren Elektrons" (1909), 1.

primarily to reinforce the point that relativity did not permit the propagation of effects with a velocity greater than that of light, and noted, "If relativistic electrodynamics is correct, we are still far from possessing a dynamics of the parallel translation of a rigid body."[37] While Einstein expressed caution in print, in correspondence with Sommerfeld that we consider more fully in the following chapter, he made it clear firstly that he did not think relativity provided a definitive treatment of the mechanics of the electron, and secondly that a more elementary foundation to electrodynamic theory and understandings of electricity and magnetism was required. As Russell McCormmach has shown, Einstein undertook strenuous work to extend electron theory through 1909, without reaching a point at which he felt able to publish his efforts.[38]

Thus previous discussions of rigidity had highlighted the centrality of the concept, but offered reasons for caution. There were two significant relativistic treatments of acceleration on which Born could draw. The most important conceptually had been presented in the concluding sections of the major research review on relativity that Einstein published at the end of 1907. There, Einstein indicated that he too was concerned with the fact that the principle of relativity applied to acceleration-free systems, being limited to drawing inferences about systems moving at uniform velocities relative to one another (based on the equivalence of the laws of physics in such systems). The point had often been raised as a criticism of the theory, but Einstein now approached it in a highly interesting and idiosyncratic way. Rather than asking with Ritz, for example, how variations of velocity could be dealt with in the Lorentz-Einstein electrodynamics, Einstein proposed that a homogenous gravitational field and a corresponding acceleration of the reference system are completely equivalent physically, thereby extending the principle of relativity to apply to uniformly *accelerated* translational motion. This conceptual extension was important because, Einstein wrote, accelerated motion was amenable to theoretical treatment (to at least some extent). Einstein then argued that at any particular time an accelerated system could be approximated by

37. Einstein, "Trägheit der Energie" (1907), 381.

38. The caution was also expressed in Einstein's major research review of relativity, in which he derived the mass-energy equivalence using a deformable envelope rather than a rigid body and pointedly left questions of the structure of the electron aside: Einstein, "Relativitätsprinzip" (1907), 431–47. The correspondence was Einstein to Sommerfeld, 14 January 1908, in Einstein, *Collected Papers*, vol. 5 (1993), 86–89. On Einstein's unpublished work on the electron, see McCormmach, "Einstein, Lorentz, and the Electron Theory" (1970).

considering instead one moving with a *constant* velocity that was momentarily at rest with respect to the system.[39] That allowed Einstein to relate inferences drawn from his earlier studies of the relativity principle to the properties of time and the path of light in a gravitational field. Ironically, at the very moment he recognized the need to go beyond what would later be called "special relativity," Einstein relied centrally upon its techniques. Born followed the same strategy in his treatment of acceleration, but used in addition the framework Minkowski had provided to represent acceleration mathematically in the context of his geometrical representation of the space-time manifold, using the concept of proper time and the distinction between timelike and spacelike vectors.[40]

Born offered a definition of rigidity that was based on a set of differential equations that described a body whose shape remained unchanged in the frame of reference momentarily at rest in regard to the body. Faced with great technical difficulties, he could not integrate the condition for rigidity for the general case, but the fact that rigidity must be identical with incompressibility for the specific case of straight line translation gave both a criterion for showing that Born's definition of rigidity made sense and a method of integrating the differential conditions of rigidity for the simplest instance of that straight-line translation. Similarly, Born did not study the general case of arbitrary accelerations, but considered in detail only the simplest case of translation with a constant acceleration, showing that in this case the motion described a hyperbolic trajectory in

39. Einstein, "Relativitätsprinzip" (1907), part V, "Relativitätsprinzip und Gravitation," 454–62. Among the assumptions that allowed this treatment, Einstein considered how the shape of a body with a certain acceleration but not velocity relative to a nonaccelerated reference frame would be influenced by that acceleration. For reasons of symmetry, such change would have to be a dilation of constant proportion in the direction of the acceleration γ, as well as possibly in the directions perpendicular to this. Further, any such dilation, "if it exists at all," would have to be in even functions of γ, and could be neglected if γ was sufficiently small to neglect terms of the second and higher powers.

40. See Minkowski, "Space and Time" (n.d.), 83–86. There, Minkowski offered the following way of dealing mathematically with the acceleration of a world line. By analogy with the circle of curvature that enables the form of a curve to be approximated more accurately than the tangent at a point, there is a definite hyperbola with three infinitely proximate points in common with a world line at P. The asymptotes of this hyperbola generate a front cone and a back cone. If M is the center of curvature, the hyperbola is an internal one with center M. If ρ is the magnitude of the vector MP, the acceleration vector at P is the vector in the direction MP with the magnitude c^2/ρ. Minkowski's work therefore integrated acceleration into his four-dimensional representation in such a way that its mathematical description became possible, and accelerated motions formed the closest approximation to arbitrary motion of a world point.

four-dimensional space-time, thus establishing a direct connection with the hyperbola of curvature Minkowski had described as a means of approximating arbitrary motion.[41]

With the support of this kinematic foundation Born went on to establish the dynamics of hyperbolic motion in the second and third chapters of the paper, seeking to determine the self force propagated by an electrically charged rigid body. The results he obtained could then allow approximate conclusions to be drawn for all motions for which the sum of the acceleration vector changed only a little, and Born recovered the results that had earlier been obtained for the Lorentz theory of the electron. In the final part of the paper Born raised his major criticisms of earlier work in electrodynamics. For Ehrenfest, Born identified these as centering on the concepts of *resultant* force, *resultant* moment, *resultant* energy (total energy), and the volume integrations used to obtain these.[42] It was here that Born dealt explicitly with the query that Planck and Einstein had failed to address, arguing that Abraham's argument depended simply on applying the principle of relativity inconsistently in the relevant calculations.[43] Throughout the paper Born followed the same general strategy.

41. Born, "Theorie des starren Elektrons" (1909), 6–16 (sections on the rigid body in the "Old Mechanics," and differential conditions), 16–19 (incompressible flow), 19–25 (rectilinear translation), and 26.

42. Born to Ehrenfest, 5 July 1909.

43. Born argued that in the Abraham and Lorentz theories of the electron, the equations of motion of the electron (without ordinary mass) are formed through a common procedure. The resultant ponderomotive force of the external field, and the field due to the electron itself, are formed through integration over the space filled by the electron at a specific time, and these resultant forces are then set equal to zero. In the Abraham theory, where the ether is assumed as a reference system at absolute rest and the electron is rigid in the old sense, the procedure has its justification. In the Lorentzian theory, however, if one defines the resultant force and so forth as the integral at a given time, different values are obtained according to the reference system chosen. The basic equations of electromagnetism are certainly invariant against the Lorentz transformation, but only if the components of the Maxwell stresses, the Poynting vector, and the energy density are also appropriately transformed. Born argued that integration of these concepts at a given time had no direct meaning in the kinematics of the principle of relativity, and insufficient observance of this circumstance lay behind the apparent appearance of an energy and momentum of deformation, on which Planck and Abraham had remarked. Rather than using the laws of the conservation of energy and momentum in this way, Born defined the resultant force of a force field as the integral of the product of the rest charge and rest force over the rest form of the electron (but left to one side the question of how to define the resultant moment when rotations were permitted). The equations of motion of a rigid electron could then be expressed such that: "The rigid electron moves in such a way that the resultant of its own field is equal and opposite to the resultant of the

He displayed concepts from the old kinematics and dynamics and showed that they are not Lorentz invariant. He then selected forms of the concepts that could be made Lorentz invariant through suitable modification and constructed an analogue to the concept employed in the old kinematics. It is a careful, pedagogic approach. Born made the links between the new physics and the old explicit by showing the steps through which the new could be achieved. His final step was to confirm Lorentz's result for the mass of the electron, and emphasize two features of his approach that he thought established a strong link between atomism and the principles of dynamics.[44]

We have noted that at two important points Born chose to study only one particular limiting case of the general conditions he was able to derive. He integrated the differential conditions of rigidity only for the simplest case of straight-line motion (among all possible accelerations of a rigid body), and studied in detail only constant acceleration. Choices of this kind made the task of constructing an electrodynamics of the rigid body manageable, and since the hyperbolic motion Born did discuss allowed arbitrary accelerated motions to be approximated, he could rightly consider the limitations reasonable.[45] Nevertheless, they proved to be the Achilles' heel of his endeavor.

---

external field." The energy law then appears as the statement, depending on the three equations of motion, that the sum of the work of the external and internal fields is always equal to zero. See Born, "Theorie des starren Elektrons" (1909), 47–48. Ebenezer Cunningham had offered a similar criticism of Abraham's argument in Cunningham, "Electromagnetic Mass of a Moving Electron" (1907).

44. The first was purely kinematical: for given dimensions Born found the world line could not exceed a certain limit of curvature, and the greater the acceleration, the smaller the body would appear. The second was the conclusion that the electron had to have one middle point and its charge had to be distributed in concentric spheres. From the empirical fact that no external lateral force was required to maintain quasi-stationary motion of an electron in cathode or Becquerel rays, Born concluded that for vanishingly small accelerations the electron itself did not propagate any lateral force. For this to be the case, in both the y and z directions (and since the direction of motion of the electron was arbitrary), the charge had to be symmetrically distributed with respect to each plane passing through a single middle point of the electron. Born, "Theorie des starren Elektrons" (1909), 25 (for the kinematic argument) and 53–54 (symmetry).

45. Early textbooks on relativity followed Born in this, and gave extensive discussions of hyperbolic motion; the theoretical tools he had used were evidently judged acceptable by their authors. See Laue, *Relativitätsprinzip* (1911), 105–16; Silberstein, *Theory of Relativity* (1914), 182–98, on 189–91.

The Paradox of Relativistic Rigidity

In the early twentieth century many physicists looked above all to electron theory for progress in their discipline. Born had addressed the conceptual relations between the concrete promise of that particle and the new principle of relativity head on. The early responses of Paul Ehrenfest and Arnold Sommerfeld indicate how important that was. Ehrenfest had to put pen to paper fast, jotting down his reaction for an (unnamed) friend:

> Dear Friend!
> Although I hope to see you later today, and this letter will perhaps only reach you later than I myself, I must *write* to you at once. The theme: Born's work.
>
> I began to read it last night—it is clear and elegant throughout—Minkowski + Born represent an enormous advance—hopefully the whole devil's business will fall apart much more quickly than I had hoped![46]

Sommerfeld was the first to rise to give his views when Born finished presenting his work to the Naturforscherversammlung in 1909: "No one can express better than I the great simplification and improvement of the theory which is given through the new definition of rigidity and the happy thoughts of Born, because I myself have also been concerned with uniformly accelerated motion and been brought to extraordinarily unclear expressions on the grounds of the old definition of rigidity."[47]

Despite such first impressions, the reactions of both physicists soon took an unanticipated turn—one that denied Born's ambitions. After praising Born's happy thoughts, Arnold Sommerfeld struck a cautionary note that was soon to dominate readings of Born's theory in particular and the relationship between relativity and electron theory in general.

46. "Lieber Freund! Obwohl ich Sie heute noch zu sehen hoffe, also dieser Brief Sie vielleicht erst später treffen wird als ich selbst muss ich Sie sofort *an*schreiben. Thema: Born-Arbeit. / Ich habe sie heute Nacht zu lesen angefangen—Sie ist überaus klar und elegant—Minkowsky + Born stellen einen enormen Fortschritt dar—also wird hoffentlich viel rascher der ganze Teufels-kram aus dem Leim gehen als ich gehofft hatte!" The fragment continues: "1. Ich hatte ihnen die Schwierigkeit gezeigt: Wenn ein relativ-starrer Stab nach rechts läuft—seit—Ø mit der constant Geschwindigkeit v_0—so können Beschleunigungen . . . " Ehrenfest to unnamed friend, undated draft letter, AHQP/EHR Reel 32 m59.

47. Sommerfeld in discussion to Born, "Dynamik des Elektron" (1909), 817.

Like Einstein before him, Born had accepted the validity of a description based on a frame of reference moving at a uniform velocity instantaneously at rest with an accelerated body. Sommerfeld drew attention to the technique and argued that the principle of relativity applied only to uniform velocities. It could yield no certain conclusions in regard to accelerated motion.[48] More significant than the simple conceptual point is the judgment that theoretical endeavor should end with it. Rather than indicating a wish to follow up Born's work as a contribution to electron theory—or asking whether there were prospects for testing Born's theorems experimentally—by late 1909 Sommerfeld gave primacy to the extent to which the principle of relativity could determine the development of theory.[49] Remember that like Ehrenfest and Born, Sommerfeld had stated a strong preference for the electromagnetic worldview just three years earlier; we will consider the complex path he took toward supporting relativity in more detail in chapter 8, and there analyze its implications for our understanding of the authorship of the theory. For the moment I

48. For example, he wrote that in relation to the form of the ponderomotive force it would be possible to add to Born's theorem terms that no longer contained the velocity. Sommerfeld said he himself was in favor of neglecting such terms and he thought Born's theorems were in any case the simplest among many possible, but reiterated that the principle of relativity could not yield a compelling basis for them. To a large extent Sommerfeld's view has been accepted by commentators on Born, who have often regarded Born's electron theory as an ambitious faux pas. In Miller's view, Born did not understand that the principle of relativity could only discuss uniform motion. Goldberg writes that Born (in common with Minkowski) did not recognize the difference between Einstein's kinematical approach and Lorentz's dynamical interests, while Pyenson comments that Born (and Abraham) "found the theories [of relativity] difficult to understand." See Miller, *Einstein's Special Theory* (1981), 243–45; Goldberg, *Understanding Relativity* (1984), 193; Pyenson, "Einstein's Early Scientific Collaboration" (1985), 219. More appreciative views have been offered by historians who have argued that Born's goals and formalism, and the paradox described below, contributed to Einstein's pathway to general relativity. See Pais, *"Subtle is the Lord"* (1982), 214; Stachel, "Rigidly Rotating Disk" (2002); Maltese and Orlando, "Definition of Rigidity" (1996).

49. Born later commented that after his contribution to the 1909 Naturforscherversammlung (and therefore before the publication of Ehrenfest's paper, discussed below), he and Einstein discussed rotation, being puzzled that a body at rest can never be brought into uniform rotation. Einstein corresponded with Sommerfeld about his search for an extension of the relativity principle to uniformly rotating systems, analogous to his 1907 treatment of uniform acceleration. By January 1910, the lack of empirical data bearing on the question had discouraged him from pursuing the question actively for the present. See Born, "Definition des starren Körpers" (1910); Einstein, *Collected Papers*, vol. 5 (1993); Einstein to Sommerfeld, 29 September 1909 and 19 January 1910 (Docs. 179 and 197), 210–11 and 228–30.

want to note that his position in 1909 did not involve a complete abandonment of the importance of the electromagnetic program.[50]

Paul Ehrenfest soon sent the distilled results of considerable deliberation on Born's work to the *Physikalische Zeitschrift*. In a brief note he presented an intriguing paradox that arises in understanding the rotation of a rigid body relativistically.[51] Consider a relativistically rigid cylinder of radius R and height H. It is given a constant, arbitrary angular motion around its axis. Let R be the radius the figure shows for a rest observer. Ehrenfest showed that R must fulfill two contradictory demands. On the one hand, the periphery of the cylinder must show a contraction against the rest state, for each element of the periphery moves in its own direction with the momentary velocity $R'w$. Hence $2\pi' R < 2\pi R$. On the other hand, if one considers any element of the radius, its momentary velocity is normal to its extension. Therefore the element of the radius can exhibit no contraction, $R = R$.[52]

Einstein had earlier considered the possibility of any deformation due to acceleration and come to the conclusion that "if it exists at all," it could safely be ignored for his purposes.[53] However, Ehrenfest's concerns had focused on angular momentum and "rolling" from the first question he posed to Einstein in 1907. His letters and notes show that his thinking on relativity and rotation developed through several stages before culminating in his ability to pose a paradox. In a draft letter to Abraham Joffe most likely written between March and July 1909, Ehrenfest indicated that he wished to discuss the kinematics of "arbitrarily rotated, arbitrarily accelerated 'relativistically' rigid mechanisms and the principle of relativity."[54] (In a similar context Born would likely have written only "arbitrarily accelerated.") This was the second of six possible discussion points Ehrenfest listed for Joffe, all of which were associated with relativity, thermodynamics, or both. It is quite clear that shortly after

50. See Seth, "Quantum Theory and the Electromagnetic World View" (2004).

51. Ehrenfest, "Gleichförmige Rotation" (1909).

52. Ibid.

53. Einstein, "Relativitätsprinzip" (1907), 454–55. See note 39 above.

54. "2) Kinematik beliebig rotierender, beliebig beschleunigter 'relativ'-starrer Mechanismen und Relativ-prinzip." Paul Ehrenfest undated draft letter to Joffe-Djadja, AHQP/EHR Reel 32 m16. Apart from its general concern with rigidity and relativity, this dating is suggested by the fact that Ehrenfest refers to Planck's discussions of the entropy of moving systems, a subject he raised in correspondence with Laue in April 1909. Ehrenfest draft letter to Laue, "IV, 1909," AHQP/EHR Reel 23.

Born's first letter in March 1909, Ehrenfest subjected the foundations of the principle of relativity and the results derived from it to close analysis. Ehrenfest's fifth point indicates just the kind of work an opponent would have to undertake to challenge the theory successfully. Ehrenfest wrote: "Kohl and Hicks on the Michelson experiment / In general: unschematization of the Michelson experiment."[55] The Michelson-Morley experiment had continued to receive critical attention, now in large part because of its role in justifying relativity. In 1902 William Hicks had stated that analysis of the reflection from the moving mirrors was critical to the experiment and claimed briefly that the expected effect was the reverse of that previously supposed (as a result of an algebraic error Hicks had mistaken the direction and sign of the FitzGerald-Lorentz contraction). Writing from Vienna in an extensive paper that appeared early in 1909, Emil Kohl took up the question of reflection from moving mirrors again, once more suggesting that the customary analysis of the experiment was incorrect—a perspective that Ehrenfest thought well worth pursuing.[56] A further undated fragment indicates a more concrete step in Ehrenfest's deliberations. Headed "Separate Questions," it concerned the kinematics of a diamond sphere rotating uniformly around one and the same diameter, with the center at rest. Ehrenfest asked, given a sphere, which in the case of translation becomes an ellipsoid, "what happens to it in the case of the above rotation?" His first assumption, that it remained a sphere, he regarded as "suspicious": "For assume I mark anywhere on the equator a small cross + with equal arms at rest—If I take the sphere to be very great and the angular velocity small, then the cross moves close to translationary motion with any velocity under that of light—it is suspicious that

55. "5) Kohl und Hicks über Michelson-Experiment / Allgemein: Ent-Schematisierung des Michelson Experimentes." Ehrenfest to Joffe-Djadja.

56. Hicks, "Michelson-Morley Experiment" (1902); Hicks, "The FitzGerald-Lorentz Effect (Letter to the Editor)" (1902); Hicks, "The Michelson-Morley Experiment. To the Editors of the Philosophical Magazine" (1902); Hicks, "The Michelson-Morley Experiment. To the Editors of the Philosophical Magazine" (1902). For a discussion of Hicks's work and the rebuttals of Michelson and Morley, see Swenson, *Ethereal Aether* (1972), 142–47. Kohl's paper was addressed by Laue in 1910, who showed that Kohl had based a major part of his argument on the case of plane-parallel mirrors, rather than the mirrors at a slight angle to one another that were actually used in the experiment. Discussion of the experiment hardly abated, however. See Kohl, "Michelsonschen Versuch" (1909); Laue, "Ist der Michelsonversuch beweisend?" (1910); Budde, "Theorie des Michelsons Versuches" (1911); Budde, "Theorie des Michelsons Versuches. II." (1912); Riecke, "Theorie des Interferenzversuches von Michelson" (1911); Laue, "Theorie des Michelsonsversuches" (1912).

the cross is not deformed although it moves close to pure translation."[57] Born's quick confidence may also have stimulated Ehrenfest to consider rotation more critically, for in describing his theory Born admitted that he had only discussed rotation-free motion, allowing "perhaps you might not regard your 'surface theorem' objection as having been disproved."[58] Given Born's need to provide a precise definition of rigidity it was a small step for Ehrenfest's suspicion to become a contradiction, impossible to ignore.

Ehrenfest's publication was brief and pithy—he was quite content to leave some facets of his pathway to relativity invisible. In particular, beyond addressing Born's model he offered no hint of the important causative role we have seen Born's work to play. However, he did present his definition of rigidity as having been attained in connection with Minkowski's ideas—and made no mention of Einstein.[59] At this point Ehrenfest was far more clearly defending relativity than he was Einstein. (It was only in 1912 that he first met Einstein, who hoped Ehrenfest might succeed him in Prague. The two became close friends then.) Ehrenfest's perspective on the direction of future theory was also now more circumscribed than the intent expressed in earlier probings. Rather than stating directly what implications his contradiction held for the relationship between relativity and electron theory, Ehrenfest's conclusion commented only on what it required of a characterization of deformation.

57. "*Einzelfragen* / 1) Zur Kinematik einer rotierenden (gleichförmig um ein und denselben Diameter) Diamantkugel-Centrum ruht. / *Frage:* Es sei eine *geborene* Kugel—bei Translation wird sie ein Ellipsoid—was thut sie im Fall obiger Rotation? / *Annahme* A) Sie bleibt eine Kugel—das ist *verdächtig*—denn angenommen ich markiere irgend wo auf ihrem Aequator ein kleines Kreuz + gleicharmig im Fall der Ruhe—Wenn ich die Kugel sehr gross nehme und die Winkelgeschwindigkeit klein, so bewegt sich das Kreuz *nahezu* Translatorisch mit irgend einer Unterlichtsgeschwindigkeit—Es ist verdächtig, dass das Kreuz sich gar nicht deformiert obwohl es sich doch beinahe rein Translatorisch bewegt." Ehrenfest undated fragment, AHQP/EHR Reel 32 m16.

58. "Allerdings beschränke ich mich in dieser Arbeit auf rotationsfreie Bewegungen, und Sie werden dadurch vielleicht Ihre 'Flächensatz'-Einwürfe nicht als widerlegt ansehen." Born assured Ehrenfest that in the meantime he had got much further and already knew for certain that rotationary motions led to no contradictory conclusions: "Ich kann Sie aber versichern, daß ich inzwischen viel weiter bin und schon sicher weiß, daß auch die rotierenden Bewegungen zu keinerlei Widerspruch Anlass geben." Born to Ehrenfest, 5 July 1909.

59. Ehrenfest wrote, "im Anschluß an Minkowskis Ideen," Ehrenfest, "Gleichförmige Rotation" (1909). Klein discusses only the published papers of Einstein, Ehrenfest, and Born in his treatment of Ehrenfest's work on relativity and does not speculate on when or why Ehrenfest was persuaded of the theory. Klein, *Paul Ehrenfest* (1970), 150–54.

Two other authors responded to Born's paper with detailed studies of the kinematics of the relativistically rigid body, taking up a task Born had neglected. It was to be expected, as Born found, that a rigid body in uniform accelerated motion was completely described by the motion of one of its points. This was also the case for the rigid body of the old mechanics. But Born had not considered generally what kind of motions were possible for the body he had described; he had not investigated how close the analogy between his relativistic definition of rigidity and the customary rigid body actually was. Independently, Gustav Herglotz, a close friend of Paul Ehrenfest, and Fritz Noether, a student who had just completed his PhD at Munich (where Sommerfeld taught), both subjected the kinematics of Born's definition of a rigid body to close scrutiny.[60] Herglotz asked whether Born's rigid body possessed the six degrees of freedom, "which one would wish if his new 'rigid' body was to have the same fundamental significance in the system of the electromagnetic worldview as the rigid body has in the system of the mechanical worldview."[61] Rather, he concluded that as soon as the motion of an arbitrarily prescribed point is known, the motion of the whole body is in general completely determined, and Born's rigid body has only three degrees of freedom.[62]

Both Herglotz and Noether considered the kinematic consequences of Born's condition in detail, and neglected Born's interest in the dynamics of the electron. Born's first reaction was to stand and fight. Any assumption was hypothetical, but some assumptions were required and his had many advantages. He urged the continued usefulness of his work, arguing that as yet there were no phenomena whose explanation called on rotations of the electron, and by restricting rigidity to individual electrons one could reconcile the condition to the rotation of material bodies. The material body is no continuum, and if it is composed of atoms or electrons that are rigid in his sense, when the whole body is brought into rotation the individual electrons would describe curvilinear paths for which his kinematics would suffice. However, in a series of papers Born loosened his definition of rigidity to encompass wider classes of motion and gradually

60. Both corresponded with Ehrenfest on the kinematics of relativity, Herglotz from 9 December 1909, and Noether from 23 November 1910, AHQP/EHR Reels 21 and 24. Their papers were Herglotz, "Vom Standpunkt des Relativitätsprinzips als 'starr' zu bezeichnenden Körpe" (1909); Noether, "Kinematik des starren Körpers" (1909).

61. Herglotz, "Vom Standpunkt des Relativitätsprinzips als 'starr' zu bezeichnenden Körpe" (1909), 393.

62. There are some exceptional cases. Ibid., 404.

focused on simply establishing the kinematics of relativistic motion rather than providing a basis for a theory of the electron.[63]

More generally, following Herglotz and Noether's work, attention moved away from rigidity toward characterizing deformation. In response to Ehrenfest's paper, for example, Max Planck wrote that the task of determining the deformation of an arbitrarily accelerated motion was, in relativity as well as in ordinary mechanics, essentially a problem of elastic theory.[64] In discussion at the 1910 Naturforscherversammlung (a year after Born first presented his theory), Born and Sommerfeld responded to a paper that argued for the existence of superluminal velocities by stressing that relativity forbade signal velocities faster than the speed of light, arguing that this had consequences for the rigid body theory.[65] Max Laue made the point very clearly in 1911, demonstrating that in relativity a body must posses not three, nor six or nine, but an infinite number of degrees of freedom.[66] No action can be propagated with a velocity faster than that of light. Therefore an impulse given to a body simultaneously at *n* different places will initially produce a motion to which at least *n* degrees of freedom must be ascribed. In this year Planck demonstrated that he at least thought that the lessons of the relation between rigidity and relativity had been learned. He wrote to Wien that the *Annalen* should divert papers on relativity dealing with "formulation, illustration, definitions (rigid bodies!)" to the *Physikalische Zeitschrift* or to mathematics journals, which would leave more room in the *Annalen* for "real physical investigations."[67]

Despite Einstein's repeated insistence that the theory of relativity was based on rigid bodies, clocks and light signals, the concept of the rigid

63. For the fight, see Born, "Definition des starren Körpers" (1910). Gyeong Soon Im regards this to display a positivistic stance also evident in Born's later work on quantum mechanics, Im, "Max Born und die Quantentheorie" (1991), 17. For the relaxation, see Born, "Kinematik des starren Körpers" (1910).

64. Planck, "Gleichförmige Rotation" (1910). In a letter to Joffe, Ehrenfest displayed a similar interest in elasticity and showed that he was still concerned with the theory of the electron. He wrote of his wish to find the dynamics of the electron not on kinematic but on a "form-elastic basis" ("Dynamik des Lorentz-Elektrons auf nicht Kinematic sondern 'Form-elastischen' Grundlagen"). Ehrenfest undated draft letter to Joffe, AHQP/EHR Reel 32 m19.

65. See the discussion to Ignatowski, "Bemerkungen zum Relativitätsprinzip" (1910), at 975–76. Born noted that his conditions would aid in developing a theory of elasticity.

66. Laue, "Starrer Korper in der Relativitätstheorie" (1911). Noether had proposed a structure with 9 degrees of freedom.

67. Planck to Wien, 9 February 1911, Wien Papers, STPK, 1973.110, quoted in Jungnickel and McCormmach, *Intellectual Mastery*, vol. 2 (1986), 323.

body had been found incompatible with the theory of relativity. Just as the experimental search to provide definitive proof of theories of the electron had foundered in the difficulty of establishing absolute measures of the accuracy required, Born's aim for an electromagnetic theory of the electron compatible with relativity had dissolved in the recognition that the theoretical tools available did not give enough purchase to draw compelling conclusions.

Conclusion: The Parable of the Hedgehog and the Hare

This chapter has shown that rigidity, like concepts of relative and absolute measures, was subject to both empirical and conceptual probing. Further, rigidity proved as problematic conceptually as it might be crucial to understanding the differences between theoretical predictions and electron experiments that relied on maintaining a constant electric field between the parallel plates of a condenser. But while empirical results left the validity of relativity an open question, it is clear that in Germany at least many assessments of that theory—so often set in play by the contrast between relativity and the electromagnetic worldview—turned on the discussion of Max Born's work and the issue of rigidity. Born's research was particularly significant because it aimed to provide a relativistic theory of the electron, but in the event it proved most important for establishing limits. Simultaneously, physicists accepted limits to electron theory, and to relativity, that were closely tied to the restriction of relativity to the treatment of uniform velocities. In a fragment retained in the collection of his correspondence, Paul Ehrenfest expressed the tension this involved between certain knowledge and broader goals:

> Listen Uncle—to something you yourself have doubtless already thought out and which is too difficult for me mathematically: *Prolegomena to any future relativity-theory rigid body.*
>
> An observer B_1 and an observer B_2 (both in *uniform translation*) see an *arbitrary* moving "rigid" body H (Hare). Born says: the hare behaves in each moment such that every cell holds itself rigid.—If one shows that leads to conflicts Einstein will greet you and say: that is none of my business—am I perhaps a hare?!—I am a *uniform* observer and nature only need take note of uniform observers.

> What is now necessary is this: to show that generally *each* definition of rigidity with which *all* uniform observers are equally satisfied already leads to contradictions. I am sure that it is so!![68]

To appreciate the full bite of Ehrenfest's comparison we need to know the fable his reader might have seen in Ehrenfest's reference to the hare. According to the Brothers Grimm, early one fine day a hedgehog bet a gold coin and bottle of wine that he could outrun a hare, despite his short legs. The clever hedgehog demanded time to prepare by eating breakfast, and then met his opponent at the far end of the furrowed field. One, two, three: go! The hare set off helter-skelter, while the hedgehog took just a few steps in his furrow and relaxed. For at the other end his wife stood at the ready: "I am here already!" she called as the hare approached. The astonished hare thought to himself, "That has not been done fairly," and cried, "It must be run again, let us have it again." Once more he was off, flying like the wind in a storm. But the hedgehog's wife stayed quietly in her place. So when the hare reached the top of the field, the hedgehog himself cried out to him, "I am here already." The hare was now quite beside himself with anger and cried, "It must be run again, we must have it again." So the race continued until, having run ninety-four times, the hare collapsed in the middle of his furrow, unable to take a step further.

Ehrenfest's irony cuts two ways, against both the hare and the hedgehog of relativity, against both Born and Einstein. While Born's use of a highly limited rigid body as a model for the electron fails to capture the hare's sudden motions, Ehrenfest also mocks the idea that nature should take note only of uniform observers. After all, the hedgehog only wins by subterfuge (and Ehrenfest's note shows he wants to push difficulties with the concept of rigidity even into the realm of uniform translation).

68. "Höre Onkel—Nun noch etwas, was Du Dir sicher selber schon ausgedacht hast und was mir leider mathematisch zu schwer ist: *Prolegomena zu jeder künftigen Relativitätstheorie starrer Körper.* / Ein Beobachter B1 und ein beob. B2 (beide *gleichförmig translationiert*) schauen auf einen *beliebig* laufenden 'starren' Körper H (Hase) Born sagt: Der Hase verhält sich in jedem Moment so, dass jede Zelle in ihm sich selber starr vorkommt.—Zeigt man, dass das zu Conflicten führt so grüsst Einstein und sagt: Geht mich gar nix an—Bin ich vielleicht a Has?!—Ich bin ein *gleichförmiger* Beobachter und nur auf gleichförmige Beobachter hat die Natur Rücksicht zu nehmen. / Was jetzt notwendig ist ist dieses: zu zeigen dass überhaupt *jede* Starrheitsdefinition mit der *alle* gleichförmigen Beobachter gleichartig zufrieden sind schon zu Widersprüchen führt. Ich bin überzeugt dass es so ist!!" Ehrenfest, undated fragment "Bornstab," AHQP/EHR Reel 32 m60.

As we have seen, Einstein was as keenly aware of the limitations his theory faced as many other physicists. His efforts toward a theory of the electron remained unpublished, but in 1912 he returned to the extension of relativity to nonuniform motion. Ehrenfest's paradox helped Einstein recognize a need for non-Euclidean geometry, and Born's techniques may well have guided him to Riemannian solutions.[69]

In chapters 6 and 7, we have met the principle of relativity and Einstein largely through the perspective of the electron and three other physicists: an experimentalist who worked hard to make his procedures and results accessible to the whole community, and watched the implications of his work change before his eyes; and two theorists whose research on the electron tested the relations between the electromagnetic worldview and relativity. My intent has been to recover facets of practice in both experiment and theory that have typically been neglected in the light of discovery narratives centered on a few major breakthroughs. Remembering these features helps us understand the variegated nature of work in physics, reflected equally in apparatus like parallel(?) plates and rigid(?) rods, and in bridging concepts like absolute space, time, and measurements. Our study of Kaufmann, Ehrenfest, and Born has also revealed important features of the process by which relativity was advanced within Germany, and through which Einstein won his reputation. We have seen that views of Einstein's work ranged from ambivalence in a supporter of relativity to something as strong as hatred in another advocate of the electromagnetic program. Despite this, we have noted that these physicists were strongly aware of Einstein's work, and—especially after the contributions of Minkowski—increasingly distinguished it from Lorentz's approach despite the formal equivalence that experimentalists like Kaufmann noted. We will now explore how common understandings of this field were developed in considerably more detail, by investigating what the earliest—participant—histories tell us about both the development of the new physics and changing understandings of its authorship.

69. This facet of the path to general relativity is described in the works of Pais, Stachel, and Maltese and Orlando referred to in note 48 above.

CHAPTER EIGHT

On the Histories of Relativity

Perspectives on the past always shape understandings of the new. It is true that in 1905 Einstein presented his research on the principle of relativity in as nearly an axiomatic form as possible, and the lack of explicit citations to previous work in his paper "On the Electrodynamics of Moving Bodies" is now a well-worn trope. It might almost seem as if Einstein wanted his contribution to be seen as one without a history. But once it had appeared, other physicists immediately undertook extensive interpretative work to integrate Einstein's research with the earlier work of their colleagues—and soon Einstein did so himself. In this chapter I argue that the accounts of the past, or "research histories," that scientists offer in key papers and review studies play a substantive role in shaping understandings of *new* theory. Leaving our critical probe of the relations between relativity and electron physics to focus more broadly on relativity, I document here the developing sense of that theory as it was conveyed in a series of significant participant histories. My account begins with the brief collocations with which physicists described a field of research as the "Lorentz-Einstein" approach around 1905, and culminates in Max Laue's elaborate textbook representation of relativity as a "historical necessity" in 1911. But I take up physicists' histories in part in order to critique our own historical understanding.

The opening lines of Einstein's 1905 paper observe that according to contemporary theory a well-known asymmetry existed in the kind of explanation given for the phenomena of electrodynamic induction. That the customary approach to the *history* of relativity leads to asymmetries that do not appear to be inherent in the phenomena is not yet well known. Consider the treatment of Einstein's work in comparison to that of his contemporaries, for example. Historians have approached Einstein with a primary interest in the inception of his now famous 1905 study. They have investigated in detail the lines of thought that might have led to the paper but have devoted comparatively little attention to the period immediately afterward, in which Einstein sought to develop and propagate relativity within the physics community. On the other hand, with some important exceptions, Einstein's contemporaries have been of interest very largely to the extent to which their work shows evidence that they understood or misunderstood Einstein's interpretation of relativity.[1] Historians would generally recognize the desirability of being able to explain the character of such responses to relativity on the basis of a deeper appreciation of other physicists' research programs on their own terms, but it has proved surprisingly difficult to follow through such a historiographical program consistently.[2] Rather, studies of other physicists' work have often gone hand in hand with an evaluative contrast with Einstein's approach. Many excellent and detailed studies have taken it for granted that the most important task for a physicist post-1905 was to assimilate Einstein's paper, rather than working with the recognition that all concerned (Einstein included) read and interacted with his work, if at all, with the significantly more open aim of developing physics further—as our study of Kaufmann, Ehrenfest, and Born will have demonstrated. Thus what might be termed inception and reception accounts have dominated historical studies, and unrecognized inconsistencies in historical scrutiny have flowed from this,

1. Stanley Goldberg's and Lewis Pyenson's important studies of Max Planck's reasons for taking up relativity, and his activities in propagating the theory, have long provided the most sensitive and wide-ranging accounts of another physicist's approach to relativity. See Goldberg, "Max Planck's Philosophy of Nature" (1976); Pyenson, "Physical Sense in Relativity" (1985); Pyenson, "The Relativity Revolution" (1987), esp. 96–101. For the 1905 paper, see Einstein, "Elektrodynamik bewegter Körper" (1905).

2. For my own attempt to pursue a program of this kind in relation to Max Born, see Staley, "Max Born and the German Physics Community" (1992). A number of studies by Andrew Warwick, Olivier Darrigol, Shaul Katzir, and Leo Corry have criticized the Einstein-focused methodology of previous work on, respectively, British responses to relativity, Henri Poincaré and German electrodynamics, and Hermann Minkowski. See Warwick, *Masters of Theory* (2003); Darrigol, "Poincaré's Criticism" (1995), on 2; Darrigol, "Electrodynamic Origins" (1996); Katzir, "Poincaré's Relativistic Physics" (2005); Corry, "Minkowski and the Postulate of Relativity" (1997).

obscuring many subtleties of the process by which relativity was developed.[3]

In order to throw light on this complex process of development, this chapter investigates the very first, participant, histories of relativity. Written by physicists in the period in which relativity was actively being elaborated, the "histories" I refer to vary from the conjunction of specific names in abbreviated citations of the theory to extended treatments presented in major research articles. While previous historians have considered some of these resources, discussion of them has often been unduly limited by an implicit inception/reception methodology. But on the whole these texts have drawn very little notice, either as histories or as contributions to the interpretation of relativity. They are significant precisely because they are both; to understand their importance we must first examine how previous historians have analyzed the complex relations between physicists' research and their collective, disciplinary memory.

Discussions of the uses of history in the sciences and related studies on the generation of discovery accounts have demonstrated that scientists' retrospective accounts elide and reinterpret the past in the light of present programs and disciplinary aims.[4] These studies have often had a broad, disciplinary focus and have shown that scientists employ historical accounts to establish a canon and shape the boundaries of a discipline by rereading the past from the end of science.[5] Particularly from the announcement of eclipse observations confirming Einstein's general theory

3. Two examples of such inconsistencies, which I will discuss in detail, concern historians' selective use of the term *Lorentz-Einstein* as an indication of physicists' (mis)understanding of relativity and historians' neglect of Einstein's 1907 review of relativity as a source for understandings of the theory and its past. My argument is not against inception and reception studies as such, but against some of the assumptions that can limit their approach. Given my present focus on histories of relativity alone, this chapter is itself a reception study.

4. On the nature of retrospective reportage and participant histories, see Kuhn, "Revisiting Planck" (1983); Gilbert and Mulkay, "Experiments Are the Key" (1984); Shapin, "Talking History" (1984), (a response to Gilbert and Mulkay). On the important genre of discovery accounts, see Woolgar, "Writing an Intellectual History of Scientific Development" (1976); Brannigan, *The Social Basis of Scientific Discoveries* (1981). For disciplinary histories see, in particular, Graham, Lepenies, and Weingart, eds., *Functions and Uses of Disciplinary Histories* (1983); Shapin, "Discipline and Bounding" (1992); Laudan, "Histories of the Sciences and Their Uses" (1993). For a recent historical study of disciplinary histories of physical chemistry, see Barkan, "A Usable Past" (1992).

5. Peter Galison used this expression to describe the process by which Maxwell's equations have been successively reinterpreted from the point of view of relativity and grand unified field theories. Galison, "Re-Reading the Past from the End of Physics" (1983).

in 1919, relativity (and its histories) came to represent a defining achievement of the new subdiscipline of theoretical physics, which was highly visible both within the physics discipline and to the general public. However, my initial focus is different and more fine grained than that commonly explored in accounts of disciplinary histories. Here I investigate the role of protohistories and historical reasoning in articulating, clarifying, and stabilizing particular understandings of the physical content and broader implications of relativity (and its history) in the period up to 1911, while the theory was under active development, at a time when the concepts and achievements concerned were insecure and highly contested. But my study will suggest many continuities between work of this nature and the place of relativity in the disciplinary profile of theoretical physics and mathematics, especially, for example, in Max Planck's and Hermann Minkowski's treatments of disciplinary politics.

Since innovative work in physics depends on the ability to draw on and depart from the contributions of previous practitioners, and the subject is in a significant sense archival, it is not surprising that various forms of historical reference and reasoning should be important in the practice of physics—whether this is implicit or explicit, carried out consciously or not in different instances.[6] In the following I am largely but not exclusively concerned with what can be described as a genre of "research histories," explicitly historical but highly selective accounts of the emergence and implications of a theory, experiment, or discovery that are presented in major research or research review papers. Other kinds of historical reference to be discussed range from the judgments that may be implied by citing a body of work by specific names (such as "Lorentz-Einstein" or "relativity" theory) through to Minkowski's proposal of a counterfactual history and Max Laue's elaborate use of a concept of "historical necessity" in the first textbook on relativity in 1911. I will not attempt to delineate all the forms of historical reasoning active in physics, but it should be stated that the techniques I point out here are far from the only ways physicists work with history.[7] The significance of textual material of this

6. For a study that pays attention to a form of historical discussion in physics, see Darrigol's treatment of Bohr's use of an analogy with classical physics in the development of quantum theory: Darrigol, *From c-Numbers to q-Numbers* (1992). It is possible that the work discussed here provided some kind of model for Bohr's research through its establishment of a productive contrast between classical and relativistic physics.

7. Without attempting an exhaustive account Collins and Pinch have helpfully described six different types of enterprise in the history of science: textbook history, official history,

nature has often escaped the attention of historians of relativity owing to the rather narrow focus on novel achievement and correct (Einsteinian) interpretation in many inception/reception accounts. Here I will not take it for granted that the proper interpretation—or a fully constituted theory of relativity—can be taken to inhere essentially and self-evidently in Einstein's 1905 paper alone. Rather, building on our study of the relations between relativity and the electron in chapters 6 and 7, and considering a period in which research in the field was both controversial and uncertain in its implications and scope, this study will provide a means of charting more precisely the gradual process through which Einstein's work in particular came to be regarded as distinctively foundational of an emerging new theory, and relativity eclipsed electron theory as a major framework for theoretical development.

My argument is that numerous historical resources, now largely forgotten, constituted an important component of the work involved in forming, clarifying, and propagating particular interpretations of relativity in the German physics community. My approach here has many similarities with insightful sociological studies of the kind of interpretative work involved in reading data at the laboratory bench. It stems from the view that physicists' approaches to new theory are strongly formed through drawing contrasts and comparisons with related and historically antecedent work and that these comparisons, and the understandings that result, are often expressed in narrative form in research histories. Accordingly, I explore the sense in which the present is shaped in the light of the past at least as much as the past is rewritten in the light of the present. The protohistories I discuss were not the only factor shaping the development of theoretical understanding; but in the historiographical task of widening our accounts of theoretical physics as a practice, they should certainly be considered alongside both the new applications and conceptual extensions of a theory most commonly discussed by historians and the routine calculational work consolidating theoretical technologies that has been highlighted by Andrew Warwick's studies of British electrodynamics in this period.[8]

---

reviewers' history, reflective history, analytic history, and interpretative history. Scientists contribute to each of the first four categories, while the latter two involve authors rendering science for other disciplines or attempting to put readers in the shoes of scientists of the period. Collins and Pinch, *The Golem*, 2nd ed. (1998), 165–67.

8. See Warwick, "Cambridge Mathematics and Cavendish Physics, Part 1" (1992); Warwick, "Cambridge Mathematics and Cavendish Physics, Part 2" (1993); Warwick, *Masters of*

Without attempting to define the precise role such histories play, this chapter will argue that they provide a form of conceptual orientation integral to research practices and the articulation of meaning in physics. What I mean by this is best conveyed through the early accounts I consider from Kaufmann, Einstein, and Minkowski. In each case the author discusses the physical content of a common set of equations (the Lorentz transformations) in association with evaluations of the significance, achievements, and broader implications of recent work from Lorentz and Einstein. Thus, most often through contrast and comparison, the authors assert particular relations between the complex treatment of equations, calculations, and the like in the original papers and meaning.[9] Their judgments of meaning and significance offer a selective guide through recent (and controversial) physics, ranging from issues to do with the interpretation of specific equations through to broad implications (such as whether relativity requires the existence of the ether). That is, they discuss a spectrum of interrelated levels of meaning associated with relativity, raising just those questions that historians have regarded as crucial for the formation of a clear understanding of the emerging theory and, especially, of the differences between Einstein's and other work. Our primary concern will be to examine this interpretative discussion in its own right, but it should be clear that it is rather closely related to the research practices it stems from and refers

---

Theory (2003). For sociological studies, see, e.g., Woolgar, "Writing an Intellectual History of Scientific Development" (1976); Lynch, *Art and Artifact in Laboratory Science* (1985).

9. While it does not play a large role in this chapter, a (flexible) distinction of some kind between practices and meaning or beliefs (here described variously as interpretations, understandings, or readings) is one that has particular point in the case of relativity, where physicists such as Lorentz and Einstein ascribed different interpretations to sets of equations that were regarded as formally equivalent and also developed distinctively different approaches to the solution of problems (thus exhibiting a complex relation between beliefs, formalism, and practices). For an example of one physicist's approach to these issues, discussed below, see Lorentz's 1910 distinction between the "epistemological" and "physical" sides to relativity, in which he regarded himself as differing from Einstein and Minkowski and as having offered an equivalent system. My use of the terms *interpretation*, *understanding*, and *reading* incorporates but is somewhat wider than the epistemological dimension Lorentz referred to. With *interpretation* I refer mostly to views on physical content and physical implications, while I sometimes use *understanding* or *reading* in a broader sense to also include assessments of the relations between particular approaches and other areas of physics such as electron theory and views on the disciplinary implications of approaches. What I have in mind as practices are more differentiated subsets of what Lorentz describes as the "physical" side of relativity, in order to take account of the level of techniques and procedures at which Lorentz and others recognized that, despite a formal equivalence, Lorentz's, Einstein's, and Minkowski's approaches each involved significantly different ways of solving problems.

to in different instances.[10] While a central part of Kaufmann's discussion is framed solely in terms of a conceptual contrast, the accounts of Einstein and Minkowski give a narrative and historical cast to the comparisons at issue and will thereby help me demonstrate the distinctive nature of research *histories*. All share important features but also convey somewhat different views of relativity and its implications; such accounts certainly served pedagogical purposes, but in addition to recording they helped to shape the landscape they purported to represent.

My central concern is therefore with the way participant histories reflected and shaped interpretations of the content and implications of relativity. But they also informed understandings of its history in important ways. I will argue here that Einstein, in the very first history of relativity published, was responsible for what has become known as the "myth" of the centrality of the Michelson-Morley experiment in the development of relativity, and that these participant accounts disclose a tension between viewing relativity as the product of a community and sharply individualized attributions of discovery—a tension that still runs through present historiography.

In summary, this chapter will highlight the significance of participant histories as one among several means of developing a new set of practices and understandings. Recovering this dimension of research in physics by investigating the changing nature of protagonist accounts from 1905 will show a gradual elaboration of the kinds of historical resources deployed and help me to characterize more closely the different ways that history was used in scientific research. But the accounts we consider—from Einstein, Kaufmann, Planck, Minkowski, Bucherer, Lorentz, Sommerfeld, and Laue—will also provide a means of surveying the development of relativity. These sources confirm in outline the narrative established by previous historians, which sees the German physics community as having assimilated relativity by about 1911 after an initial plurality of approaches and mixed acceptance and rejection of Einstein's 1905 contribution. Readers will have seen the skeleton of this narrative emerging in chapters 6 and 7. There, moving between experiment and theory while focusing on the electron brought to light the connective tissue of

10. Two examples of historical accounts that reflect specific research concerns include Minkowski's emphasis on a mathematical pathway and Planck's 1910 account relating relativity to the principle of least action. The penultimate section will also suggest two instances, in the work of Max Laue and Max Born, in which particular historical (and conceptual) understandings of relativity shaped the formation of research goals and strategies.

long-term concerns and the unexpected play of common themes in different contexts—recovering facets of relativity physics that have usually remained just below the surface of historians' accounts of relativity as theory. Similarly, by suspending evaluative comparisons with Einstein's work and following more closely the character and dynamic interactions of different contributions, the fine-grained chronology of my approach in the present chapter will offer a new perspective—one that reveals unexpected features of the relationships among the work of Lorentz, Einstein, and Minkowski (e.g., the fact that Einstein initially stressed continuities between his work and Lorentz's, while Minkowski and Lorentz emphasized distinctions). This study thereby provides a means of tracing the complex process through which in Germany a plurality of approaches—many relativities with many histories—could become singular—one theory, one history—and through which the work of Einstein came to be sharply distinguished from that of others. It explores a productive tension between individualized interpretations and generalized, representative voices in the work of different physicists. In addition, the material discussed will draw out pedagogical dimensions of Einstein's scientific work post-1905 that have commonly been neglected in earlier studies. Finally, I suggest that in these writings we see the hitherto-unrecognized sources of historical accounts still current today; these participant accounts, although often brief and histories only in outline, are also protean in the sense of providing the original instances of a history of relativity.

"The Theory of Lorentz and Einstein"

Einstein's 1905 paper has often drawn comment for the paucity of its references and acknowledgments.[11] While presenting a new approach to electrodynamics and uniting his work with the principal outlines of the electron theory of H. A. Lorentz, Einstein engages in very little explicit discussion of the relation his work bears either to previous contributions or to alternative approaches. The paper presents its particular understanding of the transformation equations Einstein derived (later named the "Lorentz" transformations) primarily through outlining simple

11. See, e.g., Gerald Holton's discussion of the voices of different physicists embedded in Einstein's text; these are often represented without being named but are nevertheless partially recognizable. Holton, "Quanta, Relativity, and Rhetoric" (1993), on 90–93, esp. 91–92.

empirical relationships and establishing an instrumental-philosophical discussion of the measurement of time and distance, without explicitly situating itself within a lineage of theoretical development. Thus Einstein implies a historical relation to earlier studies only through general references to a number of sharp contrasts with contemporary theory (in relation to conceptions of induction and the ether, for example) and a number of points of concordance (such as stating the agreement of the electrodynamic foundation of Lorentz's theory of the electrodynamics of moving bodies with the principle of relativity).

Perhaps partly in response to this situation, it has been a major concern of historians and philosophers of science to reconstruct a history for the paper: to establish the precise material Einstein read, to conjecture as to the most likely sources for particular insights or approaches employed, and, particularly, to distinguish Einstein's approach from the earlier work of August Föppl, Ernst Mach, Lorentz, and Henri Poincaré.[12] Because of the centrality of the Michelson-Morley experiment in participant accounts discussed here, it is important to state at this point that historical studies of this nature have taken it as a major issue to understand whether Einstein drew on the Michelson-Morley ether-drift experiment. After considerable careful scholarship investigating the prehistory of the 1905 paper and scrutinizing Einstein's testimony from later years, Gerald Holton established that while Einstein may have known of the experiment in general terms, he was not particularly influenced by it; he had ample reasons for doubting the existence of an ether wind in 1905. Arguing this

12. Miller's study, *Einstein's Special Theory of Relativity*, presents the most detailed account of this kind. See also the critical apparatus of *The Collected Papers of Albert Einstein* and, in particular, both the editorial note "Einstein on the Theory of Relativity" and footnotes provided in association with the reprinting of the 1905 paper, in Einstein, *Collected Papers*, vol. 2 (1989), 253–310. Important earlier studies include Holton, *Thematic Origins* (1988); Goldberg, "The Lorentz Theory of Electrons and Einstein's Theory of Relativity" (1969); Goldberg, "Poincaré's Silence and Einstein's Relativity: The Role of Theory and Experiment in Poincaré's Physics" (1970); Goldberg, "Henri Poincaré and Einstein's Theory of Relativity" (1979); Hirosige, "Origins of Lorentz' Theory of Electrons and the Concept of the Electromagnetic Field" (1969); Hirosige, "The Ether Problem" (1976); McCormmach, "Lorentz and the Electromagnetic View of Nature" (1970); McCormmach, "Einstein, Lorentz, and the Electron Theory" (1970). Darrigol maintains that previous work in this field has been teleologically concerned with Einstein. See Darrigol, "Electrodynamic Origins" (1996). More recent scholarship has focused on exploring the interrelations between Einstein's different concerns. See especially Howard and Stachel, eds., *Einstein: The Formative Years* (2000); Norton, "Einstein's Investigations" (2004); Renn, "Einstein's Invention of Brownian Motion" (2005); Rynasiewicz and Renn, "Einstein's *Annus mirabilis*" (2006).

line quite strongly, historians have then proposed that the appearance to the contrary—that the Michelson-Morley experiment was crucial—has been promoted by the pedagogical needs of modern textbooks. Arthur Miller suggested as a source for this myth a 1913 edition of various original articles on relativity, which printed Lorentz's discussion of the ether-drift experiment in conjunction with Einstein's 1905 paper.[13] We might conjecture that one reason for such insistence on the distinction between Einstein's actual path and the route to relativity often described has been the wish strongly to distinguish the understanding gained through professional history of science from the accounts of physicists, textbooks, and popular histories of relativity. Through investigating the way contemporaries used the Michelson-Morley experiment from 1905, rather than focusing on the inception of Einstein's work, I will suggest quite a different lineage for the incorporation of the experiment in histories of relativity.

Given the paucity of Einstein's explicit references to other work in his 1905 paper—and possibly in response to this—it is interesting to note that the first major paper to comment on it did a considerable amount to locate Einstein's work very precisely within contemporary electrodynamics. As we saw in chapter 6, early in 1906 Walter Kaufmann published a lengthy review of his experiments testing current theories of the electron, and announced that while his results supported the different theories of Max Abraham and Alfred Bucherer, they were not compatible with the Lorentz-Einstein approach. Kaufmann also disagreed with that approach on theoretical grounds because it did not comply with the electromagnetic worldview.[14] Nevertheless, in the context of an overview of the history (and current conceptions) of the electron, he outlined the historical evolution of Lorentz's theory with some care and discussed its relation to Einstein's more recent contribution. Kaufmann pointed out that Lorentz's 1895 electron theory had expected an influence of the second order from the earth's motion on certain optical and electromagnetic phenomena but that all experiments to confirm such an influence had yielded a negative

13. See esp. Holton, "Einstein, Michelson, and the 'Crucial' Experiment" (1988), and the postscript on 477–80; Miller, *Einstein's Special Theory* (1981), 391–92, referring to the papers published in Lorentz et al., *Das Relativitätsprinzip*, 7th ed. (1974 [1913]). This edition has notes by Arnold Sommerfeld and a foreword by Otto Blumenthal. An English version was published as Lorentz et al., *The Principle of Relativity* (n.d.).

14. Kaufmann, "Konstitution des Elektrons" (1906); Kaufmann, "Nachtrag" (1906). As we saw in chapter 6, Kaufmann cited Max Abraham's argument that Lorentz's theory required the assumption of a nonelectromagnetic internal potential energy in order to comply with the energy law. See Kaufmann, "Konstitution des Elektrons" (1906), 493–94.

result. (Here Kaufmann did not cite any specific experiment but referred to discussions in Lorentz's papers.) He reported that Lorentz's important 1904 paper, "On Electromagnetic Phenomena in a System Moving with Any Velocity Less Than That of Light," had then attempted to cope with these difficulties by modifying certain fundamental assumptions about the electron and the molecular forces operating between material particles. We saw previously that Kaufmann reinterpreted Lorentz's paper in terms of distinctions central to both his experimental practice and the writings of Poincaré and Einstein. He wrote that in Lorentz's new electrodynamics, "absolute velocities" appeared as factors in calculations while the final result depended only on directly observable "relative velocities" (a feature Kaufmann thought raised epistemological difficulties); and the dimensions of all physical bodies, including their particular molecules and electrons, changed in a definite way with velocity. Further, molecular forces and the "masses" of mechanics changed in just the same way with velocity as electrostatic forces and the electromagnetic mass. Kaufmann described Einstein as having arrived at equivalent formulations by the quite different approach of putting the principle of relativity at the summit of physics. He explained that Einstein's approach involved a new definition of time and of the concept of simultaneity for spatially separated points, which involved considerations formally identical with the "local time" Lorentz introduced. While Lorentz's approach could not exclude the possibility of achieving the requisite results by other means, Einstein showed that the kinematics of the rigid body and Lorentzian electrodynamics followed necessarily from according a central role to the principle of relative motion.[15] Thus Kaufmann presented Einstein's work as being formally and observationally equivalent but conceptually quite distinct from Lorentz's. The considerable interpretative work he undertook shows that Kaufmann felt it necessary to provide a closer characterization of the relationship between Einstein's paper and contemporary theory than Einstein had given, especially in regard to a study Einstein had not considered (or indeed read at that time)—Lorentz's 1904 paper.[16] In the process he helped render

15. Kaufmann, "Konstitution des Elektrons" (1906), 490–91 (quotation), 491–93. He is discussing Lorentz, "Electromagnetic Phenomena" (1937 [1904]) (a reprint in translation of a paper published in Dutch).

16. The themes of equivalence and conceptual distinction would both be important in later discussions of relativity, in common with other episodes in the history of science and mathematics. See, e.g., Sigurdsson, "Equivalence, Pragmatic Platonism, and Discovery of Calculus" (1992). A footnote added to the reprint of Einstein's "On the Electrodynamics of

them part of a common endeavor stressing the principle of relativity and distinctions between absolute and relative measures.

As we saw in chapter 6, Kaufmann's article on electron theory attracted a great deal of attention and sparked vociferous public debate in the journals and conference halls of Germany. It was therefore extremely important for the fortunes of relativity that in two papers, one of which was presented to the Naturforscherversammlung in Stuttgart in 1906, Max Planck countered Kaufmann's experimental objections by openly supporting Einstein's approach and arguing that the experiments were not sufficiently precise to distinguish between the different electron theories on offer. Enthusiastic but caught between appreciation and a wish for caution, Planck wrote of relativity as a physical idea of the greatest simplicity and generality that deserved to be discussed from many points of view and tested to the limit of the *ad absurdum* to see whether it would withstand serious scrutiny. As we know, in addition to discussing the basic equations of mechanics appropriate to relativity, Planck went into exhaustive detail in a critical study of Kaufmann's experiment; but he did not go as far as Kaufmann had in analyzing the relationship between Lorentz and Einstein; rather, he simply described the principle of relativity as having been recently introduced by Lorentz and, in a still more general version, by Einstein.[17]

Planck and other physicists were soon to collapse this description even further and refer to the Lorentz-Einstein theory, the first (potential) protohistory I wish to discuss. Many historians of science have noted this conjunction of names, taking it as extremely important evidence of physicists' attitudes toward relativity. Russell McCormmach comments, for example, "For several years after 1905 physicists seldom distinguished

---

Moving Bodies," in Lorentz et al., *Das Relativitätsprinzip*, 7th ed. (1974 [1913]), 26 n. 2, states that this work was not known to Einstein before he wrote the 1905 paper. The importance of Lorentz's paper in particular makes the restriction of this study to works published after 1905 somewhat artificial. However, it will suffice to observe that Lorentz began this paper with a discussion of the Michelson-Morley and Trouton-Noble (1903) second-order experiments, and that he builds his argument by entering into explicit dialogue with the earlier work of a number of physicists and mathematicians, including his own and that of Poincaré. In this sense its presentation is far more clearly historical in nature than that of Einstein's 1905 paper.

17. Planck, "Prinzip der Relativität" (1906), esp. 136, 137, and Planck, "Kaufmannschen Messungen" (1906) (the discussion following Planck's paper to the Naturforscherversammlung is reported on 759–61). The essays by Goldberg and Pyenson cited in note 1 above discuss Planck's reasons for accepting relativity and his role in the development of the theory.

between Einstein's and Lorentz's formulations of the electron theory. They referred interchangeably to the 'Lorentz,' the 'Lorentz-Einstein' and the relativity theories." Similarly, Gerald Holton has written that "basic differences in the underlying world pictures [of Lorentz's and Einstein's work] were slow to be recognized and the long persistence of the term 'Lorentz-Einstein' was an indicator of it."[18]

Thus historians (taking an inception/reception approach to the study of relativity) have regarded the interchangeable use of these terms as indicating a misunderstanding of the essential nature of Einstein's work, a lack of appreciation of the full distinctions between Lorentz's dynamical theory and the kinematic nature of Einstein's approach. While it is certainly important to take the way physicists refer to theories seriously, there are several reasons for exercising caution in interpreting this conjunction of two names. The first general reason is that without more information about the way a physicist worked with relativity, any single interpretation of his attitude is highly underdetermined. Certainly those using such references were not primarily concerned with delineating distinctions, and this may be the most important point behind historians' discussions of the practice. However, while the use of these terms could indicate that contemporaries mistakenly held the work of Lorentz and Einstein to be basically similar, it could equally well reflect a number of other possible positions (which are not mutually exclusive). For example, the conjunction might be used in a primarily historical sense or to stress the perceived formal and observational equivalence of the two approaches, or it could reflect (albeit only schematically) the fact that a physicist actually endorsed a mixture of the approach of both physicists.[19]

Consider Planck's use of such an abbreviated reference. He initially did so in response to Kaufmann's discussion. In the context of his critical studies, Planck's silence on the detailed relationship between Lorentz and Einstein could be taken to imply his assent to the conceptual distinctions—and observational equivalence—Kaufmann had identified. This is an instance where it might be profitable to recognize the nature of a

18. McCormmach, "Lorentz and the Electromagnetic View of Nature" (1970), 489, and Holton, "Quanta, Relativity, and Rhetoric" (1993), 93.

19. Darrigol suggests the possibility that German physicists took a selective ("active, filtering") approach in their reading of Einstein's work in Darrigol, "Electrodynamic Origins" (1996), 311–12. The point has been strongly argued in regard to British work in Warwick, *Masters of Theory* (2003) and in relation to German physics between 1905 and 1914 in Staley, "Max Born and the German Physics Community" (1992).

physicist's discourse as a contribution within the archive of an ongoing dialogue, written with an awareness of relationships to other contributions. But there are also reasons to believe that while Planck appreciated the simplicity and generality of Einstein's approach, he may have been wary of some of the implications Einstein drew, and hence of endorsing his work wholly or exclusively at this early stage.[20] We know that shortly after this time Planck saw electron theory and an investigation of the relationship between the ether and matter (in the context of the absorption and emission of radiation within the blackbody) to hold the most fruitful prospects for research in quantum theory, an attitude reflected in his correspondence with Einstein, Ehrenfest, and Lorentz and on which Planck's views were in conflict with Einstein's emerging publications.[21] In 1909 Planck argued strongly and publicly against Einstein's suggestion of applying quantum theory to light, advocating instead that the application of energy quantization be restricted to the oscillating resonators of his blackbody theory, without giving up Maxwell's equations in the way Einstein advocated.[22] Holding views of this kind, Planck may have understood Einstein's work on relativity quite clearly in 1906 but either not accepted or been uneasy with his explicit disavowal of the ether as carrier of the electromagnetic forces. Therefore, Planck's references to both Lorentz and Einstein could reflect a consciously selective approach, endorsing a mixture of different features of the work of each. In support of this possibility, it is worth noting that while others (such as Max Laue, Jakob Laub, and Alfred Bucherer) commented favorably on the implications Einstein drew against the ether, Planck did not raise this topic in his early papers. Further, in the very letter to Einstein in which Planck discussed the substantial differences in their approach to quantum theory,

20. Note that in 1908 Planck referred to "Einstein's theory of relativity" in the singular. Planck, "Prinzip der Aktion und Reaktion" (1908), on 729. But see my discussion of the inclusive nature of his 1910 historical account of relativity, below.

21. For Planck's correspondence with Einstein on this issue, see Planck to Einstein, 6 July 1907, in Einstein, *Collected Papers*, vol. 5 (1993), 49–50. This letter is quoted in more detail below (see note 39). On the correspondence between Planck and Ehrenfest, see Kuhn, *Black-Body Theory* (1987), 131–34; on that between Planck and Lorentz see Goldberg, "Max Planck's Philosophy of Nature" (1976), 155–57. On the connections between views on quantum theory, the theory of the electron, and relativity, see Staley, "Max Born and the German Physics Community" (1992), 161–62, 217–21.

22. Planck repeated substantially the same views outlined in his letter of 6 July 1907 in the public discussion of Einstein, "Das Wesen und die Konstitution der Strahlung" (1909), on 825.

he also stated his opinion that at a time when the supporters of relativity were few in number it was extremely important for them to be in agreement with one another. This view might have led him to silence or to express only implicitly his own differences from Einstein in regard to the broader implications of relativity.

These considerations underline the caution required in interpreting the use of such brief references, which could reflect a range of judgments about the conceptual and historical relations between the work in question. However, there is a more specific reason to question whether the close conjunction of Lorentz's and Einstein's names necessarily indicates a basic misunderstanding of Einstein—for Einstein himself employs just this association.

From 1905, Einstein supplemented and developed further his approach to relativity both theoretically (exploring in particular the extremely general relation he soon identified between inertial mass and energy) and with an eye toward experiment, proposing two experiments that might be capable of testing his work. One exploited a new arrangement to determine the ratio between the transverse and longitudinal mass of the electron; the other proposed that the line spectra of canal rays be considered as a fast-moving clock. With these papers Einstein addressed concerns central to experimentalists such as Kaufmann, Bucherer, and Adolf Bestelmeyer, who as we have seen, were deeply engaged in investigating the different electron theories. If Einstein had felt misrepresented or misunderstood by those who linked his work with Lorentz's, he could have sought to correct their views in this forum. In neither case, however, nor even in discussing electron theory, did Einstein draw any explicit distinction between his work and Lorentz's. Rather, in the first of these papers he compared without further comment the values for the ratio of transverse to longitudinal mass predicted by the theories of Bucherer and of Abraham and "the theory of Lorentz and Einstein."[23]

Regarding both the use of the singular and of Einstein's own name in this text as being "quite uncharacteristic," Holton has suggested that "for this brief piece only, Einstein adopted or aped Kaufmann's terminology, and probably did so with tongue in cheek."[24] However, there are independent reasons for thinking that Einstein was actually quite happy—at

23. Einstein, "Bestimmung des Verhältnisses der transversalen und longitudinalen Masse des Elektrons" (1906), on 586; Einstein, "Neuen Prüfung des Relativitätsprinzips" (1907).

24. Holton, "Quanta, Relativity, and Rhetoric" (1993), 107–8 n. 49.

least initially—for his work to be closely associated with Lorentz's, and that the lack of an explicit distinction, except in regard to the ether, was in fact quite general in Einstein's writing at this time.

Einstein's History

In 1907 the experimentalist Johannes Stark asked Einstein (fig. 6.1) to prepare a survey article on relativity for his journal, the *Jahrbuch der Radioaktivität und Elektronik*. The invitation indicates that despite its uncertain status relativity (and Einstein) had attracted significant attention; it may also have reflected Stark's interest in Einstein's controversial views on quantum theory. It gave Einstein (rather than, say, Lorentz or Planck) the extremely valuable opportunity of summarizing the achievements of relativity, and that task also brought several papers to his attention for the first time, among them Planck's detailed discussion of Kaufmann's findings.[25] The article now draws comment principally for the attitude Einstein reveals to Kaufmann's experiments and for the major new insight developed in its concluding sections—Einstein's first extension of relativity to encompass gravitation. But for many contemporaries this paper is likely to have provided an approach to relativity comparable in importance to the original 1905 paper. In correspondence with Stark, Einstein stated that he had written in such a way that readers could familiarize themselves with relativity and its applications "relatively easily." He indicated that he took "great care in explaining the assumptions used," introducing them one by one and tracing their consequences in the same order, ascribing more importance to the "intuitiveness and simplicity of the mathematical developments than to the unity of presentation."[26] Indicating a particular kind of pedagogical care, these comments may also represent

25. Stark was one of the few physicists to take Einstein's work on quantum theory seriously at this time, publishing in December 1907 a paper in which the quantum hypothesis was used to explain the Doppler effect of light emitted by canal rays. See the correspondence between Stark and Einstein regarding the survey article, which indicates that at the time he accepted the task Einstein knew of Lorentz's 1904 paper, papers by Emil Cohn and Kurd von Mosengeil, and two of Planck's publications—but of no other theoretical works related to the subject: Einstein to Stark, 25 September 1907, and Stark to Einstein, 4 October 1907, in Einstein, *Collected Papers*, vol. 5 (1993), 74–75, 76.

26. Einstein to Stark, 1 November 1907; ibid., 77–78. The paper itself is Einstein, "Relativitätsprinzip" (1907). For Einstein's correspondence relating to requests for reprints of or comments on this paper, see, e.g., Einstein to Sommerfeld, 5 January 1908; Laub to Einstein,

FIGURE 8.1 Albert Einstein, architect of relativity and author of its first "history" (in 1907), in the Patent Office in Bern, circa 1905. Hebrew University of Jerusalem, Albert Einstein Archives. Courtesy AIP Emilio Segrè Visual Archives.

an implicit comparison with the style in which Einstein had presented his work in 1905, where perhaps he strove after "unity of presentation."

In addition to paying close pedagogical attention to the assumptions employed, Einstein began his 1907 paper by giving relativity a history. This discussion has drawn surprisingly little attention from later historians

---

2 February 1908; Arthur Schoenflies to Einstein, 15 January 1909 (in which Schoenflies indicates that he had a particular wish to have his own copies of Einstein's papers, especially this one); and George Searle to Einstein, 20 May 1909 (thanking Einstein for sending a copy of his paper on the principle of relativity at the request of Bucherer), in Einstein, *Collected Papers*, vol. 5 (1993), 85–86, 94–95, 153, 190–91.

of science,[27] although we shall see some evidence that contemporaries read and utilized Einstein's account.

Einstein opened by presenting the transformation equations that preserve Newton's equations of motion under a uniform translation, referring to this independence from the state of motion of the system of coordinates used as "the principle of relativity." He then wrote that it appeared these transformation equations had suddenly been called into question by the brilliant confirmations of Lorentz's electrodynamics of moving bodies, whose basic equations were not preserved under them. Lorentz's theory, however, had had to be modified, through the introduction of the contraction hypothesis, in order to accommodate the negative result of the Michelson-Morley experiment. Einstein wrote that this ad hoc hypothesis seemed to be only an artificial means of saving the theory, and that it therefore seemed as if Lorentz's theory should be abandoned, to be replaced by one whose foundations corresponded to the principle of relativity. "Surprisingly, however, it turned out that a sufficiently sharpened conception of time was all that was needed to overcome the difficulty discussed." It only needed to be recognized that Lorentz's "local time" could be defined as "time" in general—the Lorentz-Fitzgerald hypothesis then appeared as a compelling consequence of the theory, and only the concept of the luminiferous ether as the carrier of the electric and magnetic forces failed to fit into the theory described. Finally, Einstein concluded this historical introduction by describing his review article as "an attempt to summarize the studies which have resulted to date from the union of the H. A. Lorentz theory and the principle of relativity."[28]

27. In his study of Einstein and Michelson, Holton refers to this paper but does not consider it in detail; he regards the account Einstein gives as implicit history, in contrast to the explicit histories that Holton uses (together with evidence from the prehistory to the 1905 paper) to evaluate whether the Michelson-Morley experiment was important in the genesis of relativity. See Holton, "Einstein, Michelson, and the 'Crucial' Experiment" (1988), 352 n. 29. It seems to be the fact that the account is followed by a research review (in which Einstein's concern is to draw the results of previous studies into a coherent whole) that leads Holton to describe it as implicit history. This feature increases its interest for my study of the relations between history and research and, in my view, does little to diminish its importance as a history. I also differ from Holton in that I regard Einstein's discussion as an explicit history of Lorentz's theory and the principle of relativity, which also gives (in a number of brief and ambiguous phrases) some implicit indications of Einstein's own role and the path he took to relativity.

28. Einstein, "Relativitätsprinzip" (1907), 411–12, 413. His initial references are to Lorentz, *Versuch* (1906 [1895]). Einstein cited Lorentz's 1904 paper and his own work as foundations for his 1907 treatment.

A number of things are worthy of note here, and a comparison with Kaufmann's account discloses particular characteristics of Einstein's discussion. First, at this stage Einstein's presentation stresses similarities between his work and Lorentz's rather than differences, and perhaps more strongly than Kaufmann's paper, with its discussion of the epistemological difficulties Einstein's approach avoided. Given both the recognized success of Lorentz's theory (particularly in regard to optical phenomena, though not in regard to the mass/velocity relations of electron theory) and Lorentz's status within the physics community, it should not be surprising that the young Einstein was ready to encourage perceptions of the links between his work and that of Lorentz.

Further, the relationship Einstein depicts is not only one of conceptual similarity but also a genetic one. For example, while Kaufmann's paper had pointed out a conceptual relationship between "local time," as formulated by Lorentz and Einstein's relativity, in Einstein's account this becomes an important causal element in the development of the new approach: "it only needed to be recognized..."[29] Differences of this nature make Einstein's account quite emphatically a history rather than simply a descriptive account of the conceptual relations between two different versions of one approach, as Kaufmann had it. The historical mode allows Einstein to suggest a close and genetic relationship and also clearly implies that the second formulation is not only subsequent but actually superior, a development of the first: Einstein pictures the Lorentz theory being *united* with the relativity principle to form a higher, single theory. Notice, too, that beginning with a discussion of what soon came to be called the Galilean transformation equations, Einstein casts the history of Lorentz's theory in relation to the principle of relativity: the principle had been challenged by Lorentz's theory; the theory might have to be abandoned in favor of one according with the principle; but finally the new transformation equations show its compatibility with the principle. This constitutes a retelling of the development of Lorentz's theory in the light of Einstein's own emphasis on the principle of relativity, just as Peter Galison has shown that Maxwell's equations have been reread or reinterpreted from the vantage of successive theoretical perspectives such as relativity and grand unified theories. Poincaré had previously

29. "Es bedurfte nur der Erkenntnis, daß man eine von H. A. Lorentz engeführte Hilfsgröße, welche er 'Ortszeit' nannte, als 'Zeit' schlechthin definieren kann." Einstein, "Relativitätsprinzip" (1907), 413.

discussed the relationship between Lorentz's theory and the principle of relativity (and interpreted Lorentz's local time as the apparent time for moving observers), and he is perhaps the only physicist apart from Einstein who could have given a reading of the development of Lorentz's approach similar in kind, though he was particularly concerned with the relationship between Lorentz's theory and a group of principles, especially the principle of reaction. (Indeed, the centrality of Poincaré's interactions with Lorentz and his early formulation of concepts important in relativity make his absence from Einstein's and many other German accounts a somewhat puzzling and interesting omission.)[30] In contrast to Einstein's retelling, Kaufmann's brief account had depicted Lorentz's theory as evolving in response to empirical challenges alone.

In describing these empirical challenges in 1907, Einstein was prepared to be more colorful (borrowing the language of "ad hoc" change that both Poincaré and Lorentz had used to describe the character of Lorentz's responses) and more specific. In contrast to both his own general reference to the failure of unspecified attempts to measure the velocity of the earth relative to the ether in 1905 and Kaufmann's footnote referring to Lorentz's discussions, in 1907 Einstein assigns the Michelson-Morley experiment—by name—a central place in the development of the theory.[31]

Given that Einstein framed his discussion in terms of a theoretical principle, it would be an overstatement to see this reference as subsuming the story in the kind of "empiricist" repertoire G. Nigel Gilbert and Michael Mulkay have identified as a major genre in participant accounts of the historical development of research areas. Indeed, the emphasis on the central role of theoretical insight common in participants' and historians'

30. Galison, "Re-Reading the Past from the End of Physics" (1983). Darrigol has discussed the principle of relativity in Poincaré's work, and Katzir argues that it was central to his endeavors. Darrigol, "Poincaré's Criticism" (1995), 16–17; Darrigol, "Einstein-Poincaré" (2004); Katzir, "Poincaré's Relativistic Physics" (2005). For a discussion of Minkowski's omission of any reference to Poincaré in his "Space and Time" lecture, see Walter, "Minkowski, Mathematicians, and the Mathematical Theory of Relativity" (1999).

31. Einstein, "Relativitätsprinzip" (1907), 412–13. John Stachel discusses Einstein's reference to the Michelson-Morley experiment in the context of justification in order to surmise its role in the context of discovery, pointing out that the experiment was commonly cited as evidence for the principle of relativity (not the principle of the constancy of the velocity of light) and arguing that once Einstein took the principle of relativity to be valid for all phenomena (and particularly electrodynamic and not just mechanical phenomena), it was a reformulation of the concept of time that enabled him to see a way of squaring the principle with all of electrodynamics. This account of Einstein's own path accords well with the temporal order of his 1907 historical overview. See Stachel, "Einstein and Michelson" (2002).

accounts of relativity, quantum theory, and, especially, Einstein's approach to physics could well license the recognition of a "theoreticist" repertoire in which theoretical insight initiates a decisive resolution or essentially clarifies a choice between different interpretations, where experiment is depicted as having no clear import or even apparently speaks against the insight concerned.[32] Nevertheless, Einstein's 1907 reference draws on an empiricist genre and constitutes only the first of many occasions on which an account of the history of relativity—incorporating the early development of Lorentz's theory—was to do so. Further, Einstein's account is a more likely original source for the view that the Michelson-Morley experiment was a prerequisite for his publication on special relativity than the particular conjunction of papers published in the 1913 collection of important articles on relativity, *The Principle of Relativity*. Edited by the mathematician Otto Blumenthal, with additional comments from Arnold Sommerfeld, that publication could well have drawn on Einstein's 1907 narrative.[33]

If other historians are correct in their view that Einstein owed little to the Michelson-Morley experiment in the formulation of relativity, we have to ask what he is doing citing it. There are good reasons, of a pedagogical and justificatory nature, for Einstein to have constructed a route

32. Gilbert and Mulkay contrast empiricist genres locating the driving force of scientific change in the compelling findings of specific experiments with contingent repertoires, which focus on social and personal bases for scientific action and belief. See Gilbert and Mulkay, "Experiments Are the Key" (1984). For a sensitive study of Einstein's relation to experiment that argues against the theoreticist orientation common to many accounts, see Hentschel, "Einstein's Attitude towards Experiments" (1992). In a study of Einstein's changing philosophical orientation, Holton cites Einstein's discussion of Kaufmann's experiment in the 1907 paper (see below) as early evidence of Einstein's hardening against the epistemological priority of experiment characteristic of his initial approach to physics, on a journey toward a rational realism for which his experience in the development of general relativity was central. See Holton, "Mach, Einstein, and the Search for Reality" (1988), (on Kaufmann, see 252–54).

33. Lorentz et al., *Das Relativitätsprinzip*, 7th ed. (1974 [1913]). The chronology of development suggested by the extracts reflects the importance of the Michelson-Morley experiment as portrayed by Einstein's narrative, while the editorial footnotes (added with the consultation of the different authors concerned, with Sommerfeld responsible for those on Minkowski's "Space and Time" lecture) reflect a view of the relations between the work of Lorentz, Einstein, and Minkowski that we will see emerges strongly in Germany only after about 1910–11. In addition to the evidence of readership of this paper in particular discussed in note 26, above, I have only indirect evidence—persuasive, but not conclusive—for the significance of Einstein's account for later histories of relativity. By 1913 there were numerous sources stressing the importance of the Michelson-Morley experiment in the formation of relativity, including in particular the accounts of Laub and Laue—physicists who were certainly well aware of Einstein's 1907 paper—referred to below.

through works central to those concerned with electrodynamics rather than basing his discussion solely on the more individualistic outline of his own path. Most physicists would have recognized this aspect of his story as an accurate description of a major point in the development of Lorentz's theory, while the discussion of that theory in relation to the principle of relativity would have been less familiar (and is more indicative of the course of Einstein's theoretical reflections). In addition, since the important experiments on the electron then told against relativity, it would have seemed especially important for Einstein to emphasize whatever positive links existed with other, well-recognized experiments. Lorentz's explanation of Michelson-Morley was known to be a significant success, one that Planck in 1907 wrote was still regarded as the only experimental support for relativity. The centrality Einstein accorded the experiment, and the fact that he placed great significance on its value for (at the very least) enhancing the perception of relativity, is confirmed very clearly in a revealing letter he wrote to Arnold Sommerfeld in 1908. There, Einstein suggested that "if the Michelson-Morley experiment had not brought us into the greatest predicament, no one would have perceived the theory of relativity as a (half) salvation."[34]

Einstein's letter is interesting on a number of other counts also. First, it indicates that by this stage Einstein had received significant personal recognition, at least through private correspondence. Sommerfeld (who we know was persuaded of relativity through Minkowski's work somewhat later) had evidently written to Einstein in a highly complimentary manner. Einstein replied in a self-consciously personal vein that "in consequence of my fortunate inspiration to have introduced the principle of relativity into physics, you (and others) overestimate my scientific capabilities extraordinarily." Perhaps one of the "others" Einstein was thinking of was Max Laue, who had dedicated a reprint of his 1907 paper on optics "To the discoverer of the relativity principle" in sending it to Einstein.[35]

34. Planck, "Dynamik bewegte Systeme" (1907), on 546, and Einstein to Sommerfeld, 14 January 1908, in Einstein, *Collected Papers*, vol. 5 (1993), 86–89. In 1910 Lorentz contended that the negative result could be explained only by the principle of relativity. Lorentz, "Alte und neue Fragen" (1910), on 1244.

35. Einstein to Sommerfeld, 14 January 1908, 86–87. See also Einstein, *Collected Papers*, vol. 5 (1993), 485 n. 10. The paper Laue sent Einstein was Laue, "Mitführung des Lichtes" (1907). In a similar vein, shortly after writing to Sommerfeld, Einstein received a letter in which Jakob Laub stated that he concerned himself greatly with the "relativity physics" Einstein had introduced. Laub to Einstein, 2 February 1908, in Einstein, *Collected Papers*, vol. 5 (1993), 94–95. Laub's phrase *relativity physics* usefully points to Einstein's introduction of a

However, Sommerfeld had not limited his remarks to compliments but also raised important questions about the scope of the theory; Einstein's reply went on to take up the issue of whether he saw relativity's treatment of (for example) the mechanics of the electron to be definitive. Here Einstein stated emphatically that this was not the case and insisted that (as Sommerfeld suggested) a satisfactory physical theory required a more elementary foundation for its structure; he went on to compare relativity with classical thermodynamics at a stage before Ludwig Boltzmann had interpreted entropy as probability.[36] We will consider the significance of Einstein's use of the term *classical thermodynamics* in the following chapter. Here it is important to note that his focus on a potentially elementary foundation explains why Einstein described relativity as only a half salvation, and just as Sommerfeld and Einstein were aware of limitations, we should also recognize that other physicists' attitudes to relativity (whether acceptance, suspension of judgment, or rejection) were importantly shaped by perceptions of its limitations. In this respect it is particularly revealing that Sommerfeld had questioned Einstein specifically in regard to the relation between relativity and electron theory. As we saw in chapters 6 and 7, in association with the electromagnetic view of nature, electron theory formed the context for many early reactions to Einstein's work; it was a field in which many physicists expected to see significant progress but were unsure of the precise implications to be drawn from Einstein's formulations.[37] Finally, Einstein's language in both this letter and his 1907 paper reveals the importance of a generalized voice for much of his writing on relativity and its history. Einstein is certainly ready to step into an individualized and personal account, whether implicitly, as in his 1907 discussion of time, or explicitly, as in the way he acknowledges his inspiration to Sommerfeld. Nevertheless, as his use of the pronoun *us* when referring to the crisis occasioned by the Michelson-Morley experiment indicates, Einstein often chooses to write from a perspective that incorporates but is not limited to his own; he writes to represent the

way of working theoretically, not just a principle; I have used the term here for still broader analytical aims.

36. Einstein to Sommerfeld, 14 January 1908, 87. These views were followed by Einstein's statement on the importance of the Michelson-Morley experiment.

37. See, e.g., the published exchange between Einstein and Ehrenfest on this point discussed in chapter 7. Sommerfeld's query may be taken to imply that Einstein's response to Ehrenfest's question about the stability of the electron had not satisfied him, just as we saw it did not satisfy Ehrenfest.

perspective of a community. This generalized voice is at least as powerful and important to the author as his own.

Einstein refrained from referring to Kaufmann's damaging experimental results in the opening section of his review article; rather, there his references to Newtonian theory and the Michelson-Morley experiment make a comparison with the past seem more significant in evaluating relativity than a comparison with contemporary rivals. Nevertheless, Einstein discussed Kaufmann's results later in the paper. After suggesting that whether Kaufmann's findings or the foundations of relativity were in error could be decided only when a great variety of observational material was at hand, Einstein stated—in a passage often recalled by historians for the strength and certainty of its theoreticist convictions—that rival electron theories "should be ascribed a rather small probability because their basic assumptions about the mass of moving electrons are not made plausible by theoretical systems that encompass a wider complex of phenomena."[38] Despite the impression sometimes conveyed by theoreticist discussions, both Einstein's several attempts to suggest other experiments capable of testing relativity and his willingness to draw on Michelson-Morley in this history indicate that he was by no means indifferent to experimental confirmation.

In addition to its historical discussion, Einstein's 1907 paper is a valuable resource for the historian for just the reason it would have been important for contemporaries: it reveals the state of research on relativity at the time extremely clearly. In doing so it also discloses certain features of the community then involved in such research, which it will be helpful to note here. Following his brief history with an introductory survey of the applications of relativity, Einstein could list the work of himself and four others on topics in optics, inferences of the energy and momentum of moving systems (the most important of these being concerned with the inertial mass of energy), and the dynamics of the material point (electron). The circle of contributors was small, and all the physicists concerned had been closely involved with the editors of the journal in which Einstein published his work. Max Laue and Kurd von Mosengeil (who died before Planck prepared his work for posthumous publication in 1907) were both Planck's students, while Jakob Laub was a student of the other editor of the *Annalen der Physik*, Wilhelm Wien. Laue and Laub had also participated in Hilbert and Minkowski's 1905 seminar on electron

38. Einstein, "Relativitätsprinzip" (1907), 439.

theory. The close personal network was crucial for stimulating interest in Einstein's work and for shaping the meaning these physicists ascribed to relativity. Einstein and Planck had corresponded on the principle of relativity from mid-1906 but first met when Einstein attended the Naturforscherversammlung in Salzburg in 1909. As Miller reports, Planck's letters show that he had been very concerned by the criticisms of Abraham and Kaufmann; moreover, Alfred Bucherer had written to Planck communicating his own principle of relativity and criticizing Planck's assertion that the principle of least action was compatible with the principle of relativity. On 6 July 1907 Planck wrote to Einstein to say that he was happy to hear that Einstein did not share Bucherer's opinion, because "so long as the supporters of relativity constitute so tiny a band as they do at present, it is doubly important that they agree with each other."[39] Both Laue and Laub wrote first on the application of relativity to optics, and each visited Einstein in Bern. Laue did so in 1906, before he had published on relativity. Laub's first publication in 1907 prompted Laue to observe in correspondence with Einstein and Laub that the article had two errors and an inconsistency in the reasoning. Einstein took the managerial concern a step further, pointing out in the introductory section of his 1907 paper that Laub's article was not error free. Shortly afterward Laub visited Einstein, and their discussions resulted in two collaborative papers.[40] Thus through correspondence and visits in Bern, Einstein, Planck, Laue, and Laub worked hard to ensure that they wrote on relativity in a similar sense, and this small constellation of figures recalls American work on velocity of light early in Michelson's career, or the network of research we saw forming around Kaufmann's experiments. The 1907 paper indicates that, far from being isolated and distant in the Swiss Patent Office, Einstein had already played a central role in guiding an ongoing research project among a small but relatively tightly knit interpretative community. Personal contact and correspondence supplemented published papers in the formation and maintenance of this community; it is the related work of these physicists that provided the foundation for the "union" of

39. Planck to Einstein, 6 July 1907 (cit. n. 21), quoted in Miller, *Einstein's Special Theory* (1981), 235. See also Jungnickel and McCormmach, *Intellectual Mastery*, vol. 2 (1986), 251.

40. Laue to Einstein, 4 September 1907, in Einstein, *Collected Papers*, vol. 5 (1993), 72–73 (Laub had interchanged group and phase velocity), and Einstein, "Relativitätsprinzip" (1907), 414. On Laub's collaboration with Einstein, see Pyenson, "Einstein's Early Scientific Collaboration" (1985), 220–25.

Lorentz's theory and the principle of relativity that Einstein was able to articulate in 1907.

Minkowski's Distinctions

But within the following year there was to be a contribution from someone outside this select band. Hermann Minkowski's major article, "The Fundamental Equations for Electromagnetic Processes in Moving Bodies," both introduced a new formalism for electrodynamics and dramatically increased interest in relativity theory (fig. 8.2). It too opened with a form of historical survey, one that may have drawn on Einstein's earlier account but sought to introduce quite a different emphasis. After stating that there were at present many differences of opinion concerning the basic equations of the electrodynamics of moving bodies, Minkowski introduced the following threefold distinction in the use of terms to designate relativity, because, he wrote explicitly, "this appears to me useful in order to be able to characterize the present position of the electrodynamics of moving bodies." The *theorem* of relativity expressed the purely mathematical fact that the electromagnetic equations were covariant against the Lorentz transformations and rested on the form of the differential equations for the propagation of waves with the velocity of light. The *postulate* of relativity involved a confidence, rather than a complete understanding, that yet-unrecognized laws relating to ponderable matter would be covariant against the Lorentz transformations. The *principle* of relativity involved the idea that Lorentz covariance held as a definite connection between genuine, observable quantities for moving bodies.[41] Having

41. Minkowski, "Grundgleichungen" (1908), on 55 (quotation), 54, the historical survey is on 53–56. Tetu Hirosige argues for the historical importance of Minkowski's work in promoting the development of relativity. Hirosige, "The Ether Problem" (1976). Many authors have tended to discount Minkowski's significance owing to the view that Minkowski did not fully understand the nature of Einstein's contribution (and hence, the assumption is made, could not lead others to a clear appreciation of Einstein's work). Leo Corry maintains that the roots and motivations of both Hilbert's and Minkowski's contributions to relativity have remained only partially analyzed because the histories of special and general relativity have most often been told from the perspective of Einstein's work. Corry, "Minkowski and the Postulate of Relativity" (1997). For earlier approaches to Minkowski, see the works of Goldberg and Pyenson: Goldberg, "The Early Response to Einstein's Special Theory" (1969), 111–47; Goldberg, *Understanding Relativity* (1984), 162–68; Pyenson,

FIGURE 8.2 Hermann Minkowski. Courtesy Niedersächsische Staats- und Universitätsbibliothek, Göttingen.

outlined these conceptual differences in the interpretation of the same set of equations, Minkowski then related them to a historical survey of

"The Göttingen Reception of Einstein's General Theory of Relativity" (1973); and Pyenson, "Hermann Minkowski and Einstein's Special Theory of Relativity," "Physics in the Shadow of Mathematics: The Göttingen Electron-Theory Seminar of 1905," "Relativity in Late Wilhelmian Germany: The Appeal to a Pre-Established Harmony between Mathematics and Physics," and "Mathematics, Education, and the Göttingen Approach to Physical Reality, 1890–1914," in Pyenson, *The Young Einstein* (1985), 80–100, 101–36, 137–57, 158–93. These authors take as a primary task the delineation of Minkowski's approach from Einstein's, with the former being characterized as "formalist" and "mathematical" and the latter as "physical." See also Galison, "Minkowski's Space-Time" (1979), which draws on manuscript sources to discuss the development over time of Minkowski's geometric formulation of earlier work. He held Lorentz to have discovered the theorem of relativity and to have created the postulate of relativity as the hypothesis that electrons and matter suffered contraction according to a definite law as a consequence of motion. Minkowski wrote that Einstein had then brought to its sharpest expression the fact that this postulate was not an artificial hypothesis but one forced by a new conception of time.[42]

Previously, Kaufmann had compared Lorentz's and Einstein's approaches conceptually. Minkowski's discussion shows how history can be used to serve a similar goal of conceptual understanding. Bringing together the citation of past work and a conceptual analysis allows each resource to be used to clarify the other: the conceptual analysis gives an interpretation of the earlier work, perhaps from a different perspective than had previously been possible, while the reader's independent knowledge of the past work (now referred to only briefly) can be drawn on to assist understanding of the conceptual analysis presented. Here, these resources are employed to clarify present work, and we can see how important this was in a context in which several interpretations of the single set of equations were current. In Minkowski's discussion, as in Einstein's earlier work, conceptual clarification and historical interpretation go hand in hand: reference to the archive of physics clarifies its current state.

That Minkowski described Einstein's achievement as centering on the conception of time could reflect his own assessment of the 1905 paper (recall that Kaufmann spoke of Einstein's definition of simultaneity and time) but may also have been suggested by the terms Einstein gave in 1907, which raises the possibility that Einstein's historical account influenced Minkowski's evaluation of the history of relativity.[43] It is also possible that Einstein's account was productive in a second respect. When he presented his geometrical formulation to the Cologne Naturforscherversammlung in his famous "Space and Time" lecture in September 1908,

relativity, and Walter, "Minkowski, Mathematicians, and the Mathematical Theory of Relativity" (1999), which discusses Minkowski's changing attributions (or suppressions) of authorship to Lorentz, Poincaré, and Einstein.

42. Minkowski, "Grundgleichungen" (1908), 55. In a section on the concept of time, Minkowski also described Einstein's 1905 paper as meeting the wish to bring home the physical meaning of the nature of the Lorentz transformations. See ibid., 68–70, esp. 70.

43. Minkowski's use of the expression *sharpest* recalls Einstein's 1907 terminology. However, while the publication dates make it possible that Minkowski drew on the 1907 paper, "Fundamental Equations" was first presented to the Mathematical and Physical section of the Göttingen Society for the Sciences at a meeting of 21 December 1907, before the *Jahrbuch* volume appeared in January 1908.

Minkowski described himself as stripping the concept of space of its absolute nature, just as Einstein before him had stripped time of an absolute nature. We might therefore conjecture that Minkowski's increasing concentration on space in developing his space-time representations came partly in creative response to the focus on time Einstein had pursued in 1905 and articulated as such in 1907. Interestingly, Minkowski expressed something close to chagrin at the complex path that had actually led to relativity and his own late arrival on the scene, raising the counterfactual possibility of an alternative history in order to suggest that the culture of pure mathematics provided unparalleled resources for work in physics. Minkowski stated that his address was intended to show "how it might be possible, setting out from the accepted mechanics of the present day, along a purely mathematical line of thought, to arrive at changed ideas of space and time." This claim was embodied through a discussion of the transformation group G and the structure of space and time appropriate to Newtonian mechanics before Minkowski introduced (purely formally) the parameter c and considered the graphical representation of the invariant expression

$$c^2t^2 - x^2 - y^2 - z^2 = 1.$$

In the limit when $c = \infty$ the group is that appropriate to Newtonian mechanics; however, Minkowski wrote, "since G_c is mathematically more intelligible than G_∞, it looks as though the thought might have struck some mathematician fancy-free, that after all, as a matter of fact, natural phenomena do not possess an invariance with the group G_∞, but rather with the group G_c, c being finite and determinate, but in ordinary units of measure, *extremely great*. Such a premonition would have been an extraordinary triumph for pure mathematics."[44]

In fact, on this occasion mathematics had only the satisfaction of being wise after the event, but Minkowski could now reveal that the value of c was the velocity of light and that his theory gave the Lorentz transformation equations known to physics. Thanks to "its happy antecedents, with its senses sharpened by an unhampered outlook to far horizons," mathematics was able "to grasp forthwith the far reaching consequences of

44. Minkowski, "Raum und Zeit" (1974); Minkowski, "Space and Time" (n.d.), on 82–83, 75, 79. For a study of the evolution of Minkowski's views over time, see Galison, "Minkowski's Space-Time" (1979). Scott Walter has documented the extraordinary impact of the address, which appeared in three periodicals, was published as a booklet, and had been translated into French and Italian within a year. Walter, "Minkowski, Mathematicians, and the Mathematical Theory of Relativity" (1999).

such a metamorphosis of our concept of nature." Minkowski concluded his address with the thought that those who found the abandonment of long-established views painful would be conciliated by "the idea of a pre-established harmony between pure mathematics and physics."[45]

The message of his earlier paper, though seldom expressed so explicitly, had been very similar. After describing the different senses in which the Lorentz transformation equations had been used and stating the significance of Einstein's contribution, Minkowski went on to articulate a claim for his own originality: "Up till now the principle of relativity in the sense in which I have characterized it has still not been formulated for the electrodynamics of moving bodies. In the present publication I obtain, through the principle as I have formulated it, the fundamental equations for moving bodies in a completely clear, unambiguous version. Thereby it is shown that none of the forms previously adopted for these equations have satisfied the principle exactly."[46] This claim had a broad scope and a very specific target. The broader implication was that Minkowski's mathematical apparatus afforded a conceptual consistency and clarity not achieved by previous authors; this was brought home by the specific observation that despite the expectation that the equations Lorentz had adopted for moving bodies would correspond to the postulate of relativity this was not actually the case, in particular, for Lorentz's treatment of magnetized bodies. In making this observation, Minkowski drew a distinction between Lorentz's electron theory and the consequences of relativity—of a kind unprecedented in public discussion of the principle. Einstein had discussed the appropriate form of equations for the microscopic theory of the electron but had not attempted to apply relativity to the macroscopic theory of electromagnetic and optical phenomena in moving material bodies (despite the generality of the title of his 1905 paper). It was here that Minkowski showed that Lorentz's derivations were in conflict with the requirements of the principle of relativity. Thus both Minkowski's fine-grained conceptual-historical analysis and his more expansive historical fantasy carry a disciplinary import: no one is better equipped to further relativity than the mathematician. Minkowski's critical analysis of past work went hand in hand with a prospective vision of the future, and his papers show a continual and productive tension

45. Minkowski, "Space and Time" (n.d.), 79, 91. On Minkowski's appeal to preestablished harmony, see, e.g., Pyenson, "Relativity in Late Wilhelmian Germany" (1985).

46. Minkowski, "Grundgleichungen" (1908), 55.

between allying his endeavor with the body of an existing theory and giving that theory a distinctively different cast. Minkowski's introduction of a new terminology for the principle of relativity in his "Space and Time" lecture expressed this very clearly. He spoke of "the postulate of the absolute world" or, briefly, "the world postulate," because "the postulate of relativity" did not seem to convey the full significance of the requirement of invariance against the Lorentz transformations, once the concept of space had been stripped of its absolute nature.[47]

A Copernican Movement

At the same congress at which Minkowski presented relativity as a bold new worldview, Max Planck, keeping his eye on the experimental situation, prefaced a discussion of the principle of action and reaction in relativity with the cautious observation that at the present time "Einstein's theory of relativity" could in no way be regarded as certain, though he thought that since its deviations from current ordinary theories were limited to very small terms, it could be taken as correct up to those deviations and considerations based on it therefore had some importance. However,

47. Ibid., see also section 9, "Die Grundgleichungen in der Theorie von Lorentz," 76–77, esp. 76, and Minkowski, "Space and Time" (n.d.), 82–83. John Stachel discusses the distinction between the microscopic theory of the electron and its application to the macroscopic equations of material media and different approaches to the macroscopic equations in his editorial note, "Einstein and Laub on the Electrodynamics of Moving Media," in Einstein, *Collected Papers*, vol. 2 (1989), 503–7. Einstein and Laub took up Minkowski's argument for a distinction between Lorentz's theory and relativity in the treatment of magnetized bodies by proposing an experimental arrangement capable of detecting a difference between the two approaches. Einstein and Laub, "Elektromagnetischen Grundgleichungen" (1908), on 536–40. However, many authors accepted that Lorentz's approach, consistently extended, was equivalent to relativity. As has often been noted, Einstein and Laub also reacted critically to the techniques (and some physical arguments) Minkowski had developed, showing how Minkowski's results could be achieved using the mathematical apparatus Einstein had employed in 1905. See ibid., and Einstein and Laub, "Ponderomotorischen Kräfte" (1908). Their response indicates that they saw a close connection between mathematical techniques and interpretation more broadly and at this stage wished "relativistic physics" to retain the cast Einstein had given it. Laub wrote to Einstein: "I am now still more skeptical towards Minkowski's paper; if your work weren't available, with Minkowski's transformation equation for time we would still at best be at the same standpoint (as far as physical interpretation is concerned) as with Lorentzian 'Local Time.'" Laub to Einstein, 18 May 1908, in Einstein, *Collected Papers*, vol. 5 (1993), 119–21. Most physicists did not follow their stance in regard to Minkowski's formalism, and there is evidence that by 1910 Einstein had begun to appreciate the four-dimensional approach (ibid., 121, editorial footnote 12).

together with Minkowski's theoretical work, new experimental studies—also announced in Cologne—soon changed the climate of discussion. As we saw in chapter 6, toward the end of 1908 Alfred Bucherer brought to conclusion his series of experiments utilizing β-rays of radium salts to test the velocity/mass relation of the electron and proclaimed the confirmation of the principle of relativity (at the expense of his own theory of the electron). Although his papers were immediately followed by responses from Adolf Bestelmeyer refuting the accuracy of his tests (to which Bucherer in turn responded vigorously), Bucherer's work was extremely important because it gave those who supported relativity a counter to Kaufmann's earlier negative experiments.[48] H. A. Lorentz, for example, wrote that "in all probability, the only objection that could be raised to the hypothesis of the deformable electron and the principle of relativity has now been removed."[49]

It is interesting that after presenting his experimental methods and detailing his results, Bucherer, like the theorists Einstein and Minkowski before him, took the opportunity to discuss differences and similarities between Lorentz's and Einstein's approaches, this time in the context of a historical overview of electromagnetism from the time of Maxwell that he said described "the epistemological history of the ether." Bucherer saw Lorentz as the founder of the principle of relativity, despite his reliance on the hypothesis of absolute motion; he held that Einstein's derivation opened the possibility of constructing a wholly new, monistic, picture of the essence of matter, the ether, and their reciprocal connections, assuming the essential equivalence of the two.[50]

The increased attention was such that by 1910 Max Planck was ready both to abandon his earlier caution about the validity of the principle of relativity and to embark on a more elaborate use of historical argument. In an address to the Naturforscherversammlung in Königsberg, in which he spoke on mechanical explanations in modern physics, Planck suggested that just as Copernicus's work had shown that the vertical direction

48. Planck, "Prinzip der Aktion und Reaktion" (1908), 729. Bucherer's work was first presented to the Naturforscherversammlung. Bucherer, "Messungen an Becquerelstrahlen" (1908). See chapter 6 for the debate that followed and note that throughout the period discussed in this chapter, controversy attended each experimental proclamation on the confirmation or refutation of relativity; none could be taken as definitive.

49. Lorentz, *The Theory of Electrons* (1909), 329.

50. Bucherer, "Bestätigung des Relativitätsprinzips" (1909), 531–34, on 531 (quotation), 533, 534. His presentation to the 1908 Naturforscherversammlung had begun with the historical overview before describing his experiment.

was not absolute but relative to position on a spherical earth, relativity theory now showed a more general relativity of time and velocity in inertial systems.[51] Quite explicitly, however, Planck offered the Copernican analogy to teach more than such features of the conceptual nature of the theory. In fact, he first introduced it as a comparison of the social dimensions of two *revolutions*. He wrote that at present there seemed to be a prospect of a final decision between the mechanical and nonmechanical conceptions of nature as a result of a recent movement—relativity—that had affected the whole of physics: "in the wake of this movement, dissensions rage among scientists which can only be compared to those which raged around the Copernican view of the universe. I will endeavour to show what has led to this revolution, and how the crisis consequent upon it can be settled."[52]

In drawing this comparison, Planck implied that opposition to relativity was to be expected, given its revolutionary nature (thus he explained the opposition away); moreover, since relativity is compared with the successful Copernican revolution, the similar triumph of the new theory is invested with an almost historical inevitability. Those who do not accept relativity are likened to children who at first find it difficult to believe that there are people walking below them on the other side of the earth. Planck evidently considered relativity to have passed the *ad absurdum* test. His general strategy is an interesting and important one that we have already seen suggested in Einstein's history: to make the comparison between relativity and the past seem more relevant than that with alternative theories. John Earman and Clark Glymour, and more recently Alistair Sponsel, have suggested that the neglect of contemporary alternatives was of considerable importance in the presentation of tests of gravitational theories in 1919.[53] We can now see that the grounds for

51. Planck, "Stellung der neueren Physik" (1910), on 928–29, trans. as Planck, "Place of Modern Physics" (1960 [1910]), on 38–39.

52. Planck, "Stellung der neueren Physik" (1910), 923 (my translation), cf. "Place of Modern Physics," 28. In a statement incorporated in the commission report (of April 1910) that nominated Einstein for his professorship in Prague, Planck expressed himself in a similar way: "the extent and profundity of the revolution in the scope of the physical world view evoked by this principle can only be compared with that brought about through the introduction of the Copernican world-system." Quoted in Einstein, *Collected Papers*, vol. 5 (1993), xxxvi, xxxviii, in note 16 (my translation).

53. Planck, "Stellung der neueren Physik" (1910), 929 ("Place of Modern Physics," 39); Earman and Glymour, "Relativity and Eclipses" (1980); Sponsel, "Constructing a 'Revolution in Science'" (2002).

such a narrowing of possibilities had been prepared in histories of special relativity (and it contrasts with Kaufmann explicitly noting that none of the available theories might prove adequate to his results).

Lewis Pyenson regards Planck as having proclaimed Einstein the "new Copernicus" in this lecture. It is certainly true that two years earlier, in Cologne, Planck had referred to "Einstein's theory"; however, it is interesting that in 1910 Planck actually compared the *movement* of relativity with the Copernican theory and, when he came to list those responsible for the new movement, gave an assessment quite similar to Minkowski's. Planck wrote: "Under the pioneers of the new terrain we have first to mention Hendrik Antoon Lorentz, who discovered the concept of relative time and introduced it into electrodynamics, without however drawing any radical conclusions from it; then Albert Einstein, who first had the boldness to proclaim the relativity of time as a universal postulate; and Hermann Minkowski, who succeeded in fashioning the theory of relativity into a consistent mathematical system."[54] In fact, Planck devoted some attention to Minkowski's contributions in the context of remarks that considered the acceptance and development of relativity in terms of the membership of different communities of physicists. Planck suggested that it was not an accident that the abstract problems of relativity had chiefly attracted mathematicians, particularly after Minkowski had shown that the mathematical methods employed "were the same as those for four dimensional geometry." However, relativity was important not only to those who could use their mathematical expertise in its development. Planck wrote that even experimentalists (at least those who were "genuine" and "unprejudiced") were not opposed to the theory of relativity. Members of this community, he said, allowed matters to proceed quietly and took their stand from what could be proved experimentally in a field that taxed the limits of instrumentation.[55]

54. "It was Planck ... who in 1910 elevated Einstein to the status of the new Copernicus." Pyenson, "The Relativity Revolution" (1987), 77 (see also 66). For the reference to "Einstein's theory," see Planck, "Prinzip der Aktion und Reaktion" (1908), 729. For the quotation, see Planck, "Stellung der neueren Physik" (1910), 929 (cf. "Place of Modern Physics," 39–40).

55. Planck, "Stellung der neueren Physik" (1910), 929–30 (cf. "Place of Modern Physics," 40). On Minkowski as the source for a "mathematical" theory of relativity, see Walter, "Minkowski, Mathematicians, and the Mathematical Theory of Relativity" (1999). Among such "mathematicians" Planck may have included Dimitri Mirimanoff and Philipp Frank, who published papers in the *Annalen* exploring the relationships between the principle of relativity and Lorentz's electron theory based on Minkowski's formalism, and Max Born, whose work on a rigid-body model of the electron was discussed in chapter 7.

Planck, then, offered a rudimentary map of the dynamics of the debate around relativity in Germany; in several features, at least, such a map would have looked quite different just a few years earlier. Remembering how hard Planck had worked to prevent relativity from being dismissed following Kaufmann's results, and noting the dispute that still attended each experimental study, it is surprising that Planck could represent the experimental community as proceeding *quietly*. A precondition of this comparative judgment was the increasing discussion of the theoretical nature of relativity that was fostered by Einstein, Planck himself, and Minkowski. In 1908 Planck spoke cautiously of the validity of the principle; only in 1910 could he describe relativity as having given rise to raging dissension. It is thus of the essence that Planck ascribed the new revolution not to the individual Einstein but to the movement of a community, and Planck clearly makes the most of the episode's potential to assert the disciplinary preeminence of theoreticians.

In October of the same year H. A. Lorentz presented a similarly broad and incrementalist picture of the development of relativity in a series of six lectures in Göttingen titled "Old and New Questions in Physics." However, unlike Bucherer and Planck, Lorentz attributed the principle to Einstein.[56] His lectures introduced relativity in the context of an account of the ether and electron theory before going on to discuss quantum theory—also recently accorded fundamental importance. Like Planck, but speaking less explicitly in disciplinary terms, Lorentz described relativity as having a multifaceted nature: mathematical, epistemological, and physical. His brief discussion of the epistemology of space and time took up the question that had been so important in shaping other physicists' views of relativity—the relation between his own and subsequent work. It is revealing that while Lorentz referred to current research discussing

56. He wrote that it was a particular pleasure to speak of Einstein's principle of relativity in the town in which Minkowski had worked. Lorentz, "Alte und neue Fragen" (1910), 1236. Lorentz had also discussed Einstein's work in his major study of electron theory in 1909, describing it as a "very interesting interpretation" of his own results (which Lorentz had developed using "effective coordinates" and "effective time"). He wrote that, "the chief difference [is] that Einstein simply postulates what we have deduced, with some difficulty and not altogether satisfactorily, from the fundamental equations of the electromagnetic field. By doing so, he may certainly take credit for making us see in the negative result of experiments like those of Michelson, Rayleigh and Brace, not a fortuitous compensation of opposing effects, but the manifestation of a general and fundamental principle." Lorentz, *The Theory of Electrons* (1909), 223–30 on 223, 230. Here he also noted that his own approach had the virtue of endowing the ether with a certain degree of substantiality (see below).

the mathematical and physical sides of relativity, he made epistemology central to the meaning of the theory but largely independent of the endeavor to create a relativistic physics. Lorentz indicated clearly that he recognized a strong reading of the implications of relativity, attributing to Einstein and Minkowski (without distinction) a denial of the existence of the ether and true time, but he stated that nevertheless his own preference was to see space and time as separate and distinct and to hold to both an ether and a conception of true time. Lorentz thus made it extremely clear that he wanted no full "union" with the new approach. Nevertheless, he was very happy to assimilate the continuing work in the field, and his discussion of the "physical side" indicates the inclusive nature of his understanding of relativistic physics, framed as it was in terms made possible by Minkowski.[57]

We have now considered a number of highly consequential instances in which different physicists integrated relativity into a historical framework in arguing for a particular understanding of its nature and consequences. In comparison to Planck's general article, the research papers I have cited related relativity to immediately antecedent studies and Newtonian mechanics in order to convey the character of particular conceptual innovations, though often with quite different evaluations and implications. Einstein, for example, stressed the evolutionary nature of relativity's development and its close association with Lorentz's theory of the electron, while Minkowski highlighted distinctions in drawing out different ways in which the principle could be conceived. Lorentz, for his part, isolated epistemological questions from other aspects in order to repudiate a complete union between his work and a strong reading of relativity. But in addition to providing orientation at this fine-grained level, all had also taken the opportunity to put the story in the context of more general messages, providing research and disciplinary guidelines on a broader scale. Einstein argued for the abandonment of the luminiferous ether and Bucherer for a monistic approach, while Minkowski held that it was important to

57. Lorentz began by outlining the transformation equations, pointing backward in time to show that in one form Woldemar Voigt had already employed them in an 1887 discussion of the Doppler effect (Minkowski had pointed this out in Minkowski, "Raum und Zeit" [1974], 58). He then went on to discuss the ways in which the equations of motion of the electron (and other fields of research) could be brought into accord with the principle. Here Lorentz described Minkowski's introduction of the concept of proper time as one of his more beautiful discoveries and cast his own discussion in terms suggested by Minkowski's work (comparing "Newtonian" with "Minkowskian" force, for example). See Lorentz, "Alte und neue Fragen" (1910), 1236–44. Lorentz makes his own preferences clear on 1236.

show that what had been achieved through a difficult historical progression in physics could be approached more directly by the royal road of pure mathematics. Drawing on the subject of his own recent research papers and his concern with unifying principles, Planck's account related relativity to what he regarded as the "chief law of physics, the pinnacle of the whole system," the principle of least action.[58]

One feature of Planck's article, then, was that it related relativity to its antecedents in developing a particular story of the progress of physics. However, like Minkowski's counterfactual history, with its combined conceptual and disciplinary import, Planck's extended use of a historical analogy involved quite a new employment of historical material to aid the propagation of relativity, one that explicitly addressed both conceptual and social dimensions. Planck's discussion is abbreviated, listing pioneers and mentioning subcommunities in general terms only, but it reflects quite a sophisticated understanding of the varied senses in which different physicists were able to respond to relativity, relating these to differences in background. This is an account of a contextual nature; it is related to the "contingent" repertoire Gilbert and Mulkay identify as a contrast to the "empiricist" repertoire.[59]

Einstein's Paper of 1905

The earliest histories of relativity show that the recognition of different contributions as foundational (or even as constituting a discovery of the principle of relativity) was an important feature of the interpretative, historical work involved in forming and conveying understandings of relativity. However, these accounts have displayed differences in their

58. Planck wrote that from the principle of least action four equally important principles radiate in four directions—corresponding to the four universal dimensions of the new theory of relativity. The threefold principle of momentum corresponds to the three space dimensions, while the principle of energy corresponds to the time dimension. It had never before been possible to follow these principles back to their common origin. See Planck, "Stellung der neueren Physik" (1910), 930 ("Place of Modern Physics," 41). I have noted Planck's early interest in the fact that the principle of least action proved to be invariant against the Lorentz transformations. Minkowski's unification of the Lorentz force and the energy conservation law for electromagnetic processes made possible a far more concise treatment of the principle of least action than Planck had been able to provide in 1907. See Miller, *Einstein's Special Theory* (1981), 368–69.

59. Gilbert and Mulkay, "Experiments Are the Key" (1984). See also note 32, above.

evaluations, even while drawing what by 1910 was a common distinction between the contributions of Lorentz and Einstein. But more significant than any variation in evaluation is their primary concern with a distribution of credit and shadings of understanding. They have in common a representation of relativity as the work of several authors.

In 1911 this catholicity was to give way in two very different participant histories, one brief and concise, the other extended and closely argued, as Arnold Sommerfeld and Max Laue signaled and sheeted home the collapse to a single discoverer most common in our present accounts. Asked to give a paper on relativity for the Naturforscherversammlung in 1911, Sommerfeld declined to do so (and spoke instead on quantum theory) because, he said, "the principle of relativity no longer belongs to the essentially current questions of physics. Although only six years old—Einstein's work appeared in 1905—it appears to have already become a certain possession of the physicist."[60]

It is fascinating and suggestive that it is at just the point that relativity is represented as a "certain possession"—and no longer current—that it is also presented as the product of a single man. Coming on the heels of the dissension Planck described, and going one step further than Lorentz's evenhanded overview by positively transferring attention to quantum theory, Sommerfeld's treatment implies not only that he holds the key to the debate but also that the door to dissent is locked. In part these remarks may be taken to reflect changes in viewpoint already fairly widely realized in the German physics community, but the plurality of earlier approaches suggests the more significant possibility that heralding one man as the discoverer was, rather, a means of *making* relativity a sure possession. Certainly the singularity of Sommerfeld's attribution requires explanation. Drawing on our findings in chapter 7, I would suggest that by 1911 physicists may have been ready to point to Einstein alone as discoverer because this was the first period in which many of those central to German physics not only appreciated the nature of his contributions but also agreed with his interpretation of relativity—and in particular with a definite view of its implications for electron theory. I want to advance the complex argument that a certain narrowing of focus—to promote Einstein as discoverer in 1905—went hand in hand with an acceptance of rather broader features of

60. In contrast, Sommerfeld described quantum theory as a field still subject to controversy and conceptual change. See Sommerfeld, "Plancksche Wirkungsquanten" (1911), on 1057.

an Einsteinian reading, which emerged only after 1905; but this focus on Einstein was accompanied by a simultaneous recognition and utilization of the techniques and contributions of others. To make this view plausible it will be helpful to explore in some detail the contexts of such individualized attributions at different times and to examine Sommerfeld's views more closely.

Kaufmann's early description indicates the potential for some kind of recognition of Einstein from the publication of his 1905 paper, and it is not surprising that later theoreticians were ready to pronounce judgments based on conceptual distinctions rather than emphasizing formal or empirical equivalence, as many experimentalists did. Laue, Laub, and Minkowski provide examples. Nevertheless, the varied and gradualist accounts we have seen indicate that it was by no means a necessary—or trivial—matter to identify relativity primarily with any single moment or founder. Further, the fact that the early (and private and personally directed) attributions of individual authorship to Einstein came from Max Laue and Jakob Laub—physicists clearly working within an Einsteinian framework of understandings and practices—suggests that an unambiguous attribution of discovery depended on substantially shared interpretations (rather than simple recognition), at least at a time when the field was recognized to be open.

Those who accepted relativity in this period did so in the face of strong arguments for its experimental disproof. By late 1908, however, the work of Minkowski and Bucherer had substantially changed the climate of discussion. A study of the gradual development of Sommerfeld's views will reveal the complex background to that physicist's public attribution of relativity to Einstein in 1911 and show the centrality of electron theory in his thinking, a feature that is likely to have been characteristic of others as well. We saw in chapter 6 that in 1906 Sommerfeld moved from teaching technical mechanics at the Technische Hochschule in Aachen to the chair of theoretical physics in Munich, largely on the basis of highly technical studies developing the Abraham model of the electron. Responding to Planck's 1906 advocacy of the Lorentz-Einstein approach, Sommerfeld confirmed his preference for the "electrodynamic postulate" over the "mechanical-relativistic postulate," which (harking back to mechanics as it did) might appeal to those over forty.[61] Three years later, however, he

61. For Sommerfeld's response, see the discussion following Planck, "Kaufmannschen Messungen" (1906), 759–61; for examples of discussions of Sommerfeld's stance, see Holton,

changed his view on relativity, possibly for a variety of reasons but certainly in large part owing to his reading of Minkowski and to changing views on electron theory. I have noted Sommerfeld's early 1908 correspondence with Einstein. Whatever he wrote to prompt Einstein's candid responses, Sommerfeld's views of Einstein's achievements and of relativity were not then dependent on an acceptance of relativity. In fact, shortly before this exchange Sommerfeld had written to Lorentz criticizing the unhealthy dogmatism of Einstein's work, which he said expressed "the abstract-conceptual manner of the Semite." Following the account of Bucherer's study at Cologne in late 1908, Sommerfeld wrote to Lorentz congratulating him on the success of the theory; six weeks later he was ready to proclaim his own conversion to relativity, writing that the systematic form and conception of Minkowski's work in particular had facilitated his understanding. It is important to note, however, that this did not lead Sommerfeld to abandon his commitment to programmatic support of the foundational nature of electromagnetic theory.[62]

Minkowski's work was important for Sommerfeld's understanding but did not prevent him from taking Einstein's contributions seriously. Rather, Sommerfeld took a selective approach to both. From 1906, Einstein's published discussions of the electron had increasingly counseled a cautious theoretical approach that was at some variance with both his private comment to Sommerfeld that the relativistic treatment of the electron should not be regarded as definitive and with his ongoing unpublished work on the subject. We saw in chapter 7 that in 1909 Sommerfeld responded to Max Born's extension of relativity to discuss accelerated motion and treat outstanding problems in electron theory, warning that

"Quanta, Relativity, and Rhetoric" (1993), 98–99; Miller, *Einstein's Special Theory* (1981), 234.

62. Sommerfeld to Lorentz, 26 December 1907, quoted in Einstein, *Collected Papers*, vol. 5 (1993), 88–89 n. 1, and Sommerfeld to Lorentz, 16 November 1908, 9 January 19[10], quoted in Walter, "Minkowski, Mathematicians, and the Mathematical Theory of Relativity" (1999), 70. Sommerfeld made contact with Einstein in early 1908 because of their common concern with the propagation of effects with superluminal velocity, but it would be interesting to know whether the character of his correspondence was also shaped in part by prior knowledge of Minkowski's "Fundamental Equations" paper (which was delivered in Göttingen in late 1907 but did not appear in print until April 1908). Both Sommerfeld and Einstein were in dialogue with Wien on the question of superluminal signal velocities. See the editorial note "Einstein on Superluminal Signal Velocities," in Einstein, *Collected Papers*, vol. 5 (1993), 56–60. For Sommerfeld's continued commitment to an electromagnetic program (supplemented, however, by the quantum theory of radiation), see Seth, "Quantum Theory and the Electromagnetic World View" (2004).

relativity's restriction to uniform velocities implied that no certain conclusions could be reached in regard to acceleration.[63] In doing so he accepted the principle of relativity as defining the theoretical possibilities in electron theory. This stance was theoretically conservative and renounced aims until then central to much work in the field, including Sommerfeld's. It involved the judgment that the tools provided by relativity offered few prospects for the further development of electron theory and that experimental work on the mass/velocity relation was likewise unable to provide further guidelines. This is the sense in which it was precisely by defining limits to enquiry that the principle of relativity could become, in the same moment, both a certain possession and no longer current—remaining a half salvation only. In addition to his advocacy of an Einsteinian approach to electron theory in late 1909, in 1910 Sommerfeld published two papers propagating Minkowski's methods in the physics community by developing a vectorial form of Minkowski's matrix approach.[64] So Sommerfeld's (possible) private recognition of Einstein in 1908 did not reflect an acceptance of relativity; his public recognition of the physicist in 1911 was accompanied by an acceptance of Einstein's views on the relations between relativity and electron theory; but this by no means represented simply partisan acceptance of either Einstein or Minkowski.[65]

Sommerfeld signaled an Einsteinian version of relativity in a few sentences. The first and authoritative textbook on relativity—and my final example of a participant history—amounted to an extended argument for an Einsteinian interpretation. Published by Max Laue in 1911, the book's very title, *Das Relativitätsprinzip*, reflects the centrality of the principle

63. Sommerfeld's 1909 warning came in discussion to Born, "Dynamik des Elektron" (1909), 817.

64. Sommerfeld, "Relativitätstheorie I" (1910); Sommerfeld, "Relativitätstheorie II" (1910). In contrast, in his collaborative papers with Laub, Einstein had asserted that while Minkowski's methods in the main agreed with his own, they were mathematically rather forbidding—and had then provided a treatment on the basis of the formal approach Einstein himself had used in 1905. (Perhaps Sommerfeld would have seen that as dogmatic also.)

65. Given the importance of Minkowski's work in persuading Sommerfeld of the value of relativity and his ongoing development of Minkowskian techniques, it is interesting that despite Minkowski's emphasis on the radical and original nature of his concern with space, Sommerfeld did not see the Göttingen mathematician as having put relativity on a fundamentally new footing. A similarly appreciative but selective approach to Minkowski was taken by convinced relativists close to Einstein, such as Planck and Laue, and by those physicists closest to Minkowski himself, such as Max Born, who used Minkowski's formalism in his development of a rigid-body model of the electron but did not propagate Minkowski's use of the term *world postulate* for the principle of relativity, for example.

FIGURE 8.3 Max von Laue. Courtesy Deutsches Museum, Munich.

common to many early discussions of relativity (fig. 8.3). Laue attributes the discovery to Einstein in the first line of the preface; and his understanding of relativity as a field of work broader than any single contribution is reflected in his incorporation of the results and methods of many figures, especially Einstein, Planck, and Minkowski.[66] But an explicitly historical methodological approach was also integral to Laue's treatment. In addition to drawing out the particular role history played in Laue's approach, a detailed study of his argumentation will show that extensive

66. Laue, *Relativitätsprinzip* (1911), v–vi.

interpretative work went into Laue's representation of Einstein's 1905 paper as distinctively foundational.

Using a framework first outlined in Einstein's 1907 paper, Laue opened his first chapter, "The Problem-Situation," by considering the principle of relativity and the transformation equations appropriate to classical mechanics (we will explore the origins and significance of this language soon). The body of the chapter then discussed a group of empirical results that had to be incorporated in any electromagnetic theory of moving bodies. These ranged from induction through the Fizeau experiment to the dynamics of the electron (which as we saw in chapter 6, Laue dismissed in only three sentences). Among them Laue accorded the Michelson-Morley ether-drift experiment a particular importance because of the second-order level of its enquiry and its role in the development of Lorentz's work.[67] There was an important point to reviewing the whole series of experiments: to emphasize that a scientific deed of the first degree would be required to explain such apparently contradictory results from one point of view. Laue's second chapter presented a highly revealing "Historical Overview" before giving an account of two principal predecessors of relativity, describing their successes and failures in relation to the different experimental findings. Only in the third chapter did Laue enter into an extended treatment of relativity.[68]

The general importance of a historical framework to Laue's text will be evident from this brief survey. His "Historical Overview" shows in detail the very deliberate—and resonant—sense in which he used both a specific account of the relation between relativity and its predecessors and a notion of the "historical" dimensions of the theory as powerful resources

67. Laue built on Laub's important earlier review, discussed in chapter 6. He described the negative result of the Michelson-Morley experiment as having pushed Lorentz to a new hypothesis already leading toward the theory of relativity. For this reason, he wrote, "the experiment became the fundamental experiment for the theory of relativity; as one also attained the derivation of the 'Lorentz transformation' almost directly from it." This justified Laue in giving a more detailed discussion of Michelson-Morley than the other experiments reviewed (although he recognized that even then he had made several idealizing simplifications). See ibid., 13.

68. Ibid., 18. In the third chapter he discussed first the derivation of the Lorentz transformations and Einstein's kinematics before considering Minkowski's geometric interpretation of the transformations and setting up the mathematical apparatus Minkowski had made available as a prelude to treating the electrodynamics of empty space and ponderable bodies—and dynamics—in accordance with the principle.

in conveying an Einsteinian interpretation. Laue began this section with the observation that the fundamental changes relativity involved meant that, perhaps more than other theories, it required a proof of its necessity. Laue allowed that all theories had to find their essential foundations only "in themselves and in their relation to the facts." However, while he was certainly confident of the internal consistency of relativity, he recognized that many others were not reconciled to its theoretical content because of the apparently paradoxical consequences of the relativity of time; he was also aware that there existed no unambiguous, unique experimental proof of the theory. Recognizing these difficulties in demonstrating the theory's necessity, Laue stated that "there is also, in this region, a form of historical necessity which lies in the failure of all other attempts to succeed in reconciling themselves to the facts." Accordingly, he discussed two earlier approaches in some detail. The first, Heinrich Hertz's theory, had sought without success to transfer the principle of relativity from mechanics to electrodynamics; the second, Lorentz's theory, had denied the principle of relativity from the start, designating an immovable ether as the basis for all electrodynamics. Laue made it quite explicit that in choosing these forebears for discussion his particular history was partial and programmatic, dictated by the end he had in view.[69] This history was also, in his view, essential to the development, acceptance, and understanding of relativity. For Laue the conceptual foundations on which these predecessors rested were far closer to ordinary conceptions than the basic ideas of relativity; no one would have proceeded to relativity theory if the previous theories had not been developed and recognized as hopeless. Thus in his account the failures of the past bear the burden of demonstrating the necessity of relativity.

However, entangled as relativity was in this historical background, Laue could not represent the theory as completely distinct from its forebears. Noting a developmental link first, Laue wrote that Lorentz's theory in particular had so prepared the ground through the introduction of certain fundamental concepts that relativity owed this predecessor much

69. "Nun kann natürlich jede physikalische Theorie ihre eigentliche Stütze nur in sich selbst und in der Bezugnahme auf die Tatsachen finden. Immerhin gibt es auch auf diesem Gebiet eine Art historische Notwendigkeit, die in dem Fehlschlagen aller anderen Versuche liegt, zu einem befriedigenden Verständnis der Tatsachen zu gelangen." Ibid., 18–19; see 19 for the purposefulness of his choices. Admissions regarding the lack of proof for the theory are made in the foreword. Ibid., v.

credit for its rapid development.[70] Then he went on to indicate the formal similarity, pointing out that in 1904 Lorentz had so modified his theory that it accounted for all observations. Indeed, Laue allowed that an essentially experimental decision between the extended Lorentzian and relativity theories was not possible.[71] Interestingly, his negotiation of similarities and differences goes so far as implicitly to reverse the temporal order of Lorentz's and Einstein's publications. Breaking the developmental link very sharply in service of Einstein's epistemological priority, Laue identifies Lorentz's theory primarily with its pre-1904 version, describing the 1904 paper as so modifying that theory as to approach relativity. The comparison he maintains is not between "Lorentz's theory" and relativity but between the "extended Lorentzian" and relativity theories. Given this complex preparation, by the end of Laue's discussion of former theories Einstein's 1905 paper could be represented as bringing a decisive solution to the riddles posed by the past in one stroke, through a thoroughgoing critique of the concept of time. Laue handled its relation to Minkowski's later work with the assertions that the principle of relativity as he expressed it, and as it was mathematically formulated in the Lorentz transformation, contained the basis for the theory of relativity in its

70. Ibid., 19. Here Laue might have been thinking of Einstein's reinterpretation of "local time" as time in general as an example. However, his own papers also show very clearly how the presence of Lorentz's earlier work could be utilized to rapidly advance research in relativity. In a paper of 1907, for example, Laue stated that the electrodynamics Einstein derived from the principle of relativity corresponded to Lorentz's (old) theory up to the first order of v/c. This implied that Fresnel's dragging coefficients could be correctly derived from the principle of relativity, as had already been achieved for Lorentz's theory. Laue's purpose in his two-page article was to show how much more easily the problem could be solved using the principle than through the other theory, even with the simplifications Lorentz had recently introduced. Thus, recognizing particular historical and conceptual relations with Lorentz's work allowed Laue to demonstrate practical differences of technique and hence to extend research in relativity. See Laue, "Mitführung des Lichtes" (1907). A similar strategy was followed in Laue, "Wellenstrahlung einer bewegten Punktladung" (1909). While Laue used the relations with Lorentzian theory, we have seen already that Max Born utilized those between Newtonian mechanics and relativity as a guideline for more speculative work on a rigid-body model of the electron. The attempt was in part unsuccessful, but the work of both Laue and Born indicates a close and productive relationship between a historical understanding of relativity and the formation of research strategies. The articulation of particular historical and conceptual relations was thus by no means merely a matter of pedagogy and justification.

71. The preeminence of relativity lay in the fact that as close as the Lorentzian theory came, "it still lacks the great, simple, general principle" that bestowed something so imposing on relativity. Laue, *Relativitätsprinzip* (1911), 19–20.

entirety, and that all the consequences he was going to draw in the body of his textbook could be derived from the principle.[72]

Laue's use of history was multifaceted, subtle, and quite conscious. He was very clear about the partial and legitimating nature of his particular account. He implicitly recognized the possibilities of other histories while attributing a form of (historical) necessity to relativity—one history. A similar point can be made in regard to relativity itself: Laue's complex negotiation of closeness and distance implicitly recognizes the possibility of distinctly different interpretations, many relativities, while attributing a form of necessity to one theory. This was certainly not the first attribution of relativity to Einstein; however, Laue's textbook constituted the fullest published discussion of the relations between Lorentz's and Einstein's work to date, and the original and productive nature of the account should be emphasized. Einstein had also referred to one theory, but to the merger of the Lorentzian theory with the principle of relativity. Minkowski had emphasized conceptual distinctions in past theory but focused on a geometrical interpretation of the "world postulate." In contrast to other discussions distributing credit, Laue undertook extensive critical work to locate the founding of relativity in Einstein's 1905 paper. Making Einstein so unambiguously the founder of the theory was part of a strategy to render one particular reading of relativity necessary.

Michelson subsequently recognized this understanding of the relations between his work, Lorentz's theory, and Einstein's theory of relativity. In 1911 he visited Göttingen as the successor to Lorentz's visit the previous year, and according to his biographer became aware that there were deep divisions between those who supported Einstein's theory and those who supported continued reference to the ethereal medium. Michelson signaled that he belonged to the latter by joining their table; but also invited Max Born to lecture on relativity in Chicago in 1912.[73] The widespread acceptance of general relativity after 1919 created a qualitatively different situation, and in the penultimate chapter to his 1927 *Studies in Optics*, Michelson finessed the issues concerned by writing that the results deduced on the basis of Lorentz's contraction hypothesis, "follow as a direct consequence of the (restricted) theory of relativity of Einstein." Because it had both explained known phenomena and made it possible to predict and discover new phenomena, Michelson thought relativity should be

72. Ibid., 33, 46.

73. Livingston, *Master of Light* (1973), 252–60.

accorded "a generous acceptance notwithstanding the many consequences which may appear paradoxical in consequence of the difficulty we find in realizing the conditions of high relative velocities." He allowed that the ether appeared inconsistent with relativity, but described his hope that the theory would be reconciled with the existence of a medium to explain the propagation of light waves.[74] While Michelson would never have put the relations so simplistically, I want to conclude by noting that later writers have sometimes presented an instrumental view of scientific progress that renders Einstein a footnote to discussions of interferometry. For example, in a section on "The Michelson Interferometer" in a handbook of scientific instruments, a description of the instrument was followed by this brief account of its experimental and then philosophical applications: "One of the most important applications of the instrument from the purely scientific point of view was in the famous Michelson-Morley experiment, which was designed to measure the velocity of the earth relative to the so-called 'Luminiferous Ether.' To everybody's astonishment, the answer was zero. As a consequence of this and other experiments which followed, the Relativity Theory of Einstein was born, which created a revolution in thought in the domain of Physics with important repercussions in Philosophy."[75] The ether-drift experiment and the theory of relativity might both be regarded as counting among the applications of a new instrument.

Conclusion

Our study of Kaufmann, Ehrenfest, and Born in chapters 6 and 7 examined the material culture of apparatus and measurement protocols on the one hand and the conceptual elaboration of critical issues on the other. The endeavor opened up specific dimensions of practice that have largely disappeared from histories of relativity even though they were integral to research in the field, well represented in its documentary record, and were in several respects of continuing significance. The present chapter has explored a further facet of scientific research that seldom receives close historical attention; in this case, physicists' memory work itself—a form of discourse in which physicists moved between research and its

74. Michelson, *Studies in Optics* (1995 [1927]), 156–66 on 157 and 161.
75. Cooper, ed., *Scientific Instruments*, 2nd ed. (1952), 75–76.

implications, and negotiated distinctions between individual endeavor and communal achievement. This survey has shown that physicists used histories of relativity in the service of the articulation of present interpretations, to argue for a variety of general messages, and to provide an understanding of the dynamics of their discipline and shape the development of a new and controversial theory. In addition to detailed narratives of change and progress drawing on both empiricist and theoreticist themes, physicists and mathematicians worked with more flexible accounts: with counterfactual history, historical analogy, and a subtle idea of historical necessity. Often expressing—and offering a resolution to—a significant tension between many relativities and a single theory, their protohistories have exhibited a fascinating dialogue between individual interpretations and the formation of generalized, communal voices and have offered both incremental accounts stressing the distribution of credit and sharply individualized attributions of discovery. In addition, the particular stories told have disclosed important aspects of the process through which relativity came to be accepted in Germany. In some respects these are at odds with the main features of present accounts, while in other respects we can recognize antecedents of the histories we currently accept. Einstein's expression of continuities indicates a more complex course to the development of distinctions between his own and Lorentz's work than is usually allowed, and his use of the Michelson-Morley experiment demonstrates his interest in establishing a generalized path to relativity. Laue's endeavor strongly to distinguish Einstein's contributions from others has been maintained by many academic historians of science, while more recent studies exploring the point of view of different physicists have tended to stress a gradualist understanding of the period akin to the very earliest participant accounts. Thus our own histories have reflected and drawn on physicists' memory work in complex ways, perhaps too often without fully recognizing the extent to which those original accounts helped shape the landscape they represented. My own attempt has been to see the propagation of relativity as the work of a community, focusing on the gradual development and stabilization of common views that emerged from the varied and sometimes divergent work of many individual physicists.

In conclusion, it is appropriate to explore the cultural and geographical limitations of the account I have just developed, based as it is on the German physics community. In speaking of "the physicist," Sommerfeld framed his remarks in general terms. However, while an Einsteinian reading of relativity had by 1911 undoubtedly become a secure possession for a small but powerful group in Germany, historical research has shown that this was not the case for the whole of the German physics community (witness Abraham's continued opposition to the theory and Lorentz's preference for an ether-based theory), and it was far less so outside the German-speaking academic network. Andrew Warwick's studies have shown, for example, that in Britain at this time Cambridge-trained mathematical physicists incorporated the Lorentz transformations in an electronic theory of matter that owed no allegiance to the two postulates underlying Einstein's understanding of relativity and took as basic the existence of a dynamical ether. Einstein's research was for the most part considered to be of little relevance.[76] For this reason alone, physicists working in these traditions developed distinctively different histories of electrodynamics, as the 1953 attribution of relativity to Poincaré and Lorentz by the British mathematical physicist and historian Edmund Whittaker demonstrates. (It was in part a reaction against this stance that led to the development of contrary accounts by physicists and academic historians who drew on German sources in asserting the centrality of Einstein's role.)[77] It is important, therefore, to recognize that at the time they were formulated the participant accounts we have reviewed each had only local—if not simply individual—validity as descriptions of the development of relativity, though clearly their very statement could also further the propagation of particular understandings.

This chapter has shown that very soon after 1905 it was possible to interpret electrodynamics in terms of a history of the principle of relativity—surely an important point in the developing presentation of relativity as a distinct subject. However, at the same time physicists and mathematicians also integrated the theory into broader narratives, advocating a range of different research and disciplinary orientations and agendas (in this respect, too, relativity was manifold). In regard to research orientations, the development of a contrast between Newtonian (and Lorentzian) and

76. For example, Warwick writes: "British mathematical physicists did not see Einstein's work of 1905 as a self-contained or self-evident 'theory,' but rather as a modest extension of work due mainly to Larmor and Lorentz. It was not until Minkowski's work of 1909 that these mathematicians became concerned that there might be alternative physical interpretations of the status of the principle of relativity, and only after 1911 that they began to respond to the claim that the 'theory of relativity' denied the existence of the ether." Warwick, "Electrodynamics of Moving Bodies" (1989), 179–80.

77. Whittaker, *A History of the Theories of Aether and Electricity*, vol. 2 (1960), chap. 2: "The Relativity Theory of Poincaré and Lorentz." See also Holton, "On the Origins of the Special Theory of Relativity" (1988), on 196–202.

relativistic physics has been shown to have been a particularly important and productive feature of early participant histories. And we have seen that by 1910, with Planck's Copernican analogy, the historical object had widened from relativity to become the story of a revolution on a world stage. Part IV will consider the process by which accounts of this kind engendered the understanding that what had been at stake in these events was the overthrow of "classical" by "modern" physics.

PART IV

From Classical to Modern Physics

CHAPTER NINE

On the Co-Creation of Classical and Modern Physics

Previous chapters have taken up various facets of the rich practice of physics that commonly escape attention. We have sought new ways of linking experiment and theory, material and conceptual dimensions, and physicists' research and their accounts of its implications, thereby extending our appreciation of the nature of relativity physics and the scope of its history at the same time. Now we turn to recovering the forgotten history of a word—*classical*—and the emergence of a conceptual contrast between classical and modern physics that we now take for granted. Recovering this history will help us link intellectual developments with sociological and cultural understandings of disciplinary change on a still-broader stage. Our study of physicists' (proto)histories showed that specific perspectives on the past were integral to discussions of the meaning of current research programs, and disclosed the complex process in the German physics community that by 1911 resulted in Einstein being widely regarded as the author of a singular theory of relativity. By reconsidering the terms in which we currently describe the period as a whole, the final two chapters of this book will widen our analytic perspective to consider the historical genesis of still more fundamental features of physicists' understanding of change. Doing so will bring us to a further major subject—the development of quantum theory.

Descriptions of the transition from *classical* to *modern* physics have long provided the most powerful framework in which to present the great transformations in physical understanding that occurred at the end of the nineteenth century. However loosely, this language links the physics of the day to its era, gesturing at a common struggle through which fields as disparate as art, literature, and technology broke free of tradition to forge key elements of the modern world. While it pervades writings across the spectrum, from physicists through historians and popular authors, the *origins* of this framework are at present largely unknown. This chapter aims to establish when and why physicists first started to think of "classical" and "modern" physics in the way we now take for granted.

Considering a representative use of these concepts from a major protagonist will indicate the subtle historiographical situation we have to deal with. In a letter to the American physicist Robert W. Wood, Max Planck gave one of his fullest and most emotive descriptions of the groundbreaking work he had undertaken on blackbody radiation. Writing in 1931, Planck described himself as peaceable and inclined to shun dubious adventures. But by 1900 he had been "wrestling with the problem of the equilibrium between radiation and matter for 6 years, without success." Having at last found an empirically accurate formula for the energy distribution in a normal spectrum in October 1900, the final step of providing a theoretical interpretation was therefore worth any price. The whole deed, he said, was "an act of desperation." Then he described the casualties the search had involved and the lifelines to which he held fast: "Classical physics was not sufficient, that was clear to me. For according to it, energy must, in the course of time, transform completely into radiation from matter. In order for it not to do so, we need a new constant that assures that the energy does not disintegrate [indefinitely]." In finding his way, Planck had clung to the two laws of thermodynamics. "Otherwise," he said, "I was ready to sacrifice any of my other physical convictions."[1]

Planck's invocation of classical physics is at once epochal in gesture and highly specific (his reference to the transfer of energy from matter into radiation is a brief account of the consequences of the equipartition theorem, a subject we shall return to). But despite the power and clarity with which Planck links the physics of an era with an emblematic principle,

1. Planck to Robert W. Wood, 7 October 1931, AHQP, Microfilm 66, 5, as translated in Hoffmann and Lemmerich, *Quantum Theory Centenary* (2000), 49–50.

careful historical research has taught us to question whether Planck won his way to modern physics in 1900.

In an extraordinary 1978 study Thomas S. Kuhn argued that despite Planck's formal use of energy elements in forging a new interpretation, Planck did not regard himself as introducing the radical new understanding of energy quanta we now associate with modern physics.[2] Planck's 1900 papers support the point. Rather than emphasizing energy quantization or the law itself, Planck highlighted both its use of two natural constants h and k (later named Planck's and Boltzmann's constants), and the quantitative links it established between electromagnetic theory and the properties of electrons and atoms. Planck had delivered new natural constants and opened glimmers of insight into the processes conditioning the largely inaccessible world of microphysics; but his writings show little evidence that he thought he had won a new kind of physics.[3]

Kuhn convincingly demonstrated that Planck did not articulate a broadly conceived quantum physics in 1900, even if the blackbody law was later taken to found one. But he held fast to the other side of the story, insisting that Planck's approach was still "fully classical," even as late as his 1906 *Lectures on the Theory of Heat Radiation*.[4] Kuhn's recourse to this overarching interpretative framework is widely shared. Allan Needell and Olivier Darrigol have both suggested that "classical physics" is an idealization formed in the 1910s, and that its application to the open situation of turn-of-the-century physics is anachronistic. Nevertheless, benchmark accounts of the period and its major figures routinely invoke the term in

2. Kuhn, *Black-Body Theory* (1987). Several features of Kuhn's work attracted controversy, and a more adequate treatment of Planck has been developed in the work of Needell, Darrigol and Gearhart. See especially Needell, "Irreversibility and the Failure of Classical Dynamics" (1980); Needell, "Introduction" (1988); Darrigol, "Statistics and Combinatorics in Early Quantum Theory" (1988); Darrigol, "Statistics and Combinatorics in Early Quantum Theory, II" (1991); Darrigol, *From c-Numbers to q-Numbers* (1992); Gearhart, "Planck, the Quantum, and the Historians" (2002).

3. Planck's key 1900 paper highlighted the simple structure of the formula he had found and its establishment of a logarithmic relationship between the entropy and energy of the ideal oscillators that he used to model the walls of the blackbody. He thought this relationship made his own blackbody radiation law look more promising than any other, except Wien's (which has that property but doesn't fit the data). Planck, "Theorie des Gesetzes der Energieverteilung im Normalspektrum" (1900), 237. See also Kuhn, *Black-Body Theory* (1987), 110–13, and the overview in Heilbron, *Dilemmas of an Upright Man* (2000), 1–28.

4. He pointed to Planck's emphasis on the parallels between his own approach and Boltzmann's. Kuhn, *Black-Body Theory* (1987), 125.

describing the conceptual environment in which physicists developed their research programs.[5] Once again noting the power and clarity of this language, we need to question whether we should understand the physics of the era as "classical."

Analytically one can offer a clear rationale for Kuhn's insistence. If quantization defines the new physics, and the classical and modern form two mutually exclusive theoretical stances, the fact that Planck makes no recourse to discrete energy values renders his approach classical by definition. Planck's 1931 letter shows that by then he held something like this analytic perspective—and applied it to his earlier struggles. Einstein's later writings offer a similar perspective on the historical grasp of classical physics, describing the physics of the nineteenth century as dominated by a "dogmatic rigidity" in regard to matters of principle, with the success of Newtonian approaches in apparently widely different fields having fostered the belief that knowing Newton's laws of motion together with the necessary masses and forces would suffice to explain all physics. In particular, Einstein noted that Boltzmann had essentially shown that the basic laws of thermodynamics could be deduced from "the statistical theory of classical mechanics"; and he described "classical thermodynamics" as the only physical theory of universal content that he thought would never be overthrown. Our present attitude toward classical physics is general and inclusive and closely follows Planck's and Einstein's depictions of a discipline bound by dogmatic certainties that extended from Newton to the turn of the century.[6] But noting that just as Planck made no explicit use of the concept of energy quantization in 1900, he then made no explicit reference to "classical physics," this chapter will address the question: when,

5. For the critical perspective, see Needell, "Introduction" (1988), xi–xliii; Darrigol, "The Meaning of Planck's Quantum" (2001), 224. For the concept in use, see Stachel's descriptions of Einstein's work on relativity and quantum theory in Einstein, *Collected Papers*, vol. 2 (1989), xvii, xxviii, 137. Büttner, Renn, and Schemmel have recently deployed the term epistemologically without investigating physicists' own understanding. McCormmach's well-known fictional physicist is imagined looking back on his career after World War I, by which time classical physics was indeed becoming a highly important category. See Büttner, Renn, and Schemmel, "Exploring the Limits of Classical Physics" (2003); McCormmach, *Night Thoughts* (1982).

6. Note especially the continuum between classical and statistical mechanics and classical thermodynamics and that equipartition is commonly described as a distinctively classical theorem. See Einstein, "Autobiographical Notes" (1949), 19, 21, 33. We will investigate the specific senses in which Einstein used the term *classical* before 1911 below.

why, and how did Planck and his community first start using concepts of "classical physics" in their work to shape new theory?

Given the widespread use of the language of the classical and modern in fields as fundamental to European nations as education, and as highly visible as art and technology, this empirical question holds the promise of establishing significant links between research physics and cultural history. It is important to note at the outset, however, that in turn-of-the-century schooling, art, literature, or science, any particular deployment of these evocative but open words was likely to meet controversy and contest. After all, the debate over the status, values, and content of classical and modern approaches to secondary and tertiary education in 1890s Germany was so bitter that contemporaries described it as the "Schulkrieg."[7] In this environment we will have to guard against reading a meaning into early uses of such terms that was in fact won only later.

I will focus in the first instance on the concept classical. Earlier chapters have suggested that invocations of modernity were already widespread in terms such as *modern science* and the *modern theory of electricity*. Charting the changing nature of specific deployments of the term *classical* will offer more purchase on the formation and propagation of new theory in physics.[8] A survey of turn-of-the-century writings in the first section will show that the word *classical* was most often used to pick out individual works of value, but that several more synthetic concepts had been invoked by 1900. Importantly they proved controversial—and were used and debated most widely in German language areas. In this early period the term was applied most concretely to mechanics, and section two will explore the relationship between a newly minted classical mechanics and the emerging theory of relativity. Recognizing clear relations with earlier theory provided physicists with strategies for developing new research in relativity; and describing mechanics as "classical" helped superimpose a

7. Albisetti, *Secondary School Reform in Imperial Germany* (1983).

8. In contrast, Herbert Mehrtens has shown that various stances to the category of modern mathematics were central to disciplinary debates in the mathematical community. Mehrtens, *Moderne—Sprache—Mathematik* (1990). A study of uses of the term *modern physics* would show a plurality of invocations in association with specific subjects (such as modern instruments, values, laboratories, studies, etc.). Its use in an epochal sense actually preceded similar treatment of classical, often referring to modern science or physics as that since Galileo, and the introduction to Part III showed that the term was frequently used of electron theory. We shall see that by the 1911 Solvay Council, H. A. Lorentz had paired "classical" and "modern" concepts around turn-of-the-century events.

very specific relationship with the past on earlier discourses of competing methods and radical changes in foundations. The third section will show how different the circumstances were in regard to quantum theory. Here we will have to explore the reasons for the *absence* of classical language, right through to 1910. My final chapter will then demonstrate that the expansion of the concept of classical to cover physics as a whole—in the way we now accept—followed the development of a new understanding of statistical mechanics, and, importantly, occurred in the context of an elite conference for a hand-picked group of twenty-one physicists, the Solvay Council held in Brussels in 1911. Analyzing the long process through which physicists extended the meaning of the word *classical* will provide the foundation for my argument that in the sense we now understand them, classical physics and modern physics were very importantly created at the same time. Despite the power of Planck's retrospective application of the term *classical* to an entire epoch, classical and modern physics were co-creations, mirror image twins of the fault line between the physics of the past and that of the future.

The "Classics" of Physics, circa 1900

First it is important to gain a quick overview of how physicists commonly used the term *classical* at the turn of the century. The significant tensions between the ranges of uses we recover here have long been lost to our present concept of classical physics; but later sections of this chapter will show how important they were to the formation of that concept. Here we chase the changing meanings of a word that designated the value of particular works and helped define broader bodies of practice but also reached into scientists' personality and their sense of disciplinary identity. In the diverse allegiances scientists summoned with particular uses of the term, however fleeting, we can read the traces of sociological and cultural histories incipient in their work to shape the intellectual development of a discipline. I will pay particular attention to the German physics community as the seat of both the earliest general uses of the term *classical physics* and the most fruitful formulations of relativity and quantum physics, but the most common use of the term *classical* was in fact widespread in the late nineteenth century. Significantly, it was particular in focus rather than asserting any synthetic conceptual common ground. Physicists in French-, English-, and German-speaking regions all used the word *classical* most often to celebrate specific works of unusual value—the classics of their discipline. They singled out highly respected works in this way without apparent regard to any underlying theoretical commitments and regardless of their specific field. The term was used of experiments and instruments as well as written papers and books. It clearly referred more importantly to the fruitfulness of a contribution than to whether it provided a finished and definitive treatment of the subject. In a typical formulation, in 1904 Carl Barus gave the Congress of Arts and Science in St. Louis an overview of the physics of the nineteenth century. There he wrote that each of the five main subdivisions of physics (dynamics, acoustics, heat, light, and electricity) possessed its own "sublime classics." He also clearly thought that any unified interpretation of the parallel lines along which physics was developing was a matter for the future, and wrote of the value of some august body formally canonizing "researches of commanding importance."[9]

In fact, a publishing venture with a similar aim had been underway in Germany for more than a decade. Ostwald's Classics of the Exact Sciences provides the most significant institutional embodiment of the concept of the classics in science, appearing from the press of Wilhelm Engelmann in Leipzig under the editorship of the physical chemist Wilhelm Ostwald from 1889. Much later Ostwald described his motivation as stemming from the need to reduce the great disparity between the extent of the published literature in the exact sciences and that portion that has a lasting importance.[10] It may have provided a way of insisting that like the humanities, the sciences had their classics too. At the time, his publisher referred in particular to the character of present-day education in the sciences with its focus on current knowledge. Engelmann described the series as responding to the often-repeated wish to remedy science students' deficiency in historical sensibility and firsthand knowledge of the great scientific works and to make scientific texts easily accessible to readers outside of libraries.[11] Ostwald was assisted by a group of specialist editors, and particular volumes often appeared under the editorial care of

9. Barus, "Progress of Physics" (1905), 353. The paper has been reprinted in Sopka and Moyer, *Physics for a New Century* (1986). "Modern" was used quite generally to describe the physics of the present era rather than more recent physics alone (353).

10. Ostwald, *Lebenslinien* (1927), 55.

11. Wilhelm Engelmann published an announcement accompanying the first volume of the series. See Dunsch and Müller, *Ein Fundament zum Gebäude der Wissenschaften* (1989), 56–59. Price and format were important in making the volumes accessible to students.

still other scientists. For example, while Arthur von Oettingen was responsible for physics (and in 1893 took over general responsibility), in 1895 and 1898 Ludwig Boltzmann edited publications of Maxwell's papers on Faraday's lines of force and on physical lines of force (vols. 69 and 102).

It would clearly be possible for such a series to develop a specific doctrinal stance to particular fields through the careful selection of appropriate works, but the profusion and diversity of the papers thus canonized resists easy classification and emphasizes openness. For example, Oettingen gave Boltzmann the opportunity to present Maxwell's work, despite Ostwald's support of energeticism as a rival to the mechanical worldview and even after Boltzmann had criticized energetics vociferously in debate at the Lübeck Naturforscherversammlung in 1895. (Just how deeply this opposition shaped concepts of the classical will be developed further below.) Further, the ground covered by the series reached very widely from mathematics to physiology. By 1900, 119 volumes had appeared with an alphabetical span from Abel (vols. 71 and 111 in 1895 and 1900) to Wöhler (a paper coauthored with Liebig, vol. 22, 1891). Barus's overview of the physics discipline offered even less interpretative guidance than Ostwald's series. Without explicitly identifying his candidates for sublimity, Barus mentioned, listed, and praised more than a thousand worthy works under thirty-six different headings. "Great" conferred significant praise and the words "classic" or "classical" were used but three times.[12] In a final example of the use of the term to convey the particular value of specific publications, in his 1900 presidential address to the British Association for the Advancement of Science, Joseph Larmor referred to Bertrand's "classical treatise" on infinitesimal calculus, spoke of the modern theory of electricity and magnetism as having received its "classical exposition" in Maxwell's treatise, and referred also to the "classical volumes" of the BAAS with their comprehensive reports on different fields.[13] Larmor regarded Maxwell's *A Treatise on Electricity and Magnetism* as "classical" despite the fact that it exhibited outstanding gaps and difficulties.

12. One example comes in his discussion of interference, where Barus writes that Billet's plate and split lens (1858) belongs to the same "classical order" as Fresnel's work, "as do also Lloyd's (1837) and Haidinger's (1849) interferences." Barus, "Progress of Physics" (1905), 368.

13. Larmor, "Address of the President" (1900), 417, 418, 420. There is a further reference to a classical exposition from Kelvin on 428.

These examples show that the word *classical* was typically invoked in a way that was importantly specific and bound to particular publications. However, two more synthetic and interpretative uses did begin to appear in the 1890s, and they indicate a more intimate sociology that was incipient in scientists' use of specific terms to frame intellectual developments. Interestingly, the terms *classical thermodynamics* and *classical mechanics* arose in a context of contest, each being deployed in the course of heated—and related—battles over the role of established methods in the future development of their field. While pinning down the very first coinages is difficult, those who used these terms in the late 1890s and early 1900s often offered some justification for the particular way they employed them, and sometimes discussed the circumstances in which the label had first been used. Further, exactly what facets of the fields concerned might qualify as "classical" proved controversial (especially in the case of mechanics), and other ways of describing similar concepts existed. As a result physicists were often unsure enough about the meaning and/or the connotations of both classical mechanics and classical thermodynamics to use them rather self-consciously.

In 1898 the mathematical physicist Georg Helm published an account of the historical development of energetics in which he included a section on "Classical Thermodynamics." There Helm argued that the thermodynamical understanding of the relations between heat and work gained up until the mid-1850s had reached such a complete state that it constituted a system in itself, one that was both generally accepted and so well established that "it can certainly be called 'classical.'" He went on to comment that this label had first been given to it by those who opposed all efforts to develop it further.[14] As Helm recognized, the body of work at issue had historically gone under the heading of the *mechanical theory of heat* or *thermodynamics*. While some physicists followed Helm's terminology, these terms clearly remained more common labels for the field as a whole well after he wrote.[15]

14. Helm, *Historical Development of Energetics* (2000), 153–60 on 153. Deltete's introduction provides an excellent overview of Helm's career and a sensitive interpretation of the goals and language employed in the book.

15. The British physicist G. H. Bryan provides another example of a self-conscious use, writing in passing of "'classical' or conventional thermodynamics," in Bryan, "Allgemeine Grundlegung der Thermodynamik" (1903), 143. (Bryan was concerned with instances in which the equilibrium conditions of the conventional thermodynamics of irreversible processes were not exactly fulfilled.)

Helm identified classical thermodynamics with the mathematical formulation of the first and second laws of thermodynamics in the work of Clausius and Thomson in particular. But he discerned two major currents shaping the development of the subject, and his analysis of the past was motivated by the need to show the lineaments of a major battle over the physical worldview. Since 1887 Helm had been a determined advocate of the view that the energy principle provided a new foundation for physical theory, from which mechanics and indeed all physical laws might in future be derived. Impressed by Mach's work, he twinned this with a phenomenological or relational approach that stressed the primacy of defined quantitative relations between phenomena rather than ontological claims. Helm's early writings won the enthusiastic support of Wilhelm Ostwald, and both were invited to address the current state of energetics at the 1895 Naturforscherversammlung. The occasion prompted heated debate, with Boltzmann and Planck subsequently expressing their determined opposition in print. The historian Robert J. Deltete has described the event as "a disaster for energetics and an unhappy occasion for Helm personally."[16] Writing three years after that battle, Helm presented his history as having emerged from struggle. A major aim was to show that energetics needed to be understood as a unified intellectual whole, and Helm's account of the history of thermodynamics sought to demonstrate that the work of his predecessors and contemporaries already exemplified his ideal energetic program to varying degrees. As Helm had it, far from a recent arrival, the aim of founding physics on the energy principle represented long-standing intellectual imperatives.

The second major historical current Helm recognized was the endeavor to establish proofs of thermodynamic relations on the basis of specific molecular models, best exemplified in the atomistic program of the kinetic theory of gases Maxwell and Boltzmann represented so strongly. While he regarded such hypothetical mechanical methods as constituting a historically necessary transitional point, Helm sought to combat the artificial prolongation of the mechanical hypothesis and especially its conflation with energetics.[17] Accordingly, Helm's comment on the origins of the term *classical thermodynamics* may suggest that physicists' had used it to mark their opposition to attempts to develop thermodynamics further by offering a mechanical and statistical foundation for general thermodynamical principles. (Another

16. Deltete, "Helm's History of Energetics" (2000), 6.
17. Helm, *Historical Development of Energetics* (2000), 192–93.

possibility is that Helm's opponents had initially invoked it responding against the aims of energetics.) Whatever the specific context in which *classical thermodynamics* was originally used, it seems likely that Helm was ready to promote the term in order to highlight the role of general principles in thermodynamics and distance that subject from the taste of mechanism contained in a label like "mechanical theory of heat." In any case, we will see below that in the early twentieth century Albert Einstein used a similar contrast between two different approaches to thermodynamics. Einstein drew a distinction between a traditional "classical" approach to thermodynamics and the statistical approach that Boltzmann had pioneered—but unlike Helm he favored the latter.

The term *classical mechanics* was first propagated widely by the writings of Ludwig Boltzmann and Henri Poincaré, some decades after traditional understandings of mechanics had started to attract criticism. It is well known that the late nineteenth century saw a critical reexamination of the foundations of mechanics, focused particularly on the concept of force and the role of absolute space and time in the philosophical foundations of the discipline. These criticisms were developed most notably by Gustav Kirchhoff (1876), Ernst Mach (1883), and Heinrich Hertz (1894), all of whom framed their discussions as critiques of *current* understandings. They directed their attention to the "customary representation" of mechanics or its "present form."[18] The alternative form Hertz gave mechanics by incorporating hidden masses in order to provide an explanation for force was particularly important, and in 1897 Boltzmann responded with the publication of his *Lectures on Mechanics*. There Boltzmann described mechanics as having been criticized for a number of logical gaps, but wrote that rather than attempting to provide the discipline with new clothing, he thought it was more important to understand the strengths of what he described as "the old, classical mechanics."[19]

Boltzmann's vision of just what was old and classical in mechanics proved controversial, however, and those authors who followed him in

18. Kirchhoff, *Mechanik* (1876), preface (Vorrede); Mach, *Mechanik in ihrer Entwickelung* (1883), 6 and 238; Hertz, *Prinzipien der Mechanik* (1894), 5.

19. Boltzmann, *Principe der Mechanik* (1897), v. Boltzmann opened his account with a defense of atomism against misguided phenomenology. His mechanics is discussed in Miller, "Origins, Methods, and Legacy of Ludwig Boltzmann's Mechanics" (1982); Jungnickel and McCormmach, *Intellectual Mastery*, vol. 2 (1986), 190–91. Blackmore points to the tension between Boltzmann's and Poincaré's early uses of "classical" and later understandings. Blackmore, *Ernst Mach*, (1992), 119–20.

commenting on or using the term *classical mechanics* often did so to highlight different aspects of the discipline. In his 1900 textbook on theoretical physics, Paul Volkmann criticized Boltzmann's emphasis on central forces and recourse to atomistic approaches. Volkmann cited Lagrange's analytical mechanics as demonstrating that Boltzmann's focus on these features did not characterize "the system of mechanics as classic, in the sense of its development from Galileo and Newton up to the present times." Volkmann offered an alternative characterization of classical mechanics as operating equally well with both forces and conditions, and stressed the need to avoid a premature introduction of atomism.[20] In other places Volkmann used in preference "ordinary mechanics" to distinguish characteristic understandings of the field from the approach Hertz had developed.[21] When Aurel Voss incorporated a section on "The System of Classical Dynamics" in his 1901 account of rational mechanics for the *Encyclopaedia of Mathematical Sciences*, he explained that he understood by that term "something like the teachings that had succeeded to general acceptance through the influence of French mathematicians in the first half of the nineteenth century."[22]

In contrast to this kind of emphasis on distilling the most important features of current approaches, Henri Poincaré offered an interpretation of classical mechanics bound more closely to a Newtonian interpretation of the three laws. We have already seen that two chapters of his widely read book *La science et l'hypothèse* stemmed from his address to the Congress of Physicists in 1900. They opened a section on Nature, but had been preceded by a section on Force. There, Poincaré began a chapter on "The Classical Mechanics" by presenting a trenchant criticism of the philosophical foundations of mechanics (criticizing concepts of absolute space and time, and simultaneity), before going on to discuss the concept of force in detail. He described the alternative formulations of mechanics that had been developed by Kirchhoff, Hertz, and his own countryman, Jules Andrade.[23] Poincaré's use of the term *classical* may in fact have been

20. Volkmann, *Theoretischen Physik* (1900), 361–63 on 362. Volkmann's pedagogical activities are discussed in Olesko, *Physics as a Calling* (1991), 439–48.

21. Volkmann, "Gewöhnliche Darstellung der Mechanik" (1901).

22. Voss, "Prinzipien der rationellen Mechanik" (1901), 49. Voss noted also that the foundations of dynamics were most often bound together with particular metaphysical assumptions concerning inertia, causation, and etc. He would attempt to develop his approach, as far as possible, in the light of more recent conceptions.

23. Poincaré's book was soon translated into German and English. Poincaré, *La science et l'hypothèse* (1902); Poincaré, *Wissenschaft und Hypothese* (1904); Poincaré, *Science and*

prompted by Andrade, who in 1898 had written of "the classical school of mechanics" to distinguish it from the "thread or relational school." This school gained its name from Ferdinand Reech's 1852 analysis of forces as if represented by a stretched thread attached to a particle and modifying its motion.[24] Andrade linked the adjective *classical* to an interpretative approach or school rather than to the subject of his study. Poincaré was more direct in writing of "classical mechanics," but devoted far more attention to outlining critical perspectives than he did to explicating what he meant by that term. However, when he discusses energetics, it becomes clear that Poincaré limits classical mechanics to Newton's laws alone, despite the apparently broad scope of his reference to the advance involved in moving from "the classical mechanics," "the classical theory," or "the classical system" to the energetic.[25] Poincaré wrote, for example, of the principle of conservation of energy and Hamilton's principle as teaching more than the fundamental principles of the classical theory. They had both extended the realm of applications of mechanics and introduced new restrictions on the kinds of motion possible.[26] Although his language leaves some room for ambiguity, classical thus provided Poincaré with a way of distinguishing a particular interpretation in a specific field rather than describing an epoch or worldview.

This brief overview should be enough to establish the essential features of the introduction of the terms *classical thermodynamics* and *classical mechanics*. Both took shape in a dialogue between those seeking to defend or criticize customary interpretations, asserting in particular the confidence of defenders in the value of the tried and true. The precise

---

Hypothesis (1905). On classical mechanics, see Poincaré, *Science and Hypothesis* (1905), 97–110; Poincaré, *La science et l'hypothèse* (1902), 111–28.

24. Andrade, *Mécanique physique* (1898), 135–44. See also the discussion in Dugas, *A History of Mechanics* (1955), 434–58.

25. The discussion occurs in chapter 8, "Energy and Thermodynamics," Poincaré, *Science and Hypothesis* (1905), 123–24 and 128. In the original French the latter two references are to "la théorie classique," and "système classique." Poincaré, *La science et l'hypothèse* (1902), 139–40 and 143.

26. Poincaré, *Science and Hypothesis* (1905), 124; Poincaré, *La science et l'hypothèse* (1902), 140. Zahar discusses the reasons for Poincaré's distinctions between Newton's theory and Hamilton's equations. Zahar, "Poincaré's Structural Realism and his Logic of Discovery" (1996), 58. Distinctions of this kind reflect Poincaré's readiness to teach structurally equivalent formulations for methodological and philosophical reasons. It would be a mistake to extrapolate too far from Poincaré's usage in this context, and we shall see that he later accepted an extended reading of "classical mechanics" that incorporated the Hamiltonian formulation; but it is as significant to note that he did not originally use the term that generally himself.

content to be understood as falling within such a label was unclear, however. Indeed, the controversy surrounding Boltzmann's use of *classical mechanics* indicates that others could regard the attribution more as an attempt to confer canonical status on a particular interpretation than as an unproblematic recognition of a communal consensus. The tension between atomistic mechanical interpretations and phenomenological energetic approaches was often implicated in such debates—which in some respects became a way of carrying further in new contexts fundamental issues that were raised by the energetics debate in Lübeck. Finally, it is important to note that there were other ways to describe similar concepts, such as "ordinary mechanics" or "conventional thermodynamics." The openness, controversy, and interpretational tension between concepts of classical mechanics and thermodynamics have all been lost to our own concept of classical physics, assuming as it does an unproblematic canonization of traditional approaches.

These two synthetic, interpretative uses were not the only more general concepts of classical raised in the period. Even the notably fluid and open area of research that was electrodynamics attracted the label from Poincaré, despite his clear recognition that that field could not yet be regarded as fully defined. Thinking of Lorentz's development of Maxwell's theory, Poincaré wrote of the classic system of electrodynamics that "is perhaps even now not quite definitive."[27] Nevertheless, classical mechanics and classical thermodynamics are certainly the two most significant formulations, and both concepts played into understandings of the field of alternatives when relativity and quantum theory began to attract serious attention, although they did so in rather different ways. Before considering those developments, however, we have a still broader concept of classical science to consider.

In 1899 Ludwig Boltzmann introduced a new use of the term *classical* in an address on the methods of theoretical physics at the Naturforscherversammlung in Munich. The Viennese physicist wanted to serve up "the digestive tablets of critical philosophy" and thought his audience in the Royal Theater would be ready for them, since they were already surfeited with science if not all initiates of theoretical physics.[28]

27. Poincaré, *Science and Hypothesis* (1905), 225; Poincaré, *La science et l'hypothèse* (1902), 227.

28. The address was Boltzmann, "Methoden der theoretischen Physik" (1905 [1899]). Translated as Boltzmann, "Methods of Theoretical Physics" (1974 [1899]). Here I will follow

Boltzmann's talk was self-consciously a "turn of the century" address, and he used the term *classical* to convey sociological insights into the dimensions of methodological change. Boltzmann described what he regarded as a widespread phenomenon. Approaches that had once looked capable of serving the development of science forever (or other disciplines like poetry, art, or music) might suddenly be revealed as exhausted, prompting attempts to find quite other, disparate methods. Then, Boltzmann wrote, followers of the old approach will find their point of view being described as outdated and outworn, while they in turn "will belittle the innovators as corrupters of true classical science."[29] For Boltzmann this process recurred across the developmental history of all branches of intellectual endeavor:

> Thus many may have thought at the time of Lessing, Schiller and Goethe, that by constant further development of the ideal modes of poetry practised by these masters dramatic literature would be provided for in perpetuity, whereas today one seeks quite different methods of dramatic poetry and the proper one may well not have been found yet.
>
> Just so, the old school of painting is confronted with impressionism, secessionism, plein-airism, and classical music with music of the future. Is not this last already out-of-date in turn? We therefore will cease to be amazed that theoretical physics is no exception to this general law of development.[30]

Boltzmann's account shows how important it is to recognize links to the cultural setting in understanding physicists' use of *classical*. A significant local dimension animated his reference to painting, for example (and the Naturforscherversammlung audience could attend galleries and

---

the English translations of Boltzmann's lectures except where noted. The description of the audience comes from the comments with which Boltzmann introduced the publication of two inaugural addresses, delivered to a very different audience of young students. Boltzmann, "Principles of Mechanics, I, II" (1974 [1900, 1902]), 129. General treatments of Boltzmann have sometimes highlighted his reference to classical physics, without, however drawing attention to the pioneering nature of his discussion or analyzing his use of the term in detail. More attention has been paid to evaluating his philosophy than to any other feature of his work. See Cercignani, *Boltzmann* (1998); Blackmore, *Boltzmann: His Later Life and Philosophy, Book One* (1995); Blackmore, *Boltzmann: His Later Life and Philosophy, Book Two* (1995); Battimelli, Ianniello, and Kresten, eds., *Symposium on Ludwig Boltzmann* (1993); Sexl and Blackmore, eds., *Boltzmann: Gesamtausgabe*, vol. 8 (1982); Broda and Thirring, *Boltzmann* (1957).

29. Boltzmann, "Methods of Theoretical Physics" (1974 [1899]), 79–80 on 79.

30. Ibid., 79–80.

exhibitions in Munich for a reduced charge). Following the lead set by Munich artists in 1894, Gustav Klimt had broken away from the established society of artists in Vienna to found the second secessionist movement two years before Boltzmann spoke. From 1898, widely attended exhibitions prompted heated discussion on the function of modern art. Revealing the costs of reworking traditions and crossing disciplines, the debate became especially pointed when Klimt's design of a ceiling painting for the new University of Vienna was exhibited in May 1900 (six months after Boltzmann's address). Faced with a depiction of "Philosophy" they thought represented inadequately the triumph of light over darkness, eighty-seven professors signed a petition that criticized Klimt's depiction of nebulous ideas in a nebulous form and requested the university to decline the painting (fig. 9.1).[31] Perhaps the most important general point to emerge from Boltzmann's lists is his implication that whatever their virtues, the new trends in art, poetry—and physics—are likely to be less reliable than tried and true methods. But the specific way in which Boltzmann applies *classical* to his own discipline is even more interesting, because he depicts physics as being already in a postclassical phase.

Boltzmann began by outlining the approach that had been fueled by the achievements of Galileo and Newton, seeking explanations along the lines of Newton's theory of gravitation (supplemented by repulsive forces). The task of physics had looked as if it might forever consist of seeking "the law of action of a force acting at a distance between any two atoms and then to integrating the equations that followed from all these interactions under appropriate initial conditions."[32] In the 1870s and 80s Maxwell and Hertz's work developing and confirming the theory of the electromagnetic field had broken through this program. One rich consequence was epistemological. Now, following Maxwell's understanding of the limitations of mechanical models and Hertz's stress on basing understanding on the equations, physicists recognized that it was not their task to say what reality truly is. Rather, they sought a picture (*Bild*) that is both as simple as possible and that represents phenomena as accurately as

31. Klimt's secessionist movement and the convergence of political and cultural dimensions in the debate over the university paintings is discussed in Schorske, *Fin-de-Siècle Vienna* (1981), 225–45. Boltzmann had already left for Leipzig, but the brothers Sigmund and Franz Serafin Exner (professors of physiology and experimental physics at the University of Vienna) engaged in a sustained attack on Klimt's work. See Coen, "A Scientific Dynasty" (2004), 178–238.

32. Boltzmann, "Methods of Theoretical Physics" (1974 [1899]), 82.

FIGURE 9.1 Gustav Klimt's sketch of "Philosophy" for the ceremonial hall of the new University of Vienna survives only in this photograph. Its repudiation by university academics indicates the stakes involved in reworking traditions circa 1900. Courtesy Galerie Welz, Salzburg.

possible. Heilbron has described this "descriptionist" stance as characteristic of turn-of-the century attitudes toward the aims of physical theory.[33]

Alongside the rise of new methods in approaching the particular field of electromagnetism, Boltzmann detailed the possibility of a nonclassical mechanics in the work of Kirchhoff and Hertz (which he regarded as still unrealized) before describing three alternative methodological approaches that had dominated recent physics. The first, energetics, had focused on energy as the primary concept in science. Boltzmann described those who adhered to the second, mathematical phenomenology, as "moderate secessionists" who held the illusory belief that they could represent nature without in any way going beyond experience.[34] His discussion of these programs was temperate, outlining virtues in the experiment of following each; but he also left no doubt that he favored the last, the promotion of molecular and atomic theories. I have mentioned that Boltzmann's address was deeply personal. Despite the respect in which he represented an as yet unfinished program, a sortie into the future, Boltzmann characterized himself as the only person who still grasped the old doctrines with unreserved enthusiasm. He was a monument of ancient scientific memories: "I regard as my life's task to help to ensure, by as clear and logically ordered an elaboration as I can give of the results of old, classical theory, that the great portion of valuable and permanently usable material that in my view is contained in it need not be rediscovered one day, which would not be the first time that such an event has happened in science."[35] This is both the earliest and most elaborate discussion of classical physics that I know. In it Boltzmann marks himself as the first and perhaps the only figure of the nineteenth century who understood himself to be a classical physicist. At least one of his students saw him that way (fig. 9.2), but it is worth noting that few others (if any) took up or commented on Boltzmann's self-definition. Perhaps this is a register of just

33. Ibid., 83–85. On Boltzmann's epistemology, see D'Agostino, "Boltzmann and Hertz" (1990), 381–84. Heilbron's interpretation of descriptionism as a modest and defensive mode enabling the physical sciences to avoid antagonizing established values and higher cultural authorities downplays the extent to which it promoted both criticism of previous certainties and a search for unity independent of particular singular foundations. It has been challenged by Porter's focus on the methodological resources descriptionism provided for the development of the social sciences. See Heilbron, "*Fin-de-Siècle* Physics" (1982); Porter, "Death of the Object" (1994).

34. Boltzmann, "Methods of Theoretical Physics" (1974 [1899]), 93.

35. My translation. Boltzmann, "Methoden der theoretischen Physik" (1905 [1899]), 205; Boltzmann, "Methods of Theoretical Physics" (1974 [1899]), 82.

FIGURE 9.2 Karl Przibram's drawing of Boltzmann as "Der Naturphilosoph." Przibram also drew Boltzmann on a bicycle. This image appeared as front material for Blackmore, *Ludwig Boltzmann*, Book 2: *The Philosopher* (1995). Courtesy Setsuko Tanaka.

how contested this terminological terrain was. Few of Boltzmann's contemporaries would have regarded his stance as conservative in any simplistic sense. They may have been uneasy to imagine him as Karl Przibram did.

Circa 1900, "classical" was a concept with a range of uses. Most often it expressed the value accorded outstanding contributions, and sometimes it designated particular traditional (but contesting) approaches to mechanics or thermodynamics. But it is very important to note that it formed only a minor part of the vocabulary with which physicists discussed the past and considered change. Although Boltzmann's periodization of physics was common, his language was idiosyncratic. By 1900 many agreed that physics had witnessed a long-standing criticism of foundations and the rise of new programs, especially following the success of Maxwell's work. Change was rapid, their field open. Boltzmann alone gave the term *classical* a highly general and importantly sociological meaning, and it is therefore in the German-speaking world that we can see the possibility emerging of speaking of classical physics or classical physicists. This view is supported by different editions of *La science et l'hypothèse*. In the original, Poincaré confines "The Classical Mechanics" to Newton's laws alone and refers in passing to the classic system of electrodynamics, incomplete as it was. Despite his readiness to use the concept of the classic in these contexts, as we saw in chapter 5 that language was notably absent from Poincaré's thinking when he came to consider the epistemological implications of rapid change and describe the play of hypothesis in present science. Significantly, however, Ferdinand Lindemann's preface to the German edition of Poincaré's book described his discussions as extending to the whole of theoretical physics, both "in its classical form as well as in its most recent development."[36] In a German reading, Poincaré's account of change could be subsumed within an epochal understanding of the past as classical. As it happens, Joseph Larmor wrote the introduction for the 1905 English translation and never used classical so broadly.[37]

But do all these fine distinctions in the use of an open and more or less general concept matter? Is this just splitting hairs? To understand how thoroughly particular uses of the concept *classical* could enter the practice of physics and shape perceptions of its periodization, we need now to consider their role in the development of two new theories. Histories of relativity and quantum theory are commonly used to draw conclusions about the nature of modern physics and the struggle required to break free of the classical past. Here I will move in just the opposite direction, analyzing uses of the concept *classical* within these theories themselves, in order to shed new light on their development. In this regard relativity and quantum theory reveal very different dynamics, incorporating concepts of classical theory in different ways and at different times. Displaying these differences will highlight the extent to which the concept of classical physics we now accept was both constructed in the light of the modern and defined by proponents of the new.

36. Poincaré, *Wissenschaft und Hypothese* (1904), iii.

37. Larmor did contrast "the close-knit theories of the classical French mathematical physicists" with the "somewhat loosely-connected corpus of ideas with which Maxwell . . . has (posthumously) recast the whole face of physical science." Poincaré wrote rather of the "old theories of mathematical physics [Les anciennes theories de la physique mathématique]." Larmor, "Introduction" (1905), xiv; Poincaré, *Science and Hypothesis* (1905), 213; Poincaré, *La science et l'hypothèse* (1902), 217.

The "Classical" at Work: Classical Mechanics and Relativity

A full account of the history of special relativity is unnecessary at this point, but it will be instructive to pick out several aspects relevant to our present inquiry. First, recall that Albert Einstein's 1905 paper emerged within an environment in which physicists developing electron theory had begun to take on the mantle of revolutionary figures, offering electromagnetic foundations for a new worldview. Earlier chapters have explored in great detail the close interplay established between theory and experiment as Walter Kaufmann, Max Abraham, and others chased the electron as an exciting alternative to previous attempts to explain nature mechanically or energetically. Theorists drew a sharp contrast between the electromagnetic program and the mechanical worldview of the past in particular. For example, taking advantage of the fact that one could replace the inertial mass of the electron by the apparent electromagnetic mass, in 1900 the German physicist Wilhelm Wien issued the clarion call:

> I have tried to pose the question whether by starting from Maxwell's theory we could not attempt to encompass mechanics, too. This would provide the opportunity of founding mechanics on electromagnetism now that Lorentz has developed a conception of the law of gravitation according to which gravitation is said to be closely related to electrostatics. We would then have to assume that matter is composed of nothing but very small positive and negative charges which are at a certain distance from each other.[38]

Similarly, as well as using the term *classical dynamics* in his 1901 survey of rational mechanics, Voss stated that fundamental changes in the formal foundations of mechanics would be demanded by the new electrical worldview. He wrote that it was at present too early to go far into the matter, however.[39] Aiming at a worldview based on Maxwell's theory by beginning with the particle that linked ether and matter through its charge, the electromagnetic program can be seen as a concrete embodiment of the central importance fin de siècle physicists gave Maxwell's work as a turning point. Boltzmann was one of many who took notice, tracking the

38. In discussion to Lorentz's contribution to the German Naturforscherversammlung in 1900, Lorentz, "Scheinbare Masse der Ionen" (1900), 79–80.

39. Voss, "Prinzipien der rationellen Mechanik" (1901), 40.

rise of the new approach in his inaugural addresses at the University of Leipzig in 1900 and in Vienna in 1902.[40]

Before the young physicist and Patent Office clerk Albert Einstein wrote in 1905, Henri Poincaré had gone further than any other physicist in merging this search for new, electrodynamic foundations with the terms of the existing philosophical critique of central concepts of mechanics. We have seen that Poincaré denied that space and time were absolute and underlined the need to approach the understanding of simultaneity critically in 1902. He also questioned the status of mechanical principles like the principle of relativity and of action and reaction in the context of the new field of electrodynamics. Poincaré finished his discussion of the principles of mathematical physics at the St. Louis Congress of Arts and Science in 1904 by suggesting that perhaps physicists should:

> construct a whole new mechanics, of which we only succeed in catching a glimpse, where inertia increasing with velocity, the velocity of light would become an impassible limit.
>
> The ordinary mechanics, more simple, would remain a first approximation, since it would be true for velocities not too great, so that we should still find the old dynamics under the new.[41]

While Einstein had certainly not read this address, he had read Poincaré's *Wissenschaft und Hypothese* avidly. Along with his reading of Mach and Hume, that book was one of several intellectual resources that contributed to Einstein's intensive engagement with understanding the virtues and limitations of Lorentz's theory of electrodynamics. In the course of his comprehensive research program on light, molecular theory, and electrodynamics (facets of which we consider in more detail later), Einstein recognized that combining the two principles of relativity and the constancy of the velocity of light—and critiquing the concept of simultaneity—offered a new way of linking the two fields of electrodynamics and mechanics. Contrary to his initial expectations Lorentz's electrodynamics could then be reconciled to the principle of relativity.[42]

40. Boltzmann, "Principles of Mechanics, I, II" (1974 [1900, 1902]).

41. Poincaré, "Principles of Mathematical Physics" (1986 [1905]), 298.

42. Darrigol discusses Poincaré's important critical approach to principles in Darrigol, "Poincaré's Criticism" (1995), 10–23. For an excellent overview of Einstein's path to relativity, see Einstein, *Collected Papers*, vol. 2 (1989), 253–74. Also reprinted in Stachel, *Einstein*

When Einstein described his draft paper on relativity in a now famous letter to his colleague Conrad Habicht in May 1905, he wrote that it employed a "modification of the theory of space and time."[43] Einstein abandoned absolute space and absolute time, demanding that concepts of time and space be understood solely in relation to measurement processes in a stated frame of reference (whether at rest or in uniform motion). However, the tools used to measure space and time remained the customary clocks and measuring rods. While motion approaching the speed of light gave unexpected consequences, such as the contraction of length and dilation of time that had also been obtained on the basis of Lorentz's theory, Einstein's approach could readily be regarded as involving a modification of the concepts integral to mechanics as it was usually understood. When he referred to the laws of mechanics in his first presentation of the new approach in 1905, Einstein wrote of "Newton's equations of mechanics."[44]

For those concerned with developing relativity theory, relations with previous theory were particularly important on two different fronts. The understanding that Einstein's approach allowed a reformulation of Newtonian mechanics in which the consequences for bodies moving near the speed of light were unexpected played an important role in physicists' work to understand and develop the new theory. Max Planck's first published response to Einstein's work was to investigate how the "ordinary Newtonian equations of motion" could be generalized in accord with the principle of relativity. He showed that equations of motion could be established in the Lagrangian and Hamiltonian form, and later went on to discuss the principle of least action in relativity.[45] Note Planck's terminology here: with Poincaré, his language will provide a sensitive indication of physicists' readiness or reluctance to bring new labels to mechanics. By late 1907 Einstein was ready to do just that in a major research review of relativity. There, explicitly following Planck while diverging from his language, Einstein showed that a specific vector played the same role in

---

from 'B' to 'Z' (2002), 191–214. On the relations between Poincaré and Einstein, see Darrigol, "Einstein-Poincaré" (2004).

43. Einstein to Habicht, 18 or 25 May 1905, Doc. 27 in Einstein, *Collected Papers*, vol. 5 (1993).

44. Einstein, "Elektrodynamik bewegter Körper" (1905), 891–92.

45. Planck, "Prinzip der Relativität" (1906), 137 and 140. See also Planck, "Dynamik bewegte Systeme" (1907); Planck, "Dynamik bewegter Systeme" (1908); Planck, "Prinzip der Aktion und Reaktion" (1908).

relativistic mechanics as the force vector in "classical mechanics." He described the reformulations of the equations of motion of material points he drew from Planck as demonstrating "so clearly," "the analogy between these equations of motion and those of classical mechanics."[46] Einstein's terminology probably stemmed from his awareness of Boltzmann's work and Poincaré's *Wissenschaft und Hypothese*, but did not retain the distinctions Poincaré had stressed.

A second critical relationship to previous theory concerned the view that Einstein's approach constituted an internally consistent reformulation of Lorentz's theory of electrodynamics on the basis of two principles. Jakob Laub and Max Laue both offered relativistic treatments of fields in optics that Lorentz had previously addressed. They highlighted the fact that Einstein denied the existence of the ether, and Laue showed that results that Lorentz had previously obtained in optics could be achieved much more efficiently with Einstein's approach.[47] In a large proportion of the few early papers that took up relativity, then, relations with past theory were critical. Physicists sought to reinterpret concepts that were valid in traditional mechanics in ways appropriate to relativistic mechanics, or to show how Einstein's techniques could be used to establish a more efficient approach to fields that Lorentz had already treated.[48]

In his 1908 paper Hermann Minkowski highlighted both these relationships particularly sharply, going to considerable lengths to clarify the relations between "classical mechanics" and the postulate of relativity as it had been developed as a basis for electrodynamics. He reported that these were often erroneously regarded as being in conflict, and devoted an extensive appendix to setting the matter straight.[49] Recall that Minkowski offered a general form of transformation equations in which the constant c appeared. He then showed that classical mechanics could be understood as involving invariance against the form that the transformation equations took when the constant c was assumed to be infinite. In contrast, according

46. Einstein, "Relativitätsprinzip" (1907), 433–36 (on force) and 414.

47. Laub, "Optik der bewegten Körper" (1907); Laue, "Mitführung des Lichtes" (1907); Laue, "Wellenstrahlung einer bewegten Punktladung" (1909).

48. Einstein was more concerned with exploring the implications of his approach for the understanding of energy and mass and with attempting to establish alternative means of testing the theory.

49. Minkowski, "Grundgleichungen" (1908). He forecasts the topic on 56, and the appendix is on 98–111. On Minkowski, see Corry, "Minkowski and the Postulate of Relativity" (1997); Walter, "Minkowski, Mathematicians, and the Mathematical Theory of Relativity" (1999).

to the new postulate of relativity the constant took a finite value: that of the velocity of light; and the traditional assumption could be regarded as an approximation to the exact invariance of natural laws for this finite value. The last section of the paper showed that replacing infinity with the finite constant *c* allowed the axiomatic structure of mechanics to be brought to substantial completion.[50] As we saw in chapter 7, Minkowski's work quickly became central to understandings of the new field. Almost all those who wrote on relativity in Germany after the appearance of his paper did so in part in response to his presentations, and the widely perceived elegance of his approach helped persuade several previous opponents like Paul Ehrenfest and Arnold Sommerfeld to take the theory of relativity seriously. Interestingly, in his celebrated space and time lecture to the broader audience of the Naturforscherversammlung in Cologne in September 1908, Minkowski did not use the language of the classical but wrote instead of Newtonian mechanics.[51]

Earlier chapters have shown that together with papers from Lorentz, Kaufmann, and Planck, the work of Einstein and Minkowski provided the most important original source for many physicists' understanding of the emerging field of relativity theory. Einstein and Minkowski's discussions assume the transparency of "classical mechanics," and are unlikely to have led to serious confusion. Nevertheless, it is worth focusing on the term for a moment. As we saw, the controversy attending Boltzmann's earlier writings and range of formulations in play shows that just what should be understood as *classical* in mechanics was open to debate. Planck's example indicates that physicists were able to recognize and elaborate the relationship that Einstein and Minkowski described as holding between an old and new mechanics, without using the adjective *classical*. Even Minkowski's protégé Max Born declined to follow his teacher's 1908 usage. Born echoed Minkowski's use of the term when referring directly to Minkowski's work, but wrote more often of the "old mechanics," perhaps intending to stress the contrast between past and present formulations even more strongly.[52] The choice such physicists made to qualify

50. Minkowski, "Grundgleichungen" (1908), 99.

51. Minkowski, "Raum und Zeit" (1909). Reprinted in Lorentz et al., *Das Relativitätsprinzip*, 7th ed. (1974 [1913]).

52. The reference to "classical mechanics" comes in Born, "Träge Masse" (1909), 571. As we saw in chapter 7, Born explicitly sought to use the analogy between the old and new to develop relativistic analogues to concepts important in the customary kinematics, without complete success. In his major paper, Born wrote variously of "Newtonian," Galileo-Newtonian"

mechanics with a term other than *classical* may stem from uncertainty about the presumed content of *classical mechanics* (even though Einstein and Minkowski used it inclusively, without any explicit reference to the doctrinal issues implicated in earlier discussions). It may also reflect a reaction against the broader connotations of the term—potentially linking mechanics to cultural values in other fields, and involving a very different set of associations than either *mechanics* alone, or alternatives like *ordinary*, *customary*, *Newtonian*, or the *old* mechanics.

Just what the connotations of *classical* were was likely to be changing subtly as the word was freed of its original association with particularly important works, whatever their nature, to find more general uses (contested as these sometimes were). In his 1909 book *Great Men*, Wilhelm Ostwald expanded the meaning of the word still further to describe a characteristic type of scientist. Helmholtz, Mayer, and Faraday were all exemplars of what Ostwald called the "classical temperament." In contrast to temperamentally romantic counterparts like Davy, Liebig, and Gerhardt, they displayed a slow spiritual reaction time and a steady spiritual heartbeat. Ostwald's coinage helped him frame psychological insights into leading figures in the chemical and physical sciences, and the book went through four printings within a year.[53] Still other valences would have been imported from the way that a distinction between classical and modern approaches was invoked in fields outside physics. It is difficult to gauge the precise effect such connotations might have had, but it is worth remembering the cognitive difficulty workmen expressed when ordered to give a "classic treatment" to American displays of apparatus in mechanics and electricity at the 1900 World's Fair in Paris. Further, we cannot assume that contrasts between classical and modern were always read in favor of the new—especially early in the century. Many contemporaries, scientists among them, took a notably wary approach to modern art like cubism, and to the value of modern forms of schooling with their stress on contemporary languages and business needs, rather than classical languages and traditional educational values. In Germany, for

---

or "old" mechanics, conceptions of space and time, or kinematics; and this was contrasted with the "new" or "relativistic" mechanics or kinematics. See for example Born, "Theorie des starren Elektrons" (1909), 1–6.

53. Ostwald reworked the traditional language of temperaments within a binary framework, describing the classical type as likely to exhibit phlegmatic and melancholic features rather than sanguine and choleric. Ostwald, *Grosse Männer* (1909), 371–88. See also McCormmach, *Night Thoughts* (1982), 53–54.

example, bureaucratic reforms in 1900 decreed that the two major forms of secondary education were now to have equal privileges with regard to entrance into university studies. Similarly, technical institutes gained the ability to award doctorates in 1899. But despite formal parity and heated discussions about the appropriate content of different educational streams, contemporaries would have had little trouble identifying which forms of education were suitable for the intellectual elite. Classical certainly meant traditional: but also current and foundational.

In this period, then, the word *classical* was being employed more frequently, particularly within Germany; its range of meanings was being expanded; and, as a result of new uses, its connotations were likely to be changing. In such an environment German physicists proved increasingly ready to apply it to what they might otherwise have described as customary or ordinary mechanics, but often did so somewhat self-consciously. In lectures published in 1910, for example, Max Planck wrote of the transformation that was underway, "from the so-called classical mechanics of the mass-point, which has until now assumed to be generally valid, towards the general dynamics arising from the principle of relativity."[54] Planck's linguistic caution is revealing and his adoption of the new term illustrative. The language of Einstein and Minkowski was to be widely followed in the next decade, making its way into textbooks and popularizations. In 1911 Max Laue began the first textbook on relativity with a section on the principle of relativity as it was used in classical mechanics, followed by a discussion of the great changes that applying the principle to electrodynamics had wrought. (In effect he followed the same structure Einstein had used in his 1907 research review.)[55] Ludwig Silberstein's 1914 English-language text *The Theory of Relativity* similarly drew a contrast between the principle of relativity of classical mechanics and the modern doctrine of relativity. Silberstein's background in German was important. In contrast, Ebenezer Cunningham compared Newtonian dynamics with the more general principle of relativity. In general English authors held different understandings of relativity from those in Germany, much more strongly shaped by the work of Joseph Larmor; and the linguistic framework within which they discussed the theory was also subtly different.[56]

54. Planck, *Theoretische Physik* (1910), 7–8. He showed no such self-consciousness in describing Lorentz's studies as classic on 8–9.

55. Laue, *Relativitätsprinzip* (1911).

56. Silberstein, *Theory of Relativity* (1914). On English attitudes to the development of relativity theory in Germany, see chap. 8 of Warwick, *Masters of Theory* (2003).

The point holds for physicists of other nations too. For the widespread international use of common concepts of the classical in physics, we will have to wait on the development of quantum theory discussed below, and on the popularization of relativity from 1919. However, by the time Einstein published his popular account of special and general relativity in 1917, we can say that within the German physics community at least, the specific meaning physicists had given *classical* in their application of the term to mechanics and quantum theory was likely to overshadow the connotations it still summoned from other cultural realms.[57]

What have we seen? Initially developed to describe a contrast with several critical, nontraditional interpretations of mechanics and often enmeshed in doctrinal debates, the concept of classical mechanics was soon transferred to discussions concerning relativity, particularly in Germany. There it was used more inclusively and increasingly widely, without ever completely displacing alternative terms like the *customary* or *old* mechanics. In major texts and popularizations, however, *classical mechanics* was introduced as a foil to a new, more general dynamics that involved modifications or transformations of its key concepts. It is worth emphasizing that this involved a highly specific relationship between the old and new. Within the German physical community, the classical has now been incorporated within the modern. For those working with relativity these are not rival methods vying for attention in the way Boltzmann depicted the classical and the new engaged in a competitive struggle to claim the future. In the context of its specific uses in mechanics, the classical-modern divide has become instead a relationship of historical succession and conceptual incorporation of the old by the new. It expresses the successful reworking of the traditional (mechanics) into a new framework for knowledge.

It is important to note, however, that this is still not classical *physics*, but a more limited and defined subset of the physics of the past. To understand why this is the case, we need to remember that there were already revolutions afoot involving new foundations in the endeavor to replace the mechanical worldview with the electromagnetic worldview. Over time relativity changed the nature of this debate, in part through its adherents beginning to describe the lineaments of a worldview in which the principle

57. Einstein's development of general relativity did not change this typical argumentative structure in either research or popular contexts, with the "classical" principle of relativity preceding discussions of the generalizations required for the special and general theories. See for example, Einstein, "Grundlage der allgemeinen Relativitätstheorie" (1974 [1916]); Einstein, *Relativitätstheorie: Gemeinverständlich* (1917).

of relativity held a primary place, and in part through efforts to break through the primary distinction between mechanics and electrodynamics that animated much contemporary discussion. Einstein gave expression to both these endeavors in brief remarks in his research papers, describing the search for a physical worldview that would correspond with relativity (rather than be based on it), and referring briefly to the current electromechanical worldview as a single perspective.[58] In his 1909 address on the unity of the physical worldview, Max Planck pressed the point at some length, writing that electrodynamics and mechanics could not be sharply separated from each other. Did the emission of light belong to mechanics or electrodynamics, after all? He preferred to see the distinction between reversible and irreversible processes as being more fundamentally important to the physics of the future.[59] Apart from his work on relativity, a significant portion of Planck's thinking on the needs of the physics of the future was devoted to understanding his own theory of blackbody radiation. Now let us turn to that subject, for we need to look further than classical mechanics and relativity to reach the concept of classical physics we presently accept.

The "Classical" at Work? Equipartition and Quantum Theory, 1900–1910

I have already described the main features of Planck's orientation to his work around 1900. Whatever Planck thought he had done, and despite the strict analytical pairing evident in many treatments of the classical-modern contrast as pre- and postquantization, the fine studies of Klein, Kuhn, Needell, and Darrigol have shown that within the physics community the two sides of what we now think of as the quantum revolution were in fact developed separately, and to some extent independently. That is to say: in the decade from 1900 Planck's work was increasingly recognized as founding a new quantum physics, rather than being thought of simply as an extremely successful law of blackbody radiation. In this process

58. Einstein, "Trägheit der Energie" (1907), 371–72.

59. Planck, "Einheit des physikalischen Weltbildes" (1909), 64 and 68. Needell has highlighted the importance of Planck's somewhat idiosyncratic commitment to an absolute (rather than statistical) interpretation of irreversibility, right up to 1914. See Needell, "Introduction" (1988). We might trace the contrast between absolute and other through different fields as well as across theory and experiment.

the argument that it contradicted previous theory rather than represented an unproblematic extension of past thinking was as important as the recognition that it involved quantization of energy. In regard to the latter, historians have shown that in 1905 and 1906, James Jeans, Albert Einstein, and Paul Ehrenfest wrote papers in which they argued that the finite size of the constant *h* in Planck's formula involved a new treatment of energy. Their deep familiarity with statistical mechanics in general and Boltzmann's studies in particular enabled this understanding, and the relative rarity of both skills goes a long way toward explaining why the argument was only made so many years after Planck's interpretative papers.

The first argument, that prior theory also gave an answer to the blackbody problem—in conflict with Planck's law—depended centrally on understandings of the equipartition theorem and its applicability to matter and radiation. Detailed below, this argument was first stated at about the same time that Planck wrote in 1900, and the equipartition theorem is now uniformly identified as providing the "classical" approach to blackbody radiation (known as the Rayleigh-Jeans law and resulting in the "ultraviolet catastrophe"). Planck's 1931 letter to Wood puts the matter in these terms, for example. Despite this present linkage, I will show here that the language of the classical was brought into quantum theory in a very different way from its early incorporation in relativity physics: it came late and into far less promising ground. Indeed, the equipartition theorem had been controversial from its origins in the 1870s; was always recognized as being inadequate in its application to blackbody theory; and was first described as "classical" only in the papers and discussions surrounding the 1911 Solvay Council. The equipartition theorem, then, was *made* classical, a process that simultaneously extended the conceptual grasp of that word to cover far broader theoretical expanses than mechanics alone. This amounted to the christening of a new form of classical physics, turning on a very different epochal fulcrum from the one Boltzmann had described. Subsequently, both the specific, emblematic identification of the equipartition theorem as classical, and the broader notion of the existence of a classical form of physics in the nineteenth century, have consistently been read back into a period in which the physicists' who used the equipartition theorem originally considered both the theorem itself, and the complex of theory into which it played, in quite other terms.

Because the retrospective grasp of the concept of classical physics has played such an important role in the histories of quantum physics commonly offered by physicists, historians, and popular writers, the account to follow will be significantly more finely grained than was necessary in describing the incorporation of classical mechanics into relativity. We first discuss attitudes toward the doctrine of equipartition circa 1900, before outlining the framework in which James Jeans and Albert Einstein developed strong arguments that Planck's work contradicted equipartition from 1905. In both cases I will be able to demonstrate specific reasons why these physicists *did not* use the term *classical*. Then we turn to the presentation of quantum theory and equipartition in survey papers and other forums between 1906 and 1910. This will establish the fact that physicists were able to integrate the new theory into a number of different interpretative accounts of radiation physics that stressed sharp conflicts with current theory, without invoking the concept *classical*. In sum, the following sections will explain why classical language came late to quantum theory. My final chapter will show why when it finally arrived, it came in a veritable flood.

"The Maxwell-Boltzmann Doctrine of the Partition of Energy" circa 1900

The equipartition theorem was an important component of statistical mechanics as it had been developed by Maxwell and Boltzmann, and held that in any mechanical system at thermal equilibrium each degree of freedom will possess the same average kinetic energy.[60] Both its authors recognized problems in applying the theorem to the specific heats of gases, and from its inception it had met a determined critic in Lord Kelvin. Indeed, in his contributions to understanding the state of physics at 1900, Kelvin identified two clouds on the horizon of the dynamical theory of heat and light. His first and now most notorious cloud related to the failure to detect the earth's motion through the ether in the celebrated Michelson-Morley experiment (and Kelvin focused on this field alone in his address to Paris). But in the more extensive overview he gave to the Royal Institution in May 1900 and published a year later, Kelvin actually

60. Each component or mode of a system that is capable of independent motion constitutes a degree of freedom. For example, a material point has three degrees of freedom corresponding to its three independent translational motions. An atom with extended structure should have at least six, with three possible independent rotations in addition to translational motion. Each degree of freedom will possess an average energy of $1/2kT$, where k is Boltzmann's constant and T is the absolute temperature. In general, measurements of specific heats indicated fewer degrees of freedom than expected, while the multitude of spectral lines suggested molecules must have considerable complexity, with far more than six variables.

devoted far more attention to the second cloud, which was what he described as the "Maxwell-Boltzmann doctrine regarding the partition of energy."[61] Kelvin had "never seen validity" in Maxwell's demonstration of equipartition. He also held the doctrine extremely unlikely to be true, despite its attraction as a statement in pure mathematical dynamics. But going on to consider its implications for thermodynamics, and finding equipartition to be destructive of the kinetic theory of gases (its first and most obvious realm of empirical application), Kelvin could not help but describe it as a cloud on dynamical theory.[62] After discussing the serious discord between experimentally determined ratios of the specific heats of gases and those predicted on the basis of the doctrine of equipartition, he went on to offer a semi-empirical treatment of the probability distributions involved in numerous test cases with molecules enclosed in different containers. This was to be the last installment in his long-standing critical endeavor to establish specific instances in which the doctrine might be said to apply, and to demonstrate its truth or failure in those circumstances.

While Kelvin saw some possibility of overcoming the challenge posed by the Michelson-Morley result through FitzGerald and Lorentz's contraction hypothesis, his antidote to equipartition was rather different and devastatingly straightforward. In 1900 Lord Rayleigh had discussed difficulties applying equipartition to gases and general dynamics, remarking that a way was needed to escape "the destructive simplicity of the general conclusion relating to partition of kinetic energy," particularly in its application to systems involving large amounts of potential energy.[63] For Lord Kelvin, there was an obvious response. The simplest way to escape the conclusion was to deny it: "and so, at the beginning of the twentieth

61. Kelvin, "Nineteenth Century Clouds" (1970 [1900]), 324. His discussion of equipartition extends over almost thirty pages, 329–58. Brush discusses debate on the equipartition theorem in Brush, *The Kind of Motion We Call Heat* (1976), 356–63. Klein emphasized the profile of equipartition and Planck's wary engagement with statistical mechanics in explaining his lack of reference to it in 1900. Klein, "Max Planck and the Beginning of the Quantum Theory" (1962).

62. Kelvin, "Nineteenth Century Clouds" (1970 [1900]), 333.

63. Rayleigh, "The Law of Partition of Kinetic Energy" (1900). By way of example Rayleigh contrasted two situations that equipartition had to treat equivalently, applying as it does to kinetic energy without considering potential energy. The kinetic energy involved in a given motion in the line of junction between two atoms remains the same, even though in the case in which they are bound together as a diatomic molecule, significant potential energy may have to be overcome to allow such motion, in contrast to the case when only the feeblest tie relates them.

century, to lose sight of a cloud which has obscured the brilliance of the molecular theory of heat and light during the last quarter of the nineteenth century."[64] Indeed, Planck and the many others who first used his law without investigating its theoretical basis took no apparent notice of the theorem, and for several years it played only an occasional role in physicists' treatment of the blackbody. Carl Barus's 1904 overview shows that it was then still possible to treat equipartition and blackbody theory quite separately. Discussing the kinetic theory of gases, he wrote, "the difficulties relating to the partition of energy have not yet been surmounted. The subject is still under vigorous discussion, as the papers of Burbury (1899) and others testify." Without mentioning Planck, he discussed the blackbody in his section on radiation, describing it as the means for developing a new pyrometry (which was the field on which Barus reported for the 1900 congress).[65]

Despite this early separation, our present understanding of the great changes involved in turn-of-the-century physics has been deeply marked by the links that over time were forged between the theory of blackbody radiation and equipartition; and what later came to be termed *classical physics*. Lord Rayleigh became the first to apply equipartition to the radiation involved in the blackbody in a second paper in 1900 (without making any reference to classical theory). There he offered a modification of the displacement law that Wilhelm Wien had established in 1896 on thermodynamic grounds. Suspicious of the theoretical foundations of Wien's law, Rayleigh developed a formula that he thought "more probable a priori," though he wrote that his speculations were hampered "by the difficulties which attend the Boltzmann-Maxwell doctrine of the partition of energy. According to this doctrine every mode of vibration should be alike favoured; and although for some reason not yet explained the doctrine fails in general, it seems possible that it may apply to the graver modes."[66] In Rayleigh's formula equipartition applied to the energy distribution for long wavelengths, but an exponential factor prevented the effect of the doctrine from coming into play at short wavelengths or high frequencies (the ultraviolet part of the spectrum—hence later descriptions of the need to avoid the "ultraviolet catastrophe"). His note mixed

64. Kelvin, "Nineteenth Century Clouds" (1970 [1900]), 358.

65. Barus, "Progress of Physics" (1905), 364–65. Burbury was particularly important in clarifying Boltzmann's contributions to kinetic theory.

66. Rayleigh, "Remarks upon the Law of Complete Radiation" (1900).

pragmatic concerns with the conviction that equipartition had a definite, if limited, role to play in the theory of blackbody radiation. It drew little immediate comment, though in 1902 Rayleigh noted that he had intended to emphasize the form his law assumed at the low end of the spectrum, and that Planck's formula seemed best able to meet the experimental observations.[67] However, from 1905 Rayleigh's views and his interest in understanding equipartition became an important stimulant to the development of somewhat sharper statements of its consequences, in a period that saw the formulation of more potent mixes of pragmatism and conviction.

Three young theorists in particular presented the view that there was an alternative form of the blackbody law based on equipartition that, though empirically inadequate, possessed a clearer theoretical foundation than the law that Planck had developed; and this despite the fact that its unrestricted application was already so evidently problematic. However, while James Jeans, Albert Einstein, and Paul Ehrenfest all refused to lose sight of equipartition, it is important to note at the outset that they did not initially describe the consequences of that theorem as representing classical physics. This point has seldom been recognized in the historiography of their work, a fact that has had several implications for our understanding of the period. Firstly, it has undermined our appreciation of the character of the research programs they followed. This is particularly true of studies of James Jeans, who has usually been described as a classical physicist despite his incorporation of the equipartition theorem into a highly unconventional understanding of the blackbody. Secondly, identifying the equipartition theorem as classical has also obscured the status that contemporaries accorded different branches of theory (and especially statistical mechanics), thereby minimizing the interpretative and evaluative work involved in physicists' assessments of reliable methods in a new field. Finally, too early recourse to the concept of classical physics has limited our understanding of the process by which a new worldview was won from the study of the blackbody and quantum theory.

James Jeans and the True and Final State of Nature

The most dramatic need for recognizing a limitation on the application of equipartition came from the fact that as a continuous medium, even in

67. See the footnote added to the version of the paper published in Rayleigh, *Scientific Papers, 1892–1901* (1902), 484.

an enclosed space, the ether possessed an infinite number of degrees of freedom, or "modes of vibration" as Rayleigh put it. If, as equipartition demanded, equal energy went to each mode of vibration in the combined ether-matter system of the blackbody, over time all the energy would be transferred from matter, with a finite number of degrees of freedom, into the ether with its infinite degrees of freedom. This is the view Planck summarized in 1931, and the young, Cambridge-trained mathematical physicist James Jeans expressed it very clearly as early as 1901. In his first publication on the theory of gases, Jeans addressed the long-established difficulties facing equipartition by arguing that while equipartition did establish the proper distribution of energy in systems in equilibrium, the empirical failure of the law showed that gases in normal states had not yet reached equilibrium. In response Jeans built up a kinetic theory in which gases were subject to a dissipative function related both to the nature of molecular collisions and their interaction with ether.[68]

The problem of the infinite degrees of freedom in ether also underlay Jeans's considerations in 1905 when he turned explicitly to consider the enclosed and therefore nondissipative system of matter and ether as it was embodied in the blackbody (in a paper prompted in part by Lord Rayleigh). Jeans argued, however, that in this circumstance also, his calculations showed that while the general consequences of the equipartition theorem expressed the "true final" state of nature, this would in fact take millions of years to develop.[69] Jeans supposed that energy was transferred from matter to particular frequencies of radiation in the ether when the ether was set into vibration as a result of the collision of molecules. Relating the time-scale of molecular collisions to the frequency of vibration of the ether offered him a way of distinguishing between different phases in the interaction between matter in motion, and ether. In the first phase, energy would be transferred from matter relatively quickly to the lower frequencies and wavelengths of radiation in the ether. After this it would take much longer for energy to be transferred into the quicker modes of vibration.

68. See Jeans, "Distribution of Molecular Energy" (1901), 398. Jeans also addressed equipartition in key sections of his book. Jeans, *The Dynamical Theory of Gases* (1904). For a biography, see Milne, *James Jeans* (1952). Two more recent and more detailed accounts of Jeans's blackbody work also make essential use of an anachronistic concept of the classical, although Hudson has recognized difficulties in the practice. Hudson, "James Jeans and Radiation Theory" (1989); Gorham, "Planck's Principle and Jeans's Conversion" (1991); Hudson, "Classical Physics and Early Quantum Theory" (1997).

69. Jeans, "The Dynamical Theory of Gases" (1905).

Jeans regarded equipartition to indicate the ultimate state of nature but thought this state was never realized in empirical situations. An interchange with Rayleigh in *Nature* pushed both physicists to sharp formulations of the infinity problem that equipartition posed.[70] It also led them to specify not just the form but also the coefficients of the formula expressing the consequences of equipartition for radiation, which has since been known as the Rayleigh-Jeans formula for blackbody radiation (with some exceptions I will note below). In addition, Rayleigh pushed Jeans to explore the relations between Planck's approach and his own, since Rayleigh declared himself unable to follow Planck's reasoning.[71] The result was a short, critical paper in which Jeans focused on Planck's use of energy elements (whose magnitude was given by the constant h multiplied by the frequency of the radiation). These were necessarily finite, "a sort of indivisible atom of energy" that had been introduced to simplify the calculations. As far as Jeans was concerned this arbitrary element had no physical meaning, and could legitimately be removed by taking the limit in which it was equal to zero. Doing so led to an equation relating temperature to energy, which could in fact be obtained much more readily from the theorem of equipartition. For Jeans, the methods of statistical mechanics had allowed a greater specificity than the general thermodynamic reasoning that Wien had earlier followed; and stopping short of putting $h = 0$, Planck's use of statistical mechanics had simply not gone far enough.[72]

While Jeans is universally regarded as presenting the classical approach, he is unlikely to have accepted such a description in 1905. In fact, one contemporary described his approach to kinetic theory in general as a bold attempt to break away from traditional methods through his use of

70. See Rayleigh, "The Dynamical Theory of Gases" (1905); Jeans, "The Dynamical Theory of Gases" (1905).

71. On the formula, see Rayleigh, "The Dynamical Theory of Gases and of Radiation" (1905), 55 (for Rayleigh's inability to follow Planck); Jeans, "On the Partition of Energy between Matter and Aether" (1905); Rayleigh, "The Constant of Radiation as Calculated from Molecular Data" (1905).

72. He wrote: "The similarities and differences of Planck's method and my own may perhaps be best summed up by saying that the methods of both are in effect the methods of statistical mechanics and of the theorem of equipartition of energy, but that I carry the method further than Planck, since Planck stops short of putting $h = 0$. I venture to express the opinion that it is not legitimate to stop short at this point, as the hypotheses upon which Planck has worked lead to the relation $h = 0$ as a necessary consequence." Jeans, "A Comparison between Two Theories of Radiation" (1905), 294.

the notion of dissipation.[73] Similarly, while Jeans emphasized the power of statistical mechanics rather than general thermodynamical reasoning, he simultaneously denied that equilibrium conditions were actually met in experimental conditions—thereby short-circuiting the primary basis for the application of most thermodynamic and statistical arguments in the work of other theorists. Thus both Jeans and his contemporaries certainly saw his endeavors as bold, speculative approaches to new understandings of the relations between radiation and ether. In addition, his awareness of the controversial standing of equipartition suggests that Jeans would have hesitated to describe that particular theorem as classical, despite his own conviction of its ultimate purchase. Later, Jeans sought to understand the implications of the radiation laws for the motion of electrons in various orbits as a likely mechanism for the absorption and emission of radiant energy, and addressed what he saw as two particularly radical elements of Planck's approach. But despite describing Planck's law as deeply problematic for implying a "non-Newtonian mechanical system," he did not turn to the language of classicism to describe either those features that Planck's law denied, or his own approach.[74]

Albert Einstein and the Failure of the Application of Molecular-Kinetic Theory

In 1905 a second young physicist, Albert Einstein, emerged from several years' work analyzing perceived lacunae in Boltzmann's studies and developing an independent approach to the field that Willard Gibbs was to summarize in his masterful 1902 study of statistical mechanics. In the light

73. See Bryan, "Three Cambridge Mathematical Works (Book Review)" (1905), 602. Bryan, a contributor to the criticism of equipartition, responded to Jeans's arguments on that topic by suggesting that "while distributions satisfying Maxwell's law of equipartition are theoretically possible, other distributions may exist, and may, indeed, represent a normal and *persistent* state of affairs even in *conservative* systems" (Bryan's emphases). For a similar view of Jeans, see also the approach Planck took, discussed below.

74. Planck's work was radical and non-Newtonian because it necessarily involved a discontinuous, atomistic treatment of the energy of wave motions in the ether (and not just the energy exchange between the ether and the resonators that Planck had used to model matter). In addition, startlingly few degrees of freedom received any energy at all. Jeans wrote: "Planck's treatment of the radiation problem, introducing as it does the conception of an indivisible atom of energy, and consequent discontinuity of motion, has led to the consideration of types of physical processes which were until recently unthought-of and are to many still unthinkable." Jeans, "Non-Newtonian Mechanical Systems" (1910), 943. He discusses equipartition on 954.

of Gibbs's work Einstein later regarded his early publications as worthless, but he had developed powerful theoretical tools and had simultaneously been reflecting on a broad range of topics. Jürgen Renn has usefully characterized his approach as an interdisciplinary atomism, seeking to unify disparate fields of study through a consistent mathematical framework.[75] By 1905 Einstein had brought his understanding to a stage of fruitful maturity. Significantly (as noted previously), his series of papers opened with what Einstein called a "very revolutionary" argument for the quantization of light when writing to his friend Conrad Habicht. In print, Einstein described his considerations as a "heuristic point of view" that addressed a "profound formal difference" in the theoretical conceptions with which physicists approached gases and other ponderable bodies and Maxwell's theory of the electromagnetic field.[76] In Maxwell's theory, energy was considered as a continuous spatial function that could not be described by a finite number of quantities and spread continuously over a steadily continuous volume. In contrast, the energy of material bodies could not be broken up into arbitrarily many, arbitrarily small parts. Einstein saw little reason to doubt the value of Maxwell's theory in the purely optical realms in which it had initially been developed, but he looked for weak spots in phenomena in which the emission and absorption of light were involved. In the paper to follow, Einstein used an analogy between the statistical study of gases, and the form of Wien's law of blackbody radiation for short wavelengths, in order to argue that in some circumstances at least the energy of light had to be treated as having an atomic constitution. This was his revolutionary argument, and few of his contemporaries found it persuasive.

I would like to emphasize two features. First, as I stressed in chapter 7, in contrast to the faith so many contemporaries invested in an electromagnetic worldview, Einstein's considerations had already led him to recognize gaps in the grasp of electromagnetic theory. My second point is that in response he was ready to probe limitations with tools gained from other fields while holding fast to theoretical approaches within the bounds in which their validity seemed secure. Thus, Einstein evaluated the usefulness of any particular theoretical principles or models in specified contexts.

75. Renn, "Einstein's Controversy with Drude" (2000).

76. Einstein, "Erzeugung und Verwandlung des Lichtes" (1905), 132. For a recent study of the pathway to this paper, stressing the relations among different facets of Einstein's thought, see Rynasiewicz and Renn, "Einstein's *Annus mirabilis*" (2006).

I will argue below that this general methodological approach had its consequences for the language in which Einstein stated the implications of his novel arguments. I think it also made him much more likely to accept a "classical" or "current" understanding of *specific* fields than to forge a more extensive use of the concept *classical* to describe physics in general. Indeed, Einstein made breathtakingly clear arguments concerning specific formal features and offered highly concrete statements of the extent or limitations of particular principles or models. In contrast, he was less definite in stating broader perspectives, only gesturing toward worldviews. When he did the latter, he employed the terms his contemporaries used, even if he used the adjective *electromechanical* to work against their present understanding of its content. There, are then, methodological reasons why Einstein was unlikely to use the term *classical* in a general way; but as we shall see, despite his 1907 invocation of classical mechanics, Einstein's extensive use of a highly specific interpretation of classical thermodynamics is likely to have prevented him from applying the term *classical* to the equipartition theorem in particular.

Einstein combined a profound formal overview with a pragmatic respect for particular tools, and his 1905 paper considered the equipartition theorem before he went on to advance the hypothesis of light quanta. Einstein first used the equipartition of energy to establish a relation for the average energy of radiation in dynamic equilibrium with the resonators used to model the blackbody. This was an exact statement (with numerical coefficients) of the law that Rayleigh had described in 1900 and would return to a month later in his exchange with Jeans. Einstein was clear that this relation, built up using Maxwellian theory and electron theory, both failed to agree with experience and had important implications for the model that had generated it. It meant that in this model "a definite distribution of energy and the ether is out of the question" since the larger the range of frequencies, the larger the radiation energy, reaching infinity in the limit.[77] In Einstein's view the formula showed that "the greater the energy density and the wavelength of radiation, the more useful the theoretical foundations we have been using prove to be; however, these foundations fail completely in the case of small wavelengths and small radiation densities."[78] Rather than drawing on Planck's empirically successful theory, Einstein then relied on Wien's law in the specific region in

77. Ibid., 136.
78. Ibid., 137.

which it was fully confirmed by experiment (for large values of ν/T), in developing the analogy to the kinetic theory of gases that established his light quanta hypothesis.

In a paper published the following year he returned to blackbody radiation to give a new interpretation of Planck's law, which he had previously thought formed a counterpoint to his own considerations.[79] Now, like Jeans, he was ready to assert that the reason Planck's law departed from the invalid radiation formula Einstein had derived in 1905 was precisely its reliance on the assumption that the energy of a resonator could only assume values that are integral multiples of Planck's constant and the frequency. Einstein drew attention to the same features that Jeans had highlighted, but stated in contrast, "In my opinion the above considerations do not at all disprove Planck's theory of radiation: far more they seem to me to show that with his theory of radiation Mr. Planck has introduced a new hypothetical element—the hypothesis of light quanta—into physics."[80] In later papers Einstein elaborated on the unusual consequences of Planck's law. In a particularly important 1907 article, he described Planck's law as showing that molecular motions obeyed laws different from those holding in our world of sense perception. In this paper Einstein extended his use of the quantum from its first applications to light and the interchange of energy between radiation and charged resonators to develop a quantum theory of the specific heats of solids.[81] While this paper drew little immediate comment, empirical confirmation of the law of specific heats Einstein derived in 1907 was to be extremely important in attracting wider attention to quantum physics—from 1910 onward.

We have seen that Einstein referred to "classical mechanics" in his papers on relativity. A similarly specific concept of earlier and classical theory was also at play in his work on heat and molecular theory, but this probably precluded him from applying that term to equipartition.[82] In

79. Einstein, "Lichterzeugung und Lichtabsorption" (1906).

80. Ibid., 203.

81. Einstein, "Plancksche Theorie der Strahlung und die Theorie der spezifischen Wärme" (1907), 183–84. Here he stated that abandoning the mean energy value predicted by equipartition involved assuming that the application of the molecular-kinetic theory caused the conflict with experience at issue, now referring to the limiting formula as Rayleigh's. See 182.

82. This point has not yet been recognized in accounts that describe the theorem as founded in classical physics, or Einstein's paper as showing that light quanta could not be explained "on the basis of either Maxwell's theory or classical mechanics." See Einstein, *Collected Papers*, vol. 2 (1989), xvii, xxviii, 137. The formulation relies on describing statistical

two papers on the kinetic theory of heat and (Brownian) molecular motion in 1905 and 1906, Einstein drew a contrast between "classical thermodynamics" and the molecular theory of heat similar to that outlined in my discussion of Helm's work. In the first instance Einstein maintained that if confirmed his studies would show that classical thermodynamics could not be regarded as strictly valid for microscopically visible motions, while in the second he described classical thermodynamics as distinguishing in principle between heat and other forms of energy, in contrast to the molecular theory of heat.[83] Similarly, in the letter to Sommerfeld in which he described his attitude toward the need for elementary foundations in physics, Einstein compared relativity as it stood—without a more elementary understanding of electrical and magnetic processes—to "classical thermodynamics" before Boltzmann's probabilistic interpretation of entropy.[84] Even more pointedly, in 1909 Einstein discussed a weakness in Planck's use of probabilities in the calculation of entropy. Here Einstein distinguished a classical thermodynamical approach to irreversibility from the interpretation offered by the statistical theory.[85] The contrast was between an absolute thermodynamics (which indeed Planck supported), in which the state of maximum entropy was a single definite state that a system maintained once it had been reached, and the statistical approach to irreversibility in which maximum entropy was simply the most probable condition, with a system assuming a random sequence of states over time. For Einstein, statistical mechanics and the molecular theory of heat represented nonclassical thermodynamics; and the equipartition theorem was a central component of statistical mechanics.

We can now sum up our findings. Einstein clearly regarded his work on quantum theory as having opened up major contrasts. He expressed these in different ways in the three papers in which he addressed quantization

mechanics as an extension of classical mechanics. Although Boltzmann would have urged this, and Einstein's later writing provides influential examples of it, I do not know of any case in which contemporaries used such an extended understanding until 1911. For the more synthetic way in which Einstein later used the concept of classical to bridge mechanics, statistical mechanics, and thermodynamics, see note 6, above.

83. Einstein, "Molekularkinetischen Theorie der Wärme" (1905), 549; Einstein, "Brownschen Bewegung" (1906), 372. For a discussion of the contrast between statistical mechanics and general thermodynamic arguments that Einstein explored here, see Nye, *Molecular Reality* (1972), 112–18.

84. Einstein to Sommerfeld, 14 January 1908; Doc. 73 in Einstein, *Collected Papers*, vol. 5 (1993).

85. Einstein, "Gegenwärtigen Stand des Strahlungsproblems" (1909), 187.

in different contexts, without referring to classical theory. For him, the consequences of Wien's law, Planck's theory, and the implications of the latter for molecular theory showed the limitations of, respectively, continuous, field-based theories; the energy relations that could be derived on the basis of Maxwell's theory and electron theory; and molecular-kinetic theory and the laws derived from our world of sense perception more generally.[86] The first major point is that given his existing use of the concept "classical thermodynamics," he is unlikely to have thought of the particular conceptions of molecular and kinetic theory as classical.[87] Second, while he certainly aimed for unity, Einstein's critical exploration of the relations between the formal approaches followed in different fields is likely to have heightened his caution about broad-stroke characterizations of physics as a whole, either pointing to the future or the past.

With Jeans and Einstein, Paul Ehrenfest argued that Planck's law involved energy quantization in two papers published in 1905 and 1906 that emerged from a study of the relations between Boltzmann's earlier work and Planck's theory.[88] Like Jeans, on whose work he drew, Ehrenfest offered a discussion of equipartition in the course of developing his argument, without describing that theorem as classical. I will explore the sense in which Ehrenfest thought of old and new approaches to radiation theory—and Jeans's work—in the following section. For the moment it is enough to note that in 1906 Ehrenfest engaged in correspondence with Planck that is likely to have been one factor that prompted Planck to deal explicitly with the issue in his 1906 textbook on the theory of heat radiation (his first major return to blackbody radiation since 1901). After a lengthy discussion of the approach Rayleigh and Jeans had developed, and a criticism of Jeans's nonequilibrium thermodynamics, Planck stated his own view on equipartition. While Planck stressed that a physical interpretation of his new constant was still required, he contended that the difficulties raised by the Rayleigh-Jeans formula resulted from "an unjustified application of the theorem of energy equipartition to every

86. Einstein, "Erzeugung und Verwandlung des Lichtes" (1905), 132, 136–37; Einstein, "Lichterzeugung und Lichtabsorption" (1906), 199–200; Einstein, "Plancksche Theorie der Strahlung und die Theorie der spezifischen Wärme" (1907), 183–84.

87. Much later Einstein wrote that shortly after the appearance of Planck's work, "without yet having a substitute for classical mechanics," it was clear to him what kind of consequences Planck's theory held for the different fields on which he was to publish in the years from 1905 to 1907. Schilpp, ed., *Albert Einstein: Autobiographical Notes* (1979), 42–43.

88. Ehrenfest, "Physikalische Voraussetzungen der Planck'schen Theorie" (1905); Ehrenfest, "Planckschen Strahlungstheorie" (1906).

independent state variable."[89] Thus, in 1906 Planck was content to follow Kelvin's advice and ignore the second cloud on the horizon of dynamical theory.

The Quantum Theory in Overviews—and Off the Printed Page—1906–10

The discussions of Einstein, Jeans, and Ehrenfest were highly important in building a context in which Max Planck's work could be interpreted as involving energy quantization. Nevertheless, in Kuhn's account two later events were still more significant in persuading physicists to take Planck's theory and quantum discontinuity seriously. These were the discussion surrounding H. A. Lorentz's support of the equipartition theorem in 1908 and the empirical support Walther Nernst provided for Einstein's quantum theory of the specific heat of solids from 1910. In each case the key published papers made no reference to "classical." Lorentz's lecture on the partition of energy between the ether and matter was particularly significant because he argued that equipartition and the Rayleigh-Jeans law offered the only avenue available on the basis of current understandings of electron theory and kinetic theory.[90] Lorentz's paper brought a rapid response from those who had been long involved in research on black-body theory, with both Wien (privately) and Lummer and Pringsheim (very publicly) challenging the support he gave to the Jeans law. Interestingly, the latter pair characterized the formula concerned as alternatively the "Jeans-Lorentz law," or the "new law," placing it as a latecomer in the series that included laws from Stefan and Boltzmann, Wien, and Planck.[91] They described the law as completely impossible, being in crass conflict both with experiment and with everyday observation of the luminescence (or otherwise) of bodies at ordinary temperatures. Far from treating the

89. Planck stated that the equipartition theorem relied on the assumption that the probability that the energy state of a resonator falls in any particular elementary cell is simply proportional to the size of the cell, no matter how small the cell. But in the case at issue—and the justification for this was presumably empirical—the size of the cell had to be finite and was determined by the quantum of action h. Planck, *Theorie der Wärmestrahlung* (1906), 178. Gearhart discusses Planck's knowledge of and attitude toward equipartition from 1900 onward in Gearhart, "Planck, the Quantum, and the Historians" (2002), 190–93.

90. Lorentz, "Partage de l'énergie" (1934 [1908]). Nernst's papers will be discussed in chapter 10. The implications of Lorentz's stance for proponents of the electromagnetic worldview has recently been highlight by Seth, "Quantum Theory and the Electromagnetic World View" (2004). See also chapters 8 and 9 of Kuhn, *Black-Body Theory* (1987).

91. Lummer and Pringsheim, "Jeans-Lorentzsche Strahlungsformel" (1908), 449.

law as being canonical, or unusually justified in any way, they stated that it would be superfluous to discuss it were it not for the outstanding significance and authority of the two theoretical physicists who had supported it. Their intervention persuaded Lorentz that he was now bound to support Planck's theory as the only one in accord with experiment, but not without pointing out that he saw no way of preventing equipartition from applying to electrons (on current theoretical understanding).[92]

This was the rub, and it highlights the value of exploring both relativity and quantum theory—and indeed the physics of the period still more broadly—through the lens of the electron in the way I advocated in chapters 6 and 7. Current understandings of the electron were increasingly seen to be at issue in a period in which such fundamental questions as its structure and mass/velocity relations were subject to considerable controversy, and the relations between relativity and the electromagnetic worldview were a matter of active consideration. Lorentz argued that the equipartition theorem was bound to apply to any system that could be expressed in terms of the Hamilitonian equations, bringing both mechanical systems and electromagnetic systems under its sway. Nevertheless, for all his insistence that equipartition was forced given current understandings of the electron, the responses of Planck in 1906 and Lummer and Pringsheim in 1908 show that it was quite possible to assess radiation theory without drawing this conclusion, and without incorporating it in a framework of theoretical change in which equipartition should have any particular significance.

To date we have largely been considering the public record, including both research papers and the discussion sections of journals committed to carrying current scientific conversation like *Nature* and *Physikalische Zeitschrift*. Did the language of the classical enter private discussion and correspondence concerning radiation theory earlier or more fully than it appeared on the printed page? Einstein's letter to Sommerfeld offers one instance of a particularly revealing treatment of the relations between "classical thermodynamics," statistical mechanics, and other theories. A second example will show that correspondence allowed at least one other physicist to express attitudes that would have been unprintable.

92. Lorentz noted that since Drude's theory of metals had demonstrated the existence of free electrons, their contribution to the partition of energy would have to be combined with that of the resonators Planck had used to model the cavity. He favored Jeans-like strategies concerning the time involved to establish equilibrium in relation to the high frequency limit. Lorentz, "Strahlungstheorie" (1934 [1908]), 563.

With good reason, Paul Ehrenfest has long been regarded as one of the most sensitive and acute young physicists working in the early twentieth century. In chapter 7 we considered the tensions underlying his changing attitudes to relatively; but Ehrenfest is celebrated still more for his early and critical insight into Planck's radiation theory. Notes and correspondence that became available only after Martin Klein's authoritative biography of Ehrenfest and Kuhn's book on blackbody research show that Ehrenfest harbored a surprisingly strong opposition to the work he was able to critique so incisively.[93] As richly descriptive as Ehrenfest could be in print, his published writings only hint at this aspect of his views. In a fascinating letter to an unnamed friend we can identify from internal evidence as Walter Ritz, Ehrenfest drafted comments on both radiation theory and relativity, responding in particular to Minkowski's 1908 paper on relativity. I noted earlier that in contrast to what he described as Einstein's "incomprehensible" and "ununderstandable" approach, Minkowski helped Ehrenfest see a world of wonderful elegance in relativity. The first draft of the letter opened with comments that show that Ehrenfest had reacted equally strongly against the work of Einstein—and others—on radiation theory: "Ach, were it only possible to make an end to this obnoxious Bachanalia: Planck—Einstein—Jeans! If one thinks: how sullied is this theory now, which with Kirchhoff, Stefan-Boltzmann, the Wien distribution, is so wonderfully established."[94] Later drafts only strengthened the contrast Ehrenfest stated between the old and new theories of blackbody radiation. In the second and third drafts respectively, the new work had sullied the old "wretchedly" [elendiglich] and "disgustingly" [ekelhaft], while the old theory had been "established so singularly beautifully (born classically)" [so einzigartig schön (geboren-classisch) einsetzte] (second draft). In the third draft Ehrenfest ended his remarks with the statement that "God could not want the theory to conclude with any such filth."[95]

As we have seen, Ehrenfest began his studies in Vienna in 1899, where he attended Boltzmann's lectures on the mechanical theory of heat. His

93. Klein, *Paul Ehrenfest* (1970); Kuhn, *Black-Body Theory* (1987), 152–70, 188–89.

94. "Ach wäre es doch nur möglich diesem widerlichen Bacchanal: Planck—Einstein—Jeans ein Ende zu machen!—Wenn man bedenkt: Wie verschweinigelt ist jetzt diese Theorie, die mit Kirchhoff, Stefan-Boltzmann, Wien-Verschiebung so wunderbar einsetzte!—" Ehrenfest to "Dear Friend" (1), AHQP/EHR, Reel 32 m57.

95. "Gott kann es nicht wollen, dass das mit so irgend einem Dreck abschließt!—" Ehrenfest to "Dear Friend" (1), (2), and (3), AHQP/EHR, Reel 32 m57.

dissertation was completed under Boltzmann, on the motion of rigid bodies in fluids and the mechanics of Hertz. Surely Ehrenfest was better situated than anyone else to take up the language of the classical. A first puzzle, then, is that despite Boltzmann's 1899 self-description, Ehrenfest *does not* use that concept in the obituary he wrote for Boltzmann in 1906.[96] The above comments show that it enters his views of radiation theory, but we should note that Ehrenfest describes the theory as having been classically born, without identifying any particular current argument as classically founded.[97] Still more interesting is the fact that he links Planck and Einstein with Jeans. This is hard to understand in the light of our current view that Jeans championed classical physics, but I think clearly demonstrates that the nontraditional framework within which Jeans incorporated equipartition was as important to Ehrenfest as the fact that he used the theorem at all.

Besides private correspondence we might think that descriptive terms like *classical* would surface more prominently in survey papers. Planck's theory began making its way into more general overviews of physics from about 1909, there taking a place alongside electron theory and relativity as one of the most pressing problems of the day. However, papers by Einstein, Planck, Larmor, and Lorentz given between 1909 and 1910 show that in this forum also, radiation theory was described primarily as a challenge to current understandings of the electron and statistical mechanics, both notably open and fluid fields. Outlining their arguments will help establish the links being created between relativity and quantum theory; but examining the terms in which they were discussed will emphasize the scale of the changes that had to occur before they could be assimilated into the contrast between classical and modern theory that we now accept.

Einstein's address on the nature of radiation to the 1909 Naturforscherversammlung was eagerly anticipated, marking his first appearance in a public gathering of the kind.[98] He began by offering a survey of the history of conceptions of radiation, discussing both Newton's emission theory of light (which as a particle theory formed a predecessor to some features of his own approach) and the rise of the wave theory and Maxwell's theory of the electromagnetic field. This framework allowed him to

96. Ehrenfest, "Ludwig Boltzmann" (1906).

97. Without developing a full theory, Ehrenfest continued to deny the purchase of energy quantization to light—the first field to which Einstein had applied it.

98. Einstein, "Das Wesen und die Konstitution der Strahlung" (1909).

give an overview of the development of relativity, the theory for which he had become well known among German physicists. However, Einstein focused particularly on arguments for a quantized conception of radiation and for his own view that a future theory of light would involve a fusion of particle and wave perspectives. On both points the subsequent discussion showed that only Johannes Stark supported him. Einstein presented the relation between energy, mass, and the velocity of light that he had derived in 1905, specifically because it involved a modification of the basic concepts of physics, and—drawing implications his contemporaries refused to accept—he described the acceptance of Planck's theory as involving a rejection of the foundations of the current electromagnetic theory of radiation. This was, he said, because it contradicted the assumption that every imaginable distribution of energy would occur, with the finite value of the energy element restricting the number of possible complexions so that Planck's procedure did not constitute an expression for the probability of the state in Boltzmann's sense. Again, Einstein juxtaposed his own views and Planck's theory against currently accepted theories or conceptions, without invoking the concept of the classical.

In the discussion to follow Planck argued strenuously against the need to see his own theory as requiring the quantization of light energy, thereby seeking to confine the theory to the context of blackbody radiation in which he had initially developed it. He also posed the implications of his theory in terms that would have been impossible for him in 1900, but could be assimilated into the debate between the mechanical and electromagnetic worldviews. Now he argued that since mechanics and current electrodynamics did not allow discrete energy values, "we cannot produce a mechanical or electrodynamical model," of the resonators he had introduced. Physicists would simply have to get used to such a resonator, and after all had already faced similar situations in the failure of attempts to represent the luminiferous ether or current electricity mechanically.[99] On the occasions in which he spoke on quantum theory, such as the Columbia lectures he gave in 1909, Planck wrote of both Jeans and Lorentz's view of the blackbody problem, concluding that his theory showed that electron theory to date suffered a deficiency that had to be overcome, and that opinions differed widely as to how deep a modification of the basic structure of electron theory would be required.[100]

99. Ibid., 825.
100. Planck, *Theoretische Physik* (1910), 95.

In 1909 Joseph Larmor presented an account of the statistical and thermodynamical relations of radiant energy that he regarded as an expansion and generalization of ideas implied in Planck's analysis of natural radiation (which was itself an extension of ideas that Boltzmann had developed in gas theory). Most notably for our purposes, Larmor offered a highly empiricist framework, assuming that the molecular statistics of distributions of energy would be possible only in a few of the simpler cases, such as gas theory and natural radiation. Specifications of the probability of energy distributions had to remain imperfect and approximate, "to be modified and improved by each fresh addition to our knowledge of the system." Only after setting up his generalized framework did he inquire into the relation of equipartition to the schema he was presenting, there discussing the work of Jeans, Rayleigh, and Lorentz.[101]

Finally, in 1910 H. A. Lorentz gave a series of six lectures, "Old and New Questions of Physics," at the University of Göttingen. Comparing the language he used in this context with our findings in the following chapter will indicate just how much was to change in a little over a year in regard to invocations of concepts of classical theory. Lorentz opened with three lectures on relativity in which he finessed the relations between his own approach to electrodynamics and the abandonment of the ether and concepts of true time and space that Einstein and Minkowski had urged. The relativity principle involved what Lorentz described as a "new mechanics," and the contrast he drew was with ordinary or Newtonian mechanics, discussing the conditions that Newtonian forces had to satisfy if they were to fulfill the relativity principle.[102] As we saw in chapter 8, while Lorentz attributed the principle of relativity to Einstein, his approach stressed the communal basis of relativity physics. Lorentz devoted his final three lectures to building up a discussion of the quantum theory, framing his treatment by describing the equipartition theorem as

101. The central task was that of apportioning a physical system capable of containing energy of various forms into "cells" of equal opportunity, that is, cells such that the element of energy under consideration is as likely to occupy any one of them as any other. Larmor recognized that Planck's procedure depended essentially on the assumption of a discrete or atomistic constitution of energy; in his own approach a similar implication survived in the form that the ratio of the energy element to the extent of the standard unit cell is an absolute physical quantity, also determined by the observations on natural radiation. See Larmor, "Bakerian Lecture" (1909), on 85–87 (for Larmor's generalization and cells of equal opportunity), 89–90 (on Planck's energy atomism), 88 (for modifications), and 90–91 and 94–95 (for the discussion of equipartition).

102. Lorentz, "Alte und neue Fragen" (1910), 1238, 1240, 1241.

posing the main difficulty entering into the construction of a radiation formula. Now he went into more detail in rehearsing the view presented in his Rome lecture and his response to Lummer and Pringsheim. Lorentz stated that equipartition is valid for all systems that obey the laws of mechanics, or alternatively the Hamiltonian principle. Since the Hamiltonian principle can be applied to electron theory (with the electrical energy corresponding to potential energy and magnetic energy to kinetic energy), the ether is also a system in which the theorem should be valid. Outlining the absurdity to which it led, he maintained that one cannot explain a finite relation between the energy of ether and matter in equilibrium on the basis of the assumed foundations. The Hamiltonian principle cannot be applicable to radiation and a new hypothesis must be made: Planck's introduction of energy elements.[103]

Conclusion

These surveys confirm that from 1906 through to 1910, Planck's quantum theory was consistently presented as involving important departures from fundamental conceptions—in particular from the continuum approach to energy implicit in Maxwell's theory, from Boltzmann's statistical combinatorics and/or the equipartition theorem—and it was described as being in conflict with current conceptions, especially of electron theory. It was not presented in the language of classicism. Indeed, when that language was invoked, its use in relation to classical thermodynamics probably ran counter to any tendency to ascribe statistical mechanics or the equipartition theorem a classical status.

In the first section of this chapter we saw that the earliest synthetic deployments of concepts of "classical" theory were controversial. Just as importantly, "classical mechanics" and "classical thermodynamics" contested similar ground. Their formulation in the hands of Boltzmann and Helm was riven by a contrast between atomistic and phenomenological, energeticist approaches. This interpretational contrast has long been hidden by our own readiness to incorporate both classical mechanics and thermodynamics within a still more general concept of classical physics.

103. Ibid., 1248. Later he described this as a makeshift or emergency aid that could neither be theoretically grounded, nor avoided, because not enough was known about the actual processes at issue. See 1256.

Nevertheless, Einstein's writings show that it played an important role in early twentieth-century views. In conjunction with contemporary awareness of the problematic status of the equipartition theorem, the opposition between classical thermodynamics and statistical mechanics helps explain why the quantum was not initially conceived in contrast to classical theory. We can go a step further, though. My examination of the rich language in which physicists described the grounds from which the new theory departed should by now have demonstrated the very real possibility that physics might *never* have been described as classical in the major sense we now recognize. Consider, for example, the diverse stances taken by just a few of the figures studied above. Note the facts that a major supporter of mechanical theory like Lord Kelvin could argue strenuously against equipartition, while as the principal physicist upholding it, James Jeans found no use for the term *classical* in his development of a nontraditional thermodynamics; that the originator of the new blackbody law, Max Planck, first began to think in terms of the "so-called classical mechanics of the material point" in 1909; and that as the main architect of the unfolding quantum theory, Albert Einstein used distinct concepts of classical mechanics and classical thermodynamics to express facets of two very different fields, conveying in the case of the latter a very specific interpretational approach. This situation changed radically in 1911, when in the course of a small but influential conference devoted to quantum theory, six different physicists referred to classical theory in a new sense. My final chapter will endeavor to understand how this could occur so suddenly and so completely. I will argue that the Solvay Council provided the occasion for the development of a modern form of *classical physics* that built on Boltzmann's first use of the term a decade earlier, while fundamentally changing both the sense in which the term was deployed and the historical landscape it depicted.

CHAPTER TEN

The Solvay Council, 1911

The previous chapter initiated our study of the forgotten history of the word *classical* in physics and showed its early days to be run through with contest and controversy, long invisible to us. The present chapter will seek to explain why we remember a particular set of framing concepts and associate classical and modern physics with a particular historical turning point: the turn of the century. On 30 October 1911 a small group of eighteen physicists and three scientific secretaries gathered for five days of intensive work on the new quantum theory in Brussels. Many there were known for relatively recent achievements. With three Nobel Laureates among them—H. A. Lorentz, Marie Curie, and Ernest Rutherford—as well as luminaries like Emil Warburg, Max Planck, and Henri Poincaré, this was clearly an eminent group. But apart from the handful present who had worked on quantum theory previously, a casual observer would have found it hard to know why just these physicists had been invited above any number of their colleagues.[1] Leaving the conference, Albert Einstein

1. Lorentz and Curie earned Nobel Prizes in physics in 1902 and 1903. Rutherford's was the chemistry prize for 1908. Curie and Wien were soon to hear that they had been awarded the 1911 Nobel Prizes in chemistry and physics, respectively. Three physics laureates had turned down invitations to attend: Röntgen (1901), Rayleigh (1904), and J. J. Thomson (1906). Among other members of the congress both Planck and Poincaré had been seriously considered for prizes (in 1908, and 1910 and 1911 respectively). Proposals on their behalf had considerable support but fell foul of the values of particular Nobel committee members. See

wrote that he learned nothing new at what he called "the witches sabbath," but the first Solvay Council has gone down in physicists' folklore as marking the birth of the modern era. Indeed, a photograph of the gathering has become one of the most recognizable icons of physics in the early twentieth century, repeated in the sidebars of textbook after textbook.[2] Our study of Einstein's generation will conclude with the moment in which his work on relativity and quantum theory was incorporated within a common framework, presented as the modern challenge to classical physics.

The conference was convened and funded by the Belgian industrialist Ernst Solvay, but conceived and planned by the Berlin physical chemist Walther Nernst. Examining the circumstances in which it was proposed, the manner in which it was organized, and the nature of its publication will help me outline several distinctive features that together rendered the event a highly effective means of propagating quantum theory far beyond its original orbit. Thus the first aim of this chapter will be to demonstrate that Nernst could effect the rapid international dissemination of a physics initially won in Germany because he created a new kind of conference. A second goal will be to establish that the council was most original for its role in shaping a new understanding of the physics of the *past*. It saw the birth of a new, general form of classical theory, an achievement that may ultimately have been just as important as any direct stimulus it provided to research on quantum theory. Indeed, the structure of the council proceedings and the framing language that physicists began to use in 1911 were together to establish the historical narrative of the transition from classical to modern physics that we currently accept. Finally, this chapter will show that as in any instance in which complex events are summarized from a new perspective and labeled with a word or a phrase, the formulation of

---

Friedman, *Politics of Excellence* (2001), 49 (Planck) and 49–51 (Poincaré). In 1910 Poincaré had received thirty-four nominations (the most ever received for a single candidate), and Arrhenius helped mollify French disgust at the decision to pass over Poincaré by advancing Marie Curie for the 1911 prize. See 51–52.

2. Einstein assessed the meeting critically in a letter to Michele Besso, discussed in Barkan, "Witches Sabbath" (1993), 66–70 on 66. The attribution of the importance of the meeting is exemplified in, for example, Marage and Wallenborn, eds., *Solvay Councils* (1999). Previous commentators have been able to assimilate the conference to our present understanding of the classical-modern divide without recognizing the novelty of Planck's and others' use of the term *classical*. The photograph was by Benjamin Couprie, Institut International de Physique Solvay (courtesy AIP Emilio Segrè Visual Archives), but was not included in the French or German published reports of the congress.

a common view of the past had its elisions and its costs. Unlike the much more extensive and inclusive 1900 Congress of Physics, the Solvay conference has become part of physicists' collective memory. But it is as significant for what it helped physicists forget as it is for the particular views it propagated. *Einstein's Generation* will conclude with reflections on the way that the disciplinary memory cultivated in 1911 has helped single out Einstein's theoretical work as emblematic of an era—and marked that era as one unlike any other in the history of his discipline.

A New Kind of Conference

Ernst Solvay was a Belgian industrial chemist who had become extremely wealthy as a result of the international success of his process to manufacture sodium carbonate from salt, ammonia, and carbonic acid from the 1870s. Solvay combined a deep concern with the evolution of the individual and social groups with a strong interest in science. His factories instituted a social security system and pensions for workers in 1878 and an eight-hour day in 1897 (paid vacations came in 1913). At the 1900 Exposition Universelle in Paris, different branches of the firm put on what Albert Neuberger described as "magnificent" exhibits. The French affiliate at Barangeville-Dombasle showed an extremely instructive relief depicting the different stages of the Solvay process from the arrival of the raw material to the completion of the product, while the Brussels factory displayed paintings and exhibits representing the different industries in which soda had found application.[3] Solvay endowed institutes of physiology and social science at the Université Libre de Bruxelles in 1895 and 1901, together with a school of business that has borne his name since 1903, and a workman's educational center.

Solvay knew well that new knowledge might confer commercial advantages in industrial chemistry but also pursued his own less immediately pragmatic research interests in physics, developing a speculative approach to gravitation and matter. Early in the twentieth century he began expanding his business concerns in Germany. Emil Fischer and Walther Nernst helped cultivate the relationship. As Berlin's leading chemist and physical chemist, in 1909 they proposed Solvay for the Leibniz Medal of the Prussian Academy of Sciences (for his service and donations to the

3. Neuburger, "Chemie und Physik" (1900), 215–16.

FIGURE 10.1 Exhibit of the Solvay Company soda factories at the Paris Exposition Universelle in 1900. Kraemer, *XIX. Jahrhundert*, vol. 4. (1900), 215.

promotion of science). The following year Solvay and Nernst met in Brussels, where a discussion of mutual enthusiasms in fundamental physics led Nernst to suggest the possibility of holding a conference on current problems in the kinetic theory of matter and the quantum theory of radiation.[4]

For Nernst the subject to tackle was clear. Since moving to Berlin from Göttingen in 1906 he had focused on low temperature studies of the specific heat of solids as an indirect way of approaching equilibrium reactions between gases. His work was initially directed by the "heat theorem" he developed in 1906, suggesting that both the "internal" and "free" energy of solids would approach each other (in a tangential relationship) as the temperature of a solid approached absolute zero. Nernst thought experiment was best conducted with the guidance of theory, and physical chemistry should apply the methods of theoretical physics to chemical problems. On his inaugural speech to the Prussian Academy of Science he told his audience that modern Germans looked on fields of corn and ornamented factories with too much satisfaction. Too much emphasis was

4. On Nernst, see Barkan, *Walther Nernst* (1999). On the conference, see Klein, "Einstein, Specific Heats" (1965), 173, 177–79; Kuhn, *Black-Body Theory* (1987), 210–20; Mehra, *Solvay Conferences* (1975), introduction and chaps. 1 and 2.

often put on a directly practical point of view, and he pointed to Felix Klein to suggest how a strengthened scientific idealism, if united with a practical gaze, could enthuse wider circles in the best goals a learned body could set.[5]

Turning his laboratory to the problems of the very cold, Nernst worked through the literature of physics for theoretical treatments of specific heats—something few other chemists would have done.[6] Surveying the *Annalen der Physik* he found and took notice of Albert Einstein's 1907 paper on Planck's theory of radiation and the theory of specific heats—unlike any physicist before him. By February 1910 Nernst was ready to announce that his experiments confirmed the heat theorem. Meeting his initial goals, he could show close agreement between calculated and observed values of the free energy for four different reactions.[7] Chemical thermodynamics was clearly Nernst's chief concern, but in brief concluding remarks he referred to Einstein's work. Nernst commented only that the specific heats appeared to "converge to zero in accordance with Einstein's theory," but a month after submitting the paper he traveled to Zurich to meet Einstein.[8] His student and collaborator F. A. Lindemann later recalled that Nernst spent "a whole Easter vacation visiting Einstein and Sommerfeld and going for long walks with Planck discussing these matters."[9]

So the conference would be on quantum theory. Back in Berlin, Nernst wrote to Planck for support, outlining his concerns, describing the kind of event he envisaged, and listing possible participants. Planck's reply of 11 June 1910 is highly interesting. While Planck supported Nernst's aims, he wrote:

> I cannot conceal my deep concern with the possibility of carrying it out.... I would expect such a conference to be more successful if you would wait a while until more supporting evidence is available on this subject.
>
> But in my opinion another point argues even more strongly in favour of postponing such a conference for one year. The fundamental assumption for

5. Nernst, "Antrittsrede" (1906), 551.

6. And in particular those he replaced in Berlin. See Hiebert, "Nernst, Hermann Walther" (1980), 453.

7. Nernst, "Spezifische Wärme bei tiefen Temperaturen. II" (1910), 276–81.

8. Ibid., 282. See Kuhn, *Black-Body Theory* (1987), 214–15; Hiebert, "Nernst, Hermann Walther" (1980), 442.

9. Lindemann and Simon, "Walther Nernst" (1942), 104.

> calling the conference is that the present state of the theory, predicated on the radiation laws, specific heat, etc., is full of holes and utterly intolerable for any true theoretician (and that this deplorable situation demands a joint effort towards a solution?).... Now my experience has shown that this consciousness of an urgent necessity for reform is shared by fewer than half the participants envisioned by you.... I believe that out of the long list you named, only Einstein, Lorentz, Wien and Larmor are seriously interested in this matter, besides ourselves.[10]

Planck's views are revealing in several respects. First, they indicate what a small section of the international physics community he could identify as sharing a basic interest in quantum theory in June 1910. Second, Planck here associates *true theorists* with a very particular concern with the lacunae facing physical theory, driven by an urgent necessity for reform. Ideas of revolutionary reform were highly important in stimulating fresh interest in Planck's theory; and earlier chapters have shown how dominant a feature this was in discussions of the electron, the electromagnetic view of nature, and relativity before this. However, the revolutionary label was a two-edged sword, cutting against Einstein's light quanta hypothesis. Concern with revolutions is an ambiguously loose and shifting career guide for a theoretical physicist, but also one that fires interest and a thorough involvement. Indeed, as Suman Seth has argued, debate over foundations was a central feature of the emerging practice of theoretical physics as a distinct specialty.[11] Third, note that Planck questions the idea that collaborative work and joint efforts might be demanded in the search for a solution. We will see that Planck was not alone in the belief that true advances in theoretical physics would come rather from solitary work. Finally, the passive voice Planck assumes in this letter is remarkable, tending to hide the role of human agency. Planck advises Nernst to wait a while to ensure the success of the conference he intends. Let another one, or preferably two, years pass "and we will see how the crack which has developed in the theory continues to grow until all those who are now outside the problem will be drawn into it." Such a process has a normal course that Planck doubts can be accelerated.[12]

10. Planck to Nernst, 11 June 1910, quoted in Hermann, *Genesis of Quantum Theory* (1971), 136–37.

11. Seth, "Crisis and the Construction of Modern Theoretical Physics" (2007).

12. Hermann, *Genesis of Quantum Theory* (1971), 136–37.

In contrast to the passive voice of Planck's rhetoric on this occasion, we saw in the previous chapter that just a year earlier he worked vigorously to *narrow* the fissure of quantum theory and confine it to the exchange of energy between radiation and matter (and his resonators), when responding to Einstein's views on the nature of light. By now his views have changed dramatically. While Planck still calls for even more supporting evidence, Nernst's experimental confirmation of Einstein's theory of specific heats has led to a readiness to "watch the crack grow" and see quantum theory outside the narrow context of a particular interpretation of blackbody radiation. At no time previously, to my knowledge, had Planck mentioned Einstein's application of quantum theory to the specific heat of solids. In 1910 Planck returned to the quantum theory in print.[13]

In articles such as "The Unity of the Physical World Picture," Max Planck presented science as involving a process of continual de-anthropomorphization.[14] It was an idea that cleverly reworked Mach's emphasis on the historical conditions in which concepts were formed, in order to argue that building on precise measurement as it did, more recent theoretical physics actually demonstrated an increasing generality and independence from specific sensations or historical cultural frameworks. One of the clearest illustrations Planck gave for this thesis was the observation that earlier branches of physics had been strongly associated with particular sense impressions, such as acoustics and optics. Now powerful mathematical methods and simple principles had unified such disparate fields, and tone and color had made way for descriptions on the basis of wavelength and frequency.[15] Similarly, Planck had recently described relativity as showing how previously antithetical worldviews like mechanics and energetics could be unified in a new framework; and for Planck natural constants and unified laws had an absolute character transcending particular cultures. They were part of a physics that could even be recognized by Martians, with the object of research being a complete liberation of the physical world picture from the individuality of separate intellects.[16]

13. Planck, "Theorie der Wärmestrahlung" (1910). He had left it in 1906 with the publication of his textbook *Vorlesungen über die Theorie der Wärmestrahlung*.

14. Planck, "Einheit des physikalischen Weltbildes" (1909); Planck, "Unity of the Physical World Picture" (1960 [1909]).

15. Planck, "Unity of the Physical World Picture" (1960 [1909]), 2–4.

16. Ibid., 20–26, esp. 24. We should note that Planck was aware of the impossibility of a picture being independent of the painter, which he recognized as a contradiction in terms. See 25.

At specific points in their history Planck's descriptions of both relativity and quantum theory mimic this process of de-anthropomorphization. At these moments—when a direction he approves seems assured—Planck's personal role in the development of these theories disappears in favor of a focus on others and a use of geological metaphors. Two examples are his image of three pioneers mapping the new terrain of relativity, and in the present instance, of the crack of quantum theory widening until more are drawn into the problem. At these times Planck did not make much of the work he had previously undertaken in order to stimulate interest in relativity on the one hand, or to limit the meaning accorded quantum theory on the other. We will soon see still further implications of this kind of subversion of individual perspectives in the brief history that Planck was to offer the new theories in 1911.

While Nernst left the meaning of Einstein's work unspecified when referring to it in 1910, his meetings with Einstein, Sommerfeld, and Planck convinced him at least of the central importance and revolutionary nature of quantum theory—if not of its precise implications. Over the next year his concepts of both old and new theory were to sharpen rhetorically. Writing to Solvay in July 1910, Nernst argued, "we are currently in the midst of a revolutionary reformulation of the foundations of the hitherto accepted kinetic theory," and emphasized how foreign Planck's and Einstein's use of energy quanta were to previous conceptions.[17] He soon began to present his own work as a confirmation of the new physical theory. In the third article in his series on specific heats at low temperatures (published in early 1911), Nernst graphed the empirically determined values for the specific heats of a number of metals and diamond against the curves given by Einstein's "formula" for different characteristic frequencies.[18] The fact that they were in general agreement formed the second of seven conclusions (noting also that at very low temperatures the observed values fell more slowly than Einstein predicted). His fifth conclusion stated that the experiments could be seen as a "brilliant"

17. Nernst referred to previous (*bisherige*) theory on two occasions and described the changes concerned as an "umwälzenden Neugestaltung " and "weitgehende Reformation." He wrote, for example, that the new view "is so foreign to the previously used equations of motion of material points that its acceptance must doubtless be accompanied by a wide-ranging reform of our fundamental conceptions." Nernst to Solvay, 26 July 1910, as cited in Kuhn, *Black-Body Theory* (1987), 215 and 310–11, n. 25; Barkan, "Witches Sabbath" (1993), 70–71.

18. Nernst, "Spezifische Wärme bei tiefen Temperaturen. III" (1911), 308.

confirmation of the quantum "theory" of Planck and Einstein.[19] Thus, for some time Nernst separated the formula Einstein had provided from its interpretation as quantum theory. Slightly earlier, for example, he had described the formula as still only a "(very odd, even somewhat grotesque) calculational rule." Nernst echoed Planck's early response to relativity by writing that in Planck's work on radiation and Einstein's study of molecular mechanics, quantum theory had proved so fruitful that scientists had a duty to take it seriously and investigate the theory from as many sides as possible.[20] His most extensive discussion of the theoretical implications of the new approach came in 1911. Then Nernst compared it with "the old theories" and the "old conceptions of Maxwell and Boltzmann," which led both to empirically unsatisfactory laws of specific heats (discussing solids, fluids, and gases separately) and to the radiation formula of Rayleigh. He also described the ordinary kinetic theory of gases (operating in the temperature region hitherto approached) as dealing with rectilinear velocities and with values that were known from ballistics. In the new low-temperature region one could not speak of an extrapolation of the laws of mechanics, and ordinary mechanics was not sufficient.[21]

In June 1910 Planck was skeptical about the success of a conference on quantum theory, in part because he thought so few physicists were deeply concerned with the field. Just over a year later, however, Arnold Sommerfeld gave an address at the Naturforscherversammlung in Karlsruhe that suggests that by September 1911 Nernst's papers and other work on the quantum theory had already had a significant effect in stimulating interest within the German-speaking physics community.[22] Unlike Lorentz's 1910

19. Ibid., 309–10. The first conclusion was the observation that the curve of specific heats corresponded to a function $f(T/T_0)$, independent of the nature of the substance concerned, where T_0 was a characteristic constant. His third conclusion reached Nernst's original aim, describing the work as confirming his heat theorem.

20. Nernst, "Probleme der Wärmetheorie" (1911), 86. Nernst went on to write that just as Newton paved the way to the successes of theoretical physics in creating modern mechanics, and Dalton had given the atomic theory of chemistry and physics its most fruitful logical aid, so had Planck found a completely new method of scientific arithmetic (*Rechenoperationen*) in the quantum hypothesis.

21. Nernst, "Anwendung der Lehre von den Energiequanten auf physikalisch-chemische Fragen" (1911), 265, 267, 269. For the discussion of mechanics, see 274.

22. From a complete absence of papers in 1910, the German abstracting journal *Fortschritte der Physik* listed nine authors who wrote on a quantum theory of specific heats in 1911, ten in the following year, and twenty in 1913. They formed a significant proportion of those who wrote on quantum topics in general. See Kuhn, *Black-Body Theory* (1987), 207–9 and 216–17.

survey of old and new questions in physics, and no doubt partly because of the visit he had received from Nernst, Sommerfeld's address included specific heats within the purview of quantum physics in quite dramatic terms. His speech indicates the two-sided benefits that flowed from Nernst's travels south of Berlin. While Nernst had learned from leading theorists, they in turn were made aware of his work.

As I mentioned previously, Sommerfeld compared a stable and secure relativity with the very different prospects of the quantum theory. To convey just how fluid and problematic the latter was, Sommerfeld raised the specter of Planck trying to reform his original conceptions in his latest publication; of Einstein drawing the most far reaching consequences from Planck's discovery without (Sommerfeld believed) maintaining his original standpoint in all its boldness; and finally of Nernst, who had so successfully extended the empirical foundations for the theory and was also developing further the original ideas of Planck.[23] Sommerfeld stated that nothing could be more beneficial for modern physics than the clarification of views about these questions. "Here is the key to the situation," he proclaimed, "the key not only to the theory of radiation but also to the molecular constitution of matter, and indeed at present it still lies deeply concealed."[24]

Sommerfeld did not neglect to mention that the empirical foundations of quantum theory had emerged from the research of two different institutional sources. Indeed he highlighted their role: "It will always remain a glorious chapter in the history of the first decade of the Physikalisch-Technische Reichsanstalt (PTR) that they have erected one pillar of the quantum theory, the experimental foundations of cavity radiation. Perhaps the service of Nernst's institute is to be estimated just as highly, that in the regular measurements of specific heats it has yielded the other corner stone—carrying no less a load—of quantum theory."[25] However, while acknowledging these institutions, Sommerfeld interprets their value in the light of quantum physics, or something even more narrowly defined. The fields that had once been described by the broader and more inclusive labels of *radiation physics* and *chemical thermodynamics* (both referring to the research concerns of significant groups of experimental and theoretical physicists and chemists) have become the silent pillars of

23. Sommerfeld, "Plancksche Wirkungsquanten" (1911), 1057. Einstein's boldness was in transferring the quantum from the process of emission and absorption to the structure of light energy in space.

24. Ibid., 1057–58.

25. Ibid., 1060.

the new *quantum theory*. Similarly, 1905 is already becoming an annus mirabilis for Einstein in Sommerfeld's account, but he does not represent Einstein's program of research as being initially developed largely independently of Planck's law; authorship of quantum physics is invested in Planck. The character of the work and the primary goals that lay behind the initial development of quantum physics have both been made invisible. We have seen the submersion of personal activity, inclinations, and views in Planck's language in his letter to Nernst. Sommerfeld's brief history raises the question of the disappearance of much broader contexts of work; and like Planck, we should note that Sommerfeld wrote as a representative of *theoretical* physics.

Nernst had strong ideas about why a conference on quantum theory was necessary. I mentioned previously that when he wrote to Solvay early in the preparations, he depicted a "revolutionary reformulation" of "hitherto accepted kinetic theory." By the time he wrote formal invitations in June 1911 Nernst's references to earlier theory had sharpened. Now he described the challenge as one facing the principles of "the classical molecular theory and kinetic theory of matter."[26]

Nernst also had strong ideas about how the conference should be organized, limiting it to a small group of people and enjoining them to regard the event as "Confidential." That warning was handwritten in black ink across the margins of the typed invitation. Nernst clearly wanted to encourage the perception that this was an elite event.[27] But Nernst's insistence that the conference remain confidential was also designed to protect it from the crudity of normal disciplinary traffic, an aim also embodied in the size and focus of the conference. In the context in which national

26. For Nernst's 1910 language, see note 17 above. The official invitation was addressed from Brussels but asked for replies to be sent to Nernst in Berlin. The opening sentence reads "Selon toutes apparences, nous nous trouvons en ce moment au milieu d'une évolution nouvelle des principes sur lesquels était basée la théorie classique moléculaire et cinétique de la matière." Solvay to Langevin, 15 June 1911, Paul Langevin Papers, L8/01 at l'Ecole Supérieure de Physique et de Chimie Industrielles de la Ville de Paris.

27. See Solvay to Langevin, 15 June 1911, where "Confidentielle" is handwritten beside the text. In Brussels, Nernst continued to underline the elite nature of the event by making it clear that the Nobel Laureate Lord Rayleigh (his first choice for chair) was interested in the proceedings despite his choice not to attend. Nernst had attempted to persuade Rayleigh to reconsider by writing that they now stood before an extension of theoretical thinking that could only be compared with those achieved by Newton and Dalton. He was doubly interested in hearing from the person who eleven years earlier had first laid their finger "on the open wounds of theoretical physics, whose healing will concern us in Brussels." W. Nernst to Rayleigh, 24 July 1911, RA-HAFB.

associations of scientists met yearly, offering a forum for individual scientists to give reports on recent research, whatever that might be, it was unusual to meet with a particular topic in view as Nernst proposed. Perhaps the closest precedents were in the work of international committees on weights and measures and units of electrical measurement, or of British Association committees on particular fields of research or instrumentation. Such events brought together carefully selected participants, often representing diverse interests and sometimes bridging both national borders and science and industrial concerns. Whether successful or not, they usually had the aim of forming a common view on units or the key features of recent advances. Instead of setting the explicit goal of reaching formal consensus, Nernst's conference would involve many participants who had little expertise in quantum theory. It would also record diversity of opinion and highlight discussion. The invitation stated that no final decision was expected, but set the aim of clearing the way to the solution of the problems identified by a preparatory criticism. Establishing which molecular and kinetic interpretations agreed with observation and which required fundamental transformation would already be a great step in the development of atomic theory.

Both the confidentiality of the meeting and the timing of invitations were intended to carefully manage the relations between this small group of participants and the discipline as a whole. Nernst wrote to Solvay: "If the invitations are sent out six months earlier the various participants would have, on the one hand, sufficient time to prepare their reports; on the other, the interval of time [until the conference] would not be too long to give rise to unnecessarily protracted discussion about the conference amongst the scientists, something which is always better to avoid. For the same reason, it would be better to send the invitations marked 'confidential.'"[28] One final organizational feature should be noted. Nernst insisted on precirculating the papers before the meeting (if possible before the end of September).

In many ways, then, this was a new kind of forum, intermediary between the disciplinary meetings that scientists were familiar with and the committees on which they sometimes sat. It represented an intensification of the thematic approach that the 1900 congress had taken to the discipline as a whole, by focusing on just one problem; and its distinctive

28. Nernst to Solvay, 27 November 1910, as cited in Mehra, *Solvay Conferences* (1975), 7. The invitations were actually sent out in June 1911, four and a half months before the meeting took place.

nature was marked by the invitation to participate in a "scientific Council," a "Committee restricted to eminent experts [professionnels éminents]." The opening page of the published proceedings described the event as a "scientific council (a kind of private Congress)."[29]

By now it will be clear that Nernst did everything he could to highlight the unusual nature of this specific event. Among the many enthusiastic responses Solvay received from those on the final list of invitees, Wilhelm Wien wrote: "You have found a completely new way of pushing scientific progress forward." He thought it impossible for a single person to solve the difficult questions to be treated at the conference, "but the united efforts of the scientists of all nations, when they get together at such an excellent occasion for discussion, would provide the best chance of finding a way around the large number of molecular questions."[30]

The Christening of Classical Theory

The final list of participants included a handpicked group of eighteen physicists and three scientific secretaries drawn from Germany, France, Austria, Holland, Britain, and Denmark, trimmed down from the twenty-five identified in the official invitation. Expertise ranged widely, from mathematical physics through radioactivity to low temperature studies, but several physicists did not receive invitations even though their involvement in earlier theoretical and experimental discussions of statistical mechanics and blackbody radiation might have made them highly interested participants. Thus Burbury, Thiesen, Lummer, Pringsheim, Kurlbaum, and Ehrenfest knew nothing of the event, while Wien, Rubens, and the current director of the PTR, Emil Warburg, were all present. The multilingual H. A. Lorentz served as chair in lieu of Rayleigh, who had declined in part due to his poor language skills; and interestingly Lorentz began the conference by noting that advances were more likely to come from individual effort than from deliberations in a congress or more intimate gathering. Both the paper Lorentz gave and his brief welcoming address posed a contrast between "old theories" and "modern investigations," with his brief address being most pointed in its language—and extending the contrast to

29. See Solvay to Langevin, 15 June 1911 (see note 26 above) and Langevin and de Broglie, eds., *Théorie du rayonnement et les quanta* (1912).

30. As cited in Mehra, *Solvay Conferences* (1975), 9–10. The invitation listed participants in national groups.

FIGURE 10.2 Probably the most well known collective image of physicists at the turn of the century, from the Solvay Council held in Brussels in 1911. Photographie Benjamin Couprie, Institut International de Physique Solvay. Courtesy AIP Emilio Segrè Visual Archives.

one between modern studies and classical theory. Thinking particularly of approaches to fluids, weak solutions, and electron systems based on the kinetic theory of gases, he described the old theories as being "unable to penetrate the darkness that surrounds us on all sides." Lorentz feared that "perhaps even the fundamental equations of electrodynamics and our ideas about the nature of the ether will be shaken (although we may be allowed to hope that this will not be the case)"; and revealingly said that it was clear "we have no right to believe that the physical theories of the future will be subsumed under the rules of classical mechanics."[31]

When the language of the classical entered quantum theory it did so particularly through mechanics, but the Solvay conference was to see this language extended to meet the matter at hand ever more precisely and

31. Eucken, ed., *Theorie der Strahlung und der Quanten* (1913), 5–6. I will cite the German edition of the conference proceedings, which were originally published in Langevin and de Broglie, eds., *Théorie du rayonnement et les quanta* (1912).

clearly. In so doing a new use of *classical* was forged, with a more general purchase on the physics of the past. Lorentz's formal contribution was the first paper in the conference proceedings and was devoted to the application of the equipartition theorem to radiation. Lorentz described it as filling the need to convince the audience of the poverty of the "old theories" before they could move on to the matter at hand. Remembering that Planck did not consider equipartition in 1900 and dismissed its relevance in 1906, it is worth noting that by beginning with equipartition the very structure of the Brussels proceedings charted a pathway through earlier theory that differed markedly from the one through which Planck's law had been developed.

Revealingly, the language of the classical that had figured in Lorentz's welcoming comments *did not* appear in his formal paper, which of course had been written before he arrived. However, in response to a letter Rayleigh had sent to the council, Nernst spoke directly of "the classical theory of the equipartition of energy."[32] In his own paper James Jeans also wrote of "the classical theory of Maxwell and Boltzmann," doubting that this could be extended by a new hypothesis that could represent the facts as well as the equations of Planck, Nernst, and Einstein.[33] Together with three more instances I will discuss in detail shortly, these references show that *classical* has—quite suddenly—become part of the vocabulary with which several physicists considered work in quantum theory. Having shown in chapter 9 that this language was absent from earlier papers in the field, we are now in a position to recognize just how remarkable is its sudden appearance. Very likely this built on the use the term had found in relativity, but it is even more important to note the fact that the council papers were precirculated, and that one contribution to the congress offered an argument for a new use of the term.

The most significant invocation of the *classical* occurs in the paper that Max Planck delivered, "The Laws of Heat Radiation and the Hypotheses of the Elementary Quantum of Action." Boltzmann had linked the sociological use of the term *classical* to a situation in which methods that had looked as though they might serve a field forever are suddenly found wanting. Planck's opening paragraphs followed something like Boltzmann's recipe in arguing for the extension of the concept *classical* to cover this new field.

32. Eucken, ed., *Theorie der Strahlung und der Quanten* (1913), 11, 35, 43.
33. Jeans, "Kinetische Theorie der spezifischen Wärme" (1913), 59. See also 56.

> The principles of classical mechanics, fructified and extended by electrodynamics, and especially electron theory, have been so satisfactorily confirmed in all those regions of physics to which they had been applied . . . that it had looked as though even those areas which could only be approached indirectly through statistical forms of consideration would yield to the same principles without essential modification. The development of the kinetic theory of gases, which through the introduction of simple, bold conceptions shifted the scale of intuition and calculation to the order of magnitude of atoms and electrons, seemed to confirm this belief. . . .
>
> Today we must say that this hope has proved illusory, and that the framework of classical dynamics appears too narrow, even extended through the Lorentz-Einstein principle of relativity, to grasp those phenomena not directly accessible to our crude senses. The first incontestable proof of this view has come through the striking contradictions opened up between classical theory and observation in the universal laws of the radiation of black bodies.[34]

By 1911 Planck had embraced the language of the classical, and then quite consciously and deliberately extended it. Neglecting his earlier view that the application of equipartition to blackbody radiation was simply unjustified, his decision to extend the concept *classical* to statistical mechanics helped make the equipartition theorem classical. Given our awareness of the long controversy over its various applications, Planck's move could be regarded as the invention of a tradition, or at least as a new endorsement of the approach that several others had advocated, as valued and traditional. It brought the cultural value of the term *classical* into association with the field of statistical mechanics and the specific theorem of the equipartition of energy; and it is hardly to be doubted that interest in the new quantum theory was considerably heightened as a result. Here, Planck's formulation asserts, we have not just difficult choices between current theoretical possibilities, but high critical drama on an epochal stage.

Consider the implications of this brief history as a description of disciplinary change, in three interrelated realms. In terms of Planck's own development, this was a conception that clearly built on his personal struggles, but at the same time subtly altered his perspective on the nature of his previous work and the theoretical tools available. Note, for example, that in the 1890s, rather than accepting the success of the kinetic theory of gases as confirming the statistical, atomistic approach, Planck had turned

34. Planck, "Wärmestrahlung" (1913), 77.

to blackbody radiation precisely in order to circumvent that approach by developing a strict mechanical interpretation of the second law of thermodynamics. Ironically, it was only through the very events he now describes as "opening up the first striking contradiction between classical theory and observation" that Planck came to follow Boltzmann's approach—but even then he did so selectively and without abandoning his commitment to an absolute interpretation of thermodynamics.[35] It was Planck's subsequent exchanges with Einstein, Ehrenfest, Jeans, and Lorentz that had underlined equipartition and reinforced further subtleties of statistical physics. In 1911, then, Planck picked an indirect and generalized path through complex terrain, delicately negotiating tensions between his own views as an individual and those of a broader group of physicists. Rather than speaking directly of his own perspective, Planck adopts a communal voice in the same way that Einstein charted a collective path to the development of relativity when he highlighted the Michelson-Morley experiment in 1907. Thus, Planck's contribution to the Solvay Council outlined a new, collective history of physics circa 1900. Although our findings in chapter 9 indicate that his account fits precisely the views of none of the physicists centrally involved in the events, the terms of Planck's description were so seductively certain that they were later read back into the period—and into Planck's conceptual struggles in 1900, even by Planck himself. As we saw, when looking back in 1931 Planck would write, "Classical physics was not sufficient, that was clear to me," before outlining the need to avoid the consequences of equipartition and his fortunate solution: "for according to [equipartition], energy must, in the course of time, transform completely into radiation from matter. In order for it not to do so, we need a new constant that assures that the energy does not disintegrate [indefinitely]."[36] The collective path he first described in 1911 had finally been adopted as his own memory.

35. As Needell and Darrigol have shown, both the indeterminate internal structure of the resonators in Planck's model and his focus on the temporal rather than spatial nature of the disorder involved in natural radiation allowed Planck to circumvent the statistical interpretation of irreversibility.

36. Planck to Wood, 7 October 1931, AHQP, Microfilm 66, 5, as translated in Hoffmann and Lemmerich, *Quantum Theory Centenary* (2000), 49–50. In his Nobel acceptance speech in 1920 Planck had negotiated the distinction between his own views and that of the community far more clearly. There he wrote of the constant *h* blocking all attempts to fit it in the frame of classical theory, but added: "science does not owe the prompt and indubitable character of this decision to tests of the law for the energy distribution of thermal radiation, and even less to my special derivation of this law; it owes it to the unceasing progress of the researchers who have put the quantum of action to the service of their investigations. A. Einstein made

In terms of the dynamics of change within the physics community, Planck's 1911 formulation would help bring discussion of the equipartition theorem and new statistics onto the stage of worldview changes, while simultaneously shifting the base of the worldview discussion from the two previous alternatives in which it had most often been framed to a new, more complex footing. In contrast to the mechanical and electromagnetic worldviews, the very openness of the term *classical* has allowed Planck to recast the search for secure foundations to a contrast between the eras before and after quantum theory. We have seen that Planck's arguments for relativity had similarly urged physicists to cut across the distinction between mechanical and electromagnetic perspectives. Now he does so in service of a still broader program for new physics that incorporates relativity but must recognize the impotence of even that theory before the borders of microphysics and blackbody radiation. Now a new turning point was recognized for the critical contradiction between theory and experiment.

Finally, in terms of the dynamics of change in European culture more generally, Planck's move would help assert that physics was both part of its culture (involved in the struggle between classical traditions and the modern world) and could define a new framework for that worldview. Culturally, the relations between the modern and the classical had changed by 1911 also; in schooling and art, for example, modern forms were increasingly valorized at the expense of classical traditions.

Among those gathered together in Brussels, Einstein was surely most sharply aware of the novelty of the framework Planck offered. Einstein was one of the few physicists who had deployed concepts of classical in their own research, and surely recognized that Planck's extended use of "classical theory" signaled an increasingly clear commitment to major elements of Boltzmann's statistical mechanics, while allowing the sharp distinction that Einstein himself drew between Boltzmann's approach and "classical thermodynamics" to recede into the background. In the discussions that brought the Solvay Council to a close, Einstein described an extension of classical mechanics similar to Planck's discussion of classical theory, perhaps signaling his tacit assent to Planck's formulation.[37]

---

the first breakthrough in this domain." Planck, "Entstehung und bisherige Entwicklung der Quantentheorie" (1958 [1920]), 127.

37. Beginning his verbal summary of his paper, Einstein described classical mechanics as encompassing Lagrangian and Hamiltonian formulations. Having noted the limited guidance given by quantum theory, Einstein asked which general principles one could rely on. He

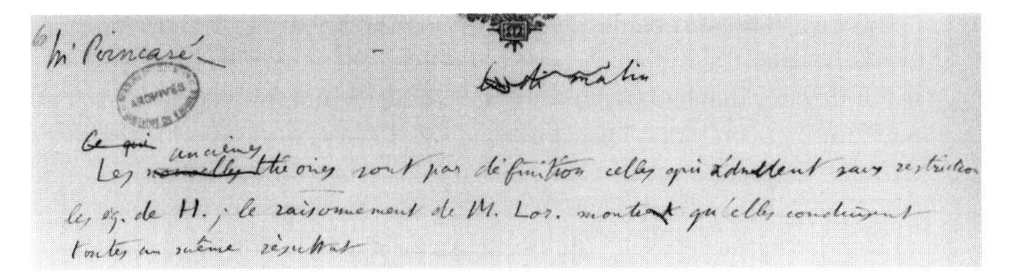

M. Poincaré

Les anciennes théories sont par définition celles qui admettent sans restriction les éq. de H.; le raisonnement de M. Lor. montre qu'elles conduisent toutes au même résultat

FIGURE 10.3 Poincaré's note in discussion to Lorentz's contribution, recording his point that previous theories had applied the Hamiltonian without restriction. Poincaré wrote first "nouvelle" and then "anciennes," revealing, perhaps, the subtle process through which his understanding of the recent past was being revised. Registre Solvay, vol. 1 of 2, p. 4, in Archives de l'Académie des sciences de l'Institut de France.

Both the tight organization of the council and its subsequent public life in print functioned as an ideal means for the propagation of new understandings of what had previously been a rather esoteric field, once largely confined to members of the German-speaking community. Nernst had succeeded in creating an organizational form that could rapidly draw together a new interpretive community, gaining adherents for quantum theory in the select ranks of a few chosen physicists from different nations and different specialties. Once the event was over, publishing its proceedings allowed Nernst to win the attention of a far broader segment of the international community, on the kind of terms he favored. The conference proceedings were zealously edited (with additional notes from individual contributors) and appeared in French and an updated German translation. James Jeans subsequently led a discussion on radiation at the British Association.[38] While the physicists present had reached little consensus on the path beyond the specific fields to which quantum theory had already been applied, Planck had helped them reach a common understanding of what the past had involved. Now others began to look back on the turn of the century in the light of a perspective and framing language first worked out in Germany.

---

thought the principle of energy conservation should stay, and stated his view that Boltzmann's definition of entropy by probability should be accepted without limitation, for "Thanks to this principle we have gained an admittedly limited enlightenment on the states of statistical equilibrium in which periodic phenomena contribute." "Discussion du rapport de M. Einstein" (1912), 436; as translated in Marage and Wallenborn, eds., *Solvay Councils* (1999), 107.

38. I have already cited both the French and German publications. The British Association discussion was published as "Discussion on Radiation" (1913). See also Jeans, *Radiation and Quantum-Theory* (1914).

No one shows this better than Henri Poincaré, new to quantum theory. Responding to Lorentz's paper on equipartition, Poincaré referred to the theories that had been shown to be unsuccessful. Jotting down his point for the secretaries, Poincaré wrote first "The new theories" before crossing "nouvelle" out to write "anciennes": "The old theories applied the Hamiltonian without limitation; Lorentz's considerations show they all lead to the same result" (fig. 10.3). As we might by now expect, the German edition took its liberties and translated "anciennes" as "klassischen."[39] Indeed, Poincaré reevaluated a great deal as a result of the conference. Soon afterward he wrote in terms that show how revealing his slip of the pen was:

> We might ask ourselves if mechanics is on the brink of a revolution. A congress in Brussels recently gathered together about twenty physicists of many nationalities. They were constantly talking about the new mechanics, as opposed to the old mechanics—but what did they mean by the old mechanics? Newton's, which was still uncontested at the end of the nineteenth century? No, they were referring to Lorentz's mechanics, and the principle of relativity, which scarcely five years ago seemed itself to be the very height of boldness.[40]

In this instance Poincaré's astonished language reflects the specific terms in which he had previously contrasted the "old" mechanics with the "new" Lorentzian form. On other occasions his writing shows that he now accepted both the periodization and terminology Planck had provided (thereby justifying the liberty his translator took). In a major 1912 paper credited with widening the hold of quantum theory in the research community, Poincaré contrasted the new to "classical mechanics," and now (in distinction to 1902) explicitly described the latter as incorporating the Hamiltonian equations. Max Born did the same in his review of the French publication of the conference proceedings.[41]

The conference in Brussels had a decisive effect in stimulating a comprehensive interest in quantum theory in France and Great Britain, which must have suggested the possibilities of creating an international agenda

39. Eucken, ed., *Theorie der Strahlung und der Quanten* (1913), 35.

40. Poincaré, "L'hypothèse des quanta" (1954 [1912]), 654.

41. Poincaré, "Théorie des quanta" (1954 [1912]), 626, 628–29. For the earlier use of old and new, see, for example, Poincaré, *Neue Mechanik* (1910). On Poincaré's participation in Brussels and response to quantum theory, see McCormmach, "Henri Poincaré and the Quantum Theory" (1967). Born's review is Born, "Rayonnement et les quanta" (1914), 166.

in physics—the Solvay conferences became a regular occurrence, continuing to bring together physicists of disparate backgrounds.[42] Planck had questioned whether the need for a joint effort was one of the fundamental assumptions behind Nernst's initiative. If he questioned either the efficacy or possibility of joint efforts in theoretical work, the effect of the first Solvay Council indicated the power of focusing the wider communities on one central problem.

Conclusion: On Monuments, Memories, and Einstein's Generation

This chapter has traced the christening of a new concept of classical theory and shown that the forum in which it was first advocated created conditions that enabled its unusually rapid propagation within a handpicked audience. I argued in chapter 5 that the organizational vigor of the German physics community enhanced their collective participation in the 1900 congress, enabling German physicists to take more away from Paris than members of other nations. Now my study of a second major conference has shown that in 1911 organizational ingenuity within the German community promoted movement in the other direction. The precisely choreographed meeting that Walther Nernst created in Brussels was able to facilitate the international diffusion of a theory that had hitherto been discussed largely in the German-speaking community and Britain. While Nernst called the Solvay Council into being in order to focus wider attention on a new problematic and a new theory, it propagated also a novel vision of the past. Although they were late to be paired in this way, henceforth quantum theory and this vision of past theory as classical physics traveled together.

Having recovered the diverse and opposing ways that concepts of *classical* were used circa 1900, and recognizing that physicists described a variety of contrasts between quantum theory and earlier approaches, we

42. Mehra, *Solvay Conferences* (1975). Born's assessment provides a good example of reactions to the first congress: "In these addresses and discussions the personalities of the speakers do not retreat behind the subjects, as is customary in scientific literature. For everyone who takes an active part in the development of science, there is a great appeal in this; for as long as the foundations of the quantum theory are still as unclear as they are today, opinion for and against the new theory depends in no small measure on personal inclinations and views." Born, "Rayonnement et les quanta" (1914), 166.

can see that despite its present ubiquity, this specific concept of the *classical* past might never have been coined. There was nothing inevitable in describing the turn of the century as Planck did in 1911. Further, had this description not come from Planck, with his authority and intimate relation to the events, and had it not been advocated in the unique environment of Nernst's "scientific Council," it may well have remained just one among several ways of describing the significance of the period. The 1900 congress had set out to document the state of a discipline and was expected to become a scientific monument. It has been forgotten despite its extraordinary scale and inclusiveness. In contrast the Solvay Council became an icon of the rise of modern physics and (making a monument of past theory) has marked the collective memory of a discipline, even though so few attended and they came without telling their colleagues. Thus Planck's description has defined an era for us, its original novelty unrecognized.

Rather than reflecting consensus views of the past, the earliest general concepts of classical theory were tendentious, deeply implicated in physicists' programs for different futures. The tensions and imperatives originally expressed in different versions of *classical mechanics* and *classical thermodynamics* conditioned the gradual formation of a new concept of past theory, but have long been invisible. Returning briefly to Boltzmann will help outline several more general findings from our interrogation of the changing meanings of classical physics. First, note that what was an importantly sociological insight in Boltzmann has now been given an epochal understanding in the German physics community, turning on a fulcrum quite different from the original one that Boltzmann had offered—on the turn of the century rather than with the success of Maxwell's work. Second, what *classical* meant has ultimately been defined not by adherents of the past traditions themselves as Boltzmann had it, but largely from the perspective of the new theories of relativity and quantum theory. In this, classical physics differs importantly from the articulation of the mechanical, energetic, or electromagnetic programs by practitioners—a point that will need to be clearly recognized in developing a more refined understanding of the process by which the new theories were won and defended.

Put rather strongly, one could say that apart from physicists who were classical by temperament in Ostwald's sense, or became classical late (like Russell McCormmach's famous night-thinker looking back after World

War I), there was only ever one "classical physicist." His name was Ludwig Boltzmann. Even given Boltzmann's voice in the wilderness, we have found that the patina of allegiance to the classical past was far more important to the formation of new theory in the twentieth century than it was to physicists of earlier times. Contemporaries had more often worked with and against theory they regarded as current, Newtonian, or old, defending stances on a continuum from phenomenological to atomistic and articulating distinctions ranging from mechanical through energetic to electromagnetic.

Having argued that we should think of Boltzmann as the only truly *classical* classical physicist, I have also shown that despite the introduction of Planck's theory of radiation in 1900 and the interplay between classical mechanics and relativity, Max Planck first christened the modern form of classical physics in 1911. The theory of relativity and the Solvay Council set the stage for the later public understanding that the fin de siècle had witnessed the overthrow of classical and the birth of modern physics. Now physics gave back to broader cultural circles a highly specific and importantly historical reading of the general term it had first begun to use in the late nineteenth century. And within the physics community the concept of classical physics—even of a classical world—would go on to become a potent heuristic foil for the development of a new quantum theory and mechanics of the atom, as Niels Bohr's work soon began to show.

Einstein's Generation has constantly sought to recover contemporary perspectives on familiar experiments, concepts, and histories, perspectives that despite their own importance have long been overshadowed by other ways of describing the past. Such neglected stories have often demonstrated significant limitations to our customary treatment of those events and achievements. Just as important distinctions between early understandings of classical mechanics and classical thermodynamics have long been forgotten, subsumed within a more general concept of classical physics, Michelson's perspective on his ether-drift experiment as the source of a new instrument has been lost in debate over whether or not the experiment was crucial to Einstein. And that debate itself neglected the considerable value Einstein gave the experiment in his own history of the rise of the principle of relativity. Similarly, the 1900 congress slipped into the shadow of the elite Solvay Council, despite its grandiose aims and extraordinary communal grasp; and the integrity of experimentalists' work to move from relative to absolute measures of space and time on a photographic plate has been forgotten in reactions to the result alone. Forgetting is integral to scientific advance, but neither our understanding of the process of science nor our appreciation of its historical development can accept the limitations imposed by such forgetfulness.

Previous chapters have pursued two very general strategies in their recovery of forgotten histories, taking up two different senses of the pun in *Einstein's Generation*. One strategy has been to pay new attention to the goals and achievements of several key contemporaries of Einstein in order to understand better the multifaceted nature of physics in the period. We have learned a great deal about the material culture of precision experimentation in the late nineteenth century by studying Michelson's endeavor to establish interferometry as a new form of instrumentation, for example. Such studies show the varied ways that physicists established links between instruments, experiments, and theory and articulated the value of their work within a complex disciplinary matrix. In particular, we highlighted the importance of standards and absolute measures in the work of Michelson and Rowland, but noted also that each pursued very different strategies, focused alternatively on establishing definitive values, or on providing key instruments, maps, and tables for spectroscopists. The generations that taught and accompanied Einstein made much more than relativity alone, and tracing the gradual propagation of Michelson's work in different contexts has complemented our already strong appreciation of the development of relativity theory with an understanding of the diverse ramifications of an instrument-based program of research—one that also owed its origins to the ether-drift experiment.

Following the work of comparatively neglected figures has helped generate a more adequate view of the compass and nature of physics in the period. But establishing critical breadth by examining the comprehensive grasp of the 1900 Congress of Physics and by exploring material, instrumental, and conceptual dimensions of *relativity physics* has also provided new grounds for understanding the significance of the events by which first relativity theory and then Einstein as an individual gained prominence in Germany. This has been our second general strategy: to pay new attention to Einstein's emergence as a central figure within his community. By 1900 both the relations between the ether and matter in general and theoretical and experimental studies of the electron in particular had attracted significant critical attention. Our study of Walter Kaufmann's research and of theorists investigating the treatment of the electron in relativity showed that experiment helped drive this field more fully than is usually appreciated and that in their different ways experimentalists and theoreticians alike pursued concerns with relative and absolute measures, and with rigidity. If these events have been absent or received short shrift in many accounts of relativity, it is in part because so many studies focus largely on origins and theoretical significance. By first exploring Einstein's reputation through the perspective of other physicists, and then examining the role of participant histories in shaping the meaning of Einstein's work after 1905, we have been able to refine our understanding of the complex process by which his community engaged with Einstein's research. In particular we saw that perceptions of Einstein's role were bound up with specific understandings of the relations between electron theory and relativity but also reflected Max Planck's use of these events to enhance the position of theoretical physics as a subspecialty in Germany.

Paying new attention to Einstein's contemporaries and to his emergence has involved the recovery of facets of the physics community that often escape accounts of physics circa 1900. Examining the origins of the broadest intellectual framework within which this period is described has similarly disclosed a forgotten history in the varied ways that physicists invoked concepts of the classical. But the specific concept of the classical past that Planck advanced in 1911 did not just elide awareness of the more complex events originally at issue; it also created a new understanding of the past. My final observation is that by focusing on 1900 as a turning point, Planck's vision has helped us pick out Einstein's era as fundamentally different from all others in the history of his discipline; for as we have it now, Einstein's era alone broke free of the classical past. While there is much to celebrate in Einstein's generation, this view relies as much on the understanding that earlier physicists were in the thrall of long-established traditions as it does on recognizing the achievements of Einstein's age in offering a more adequate understanding of the physical world. I can illustrate the form this argument takes by contrasting the appreciation we have built up of Einstein's immediate predecessors in the course of this study with Einstein's own memory.

My account of the Congress of Physics in 1900 has shown that in the year in which Einstein graduated, the physics discipline was engaged with a raft of new phenomena and had accepted the Michelson-Morley experiment as a critical challenge. Indeed, as Boltzmann's address on the methods of theoretical physics a year earlier demonstrates, the physics discipline was engaged in a search for new foundations so vigorous that Boltzmann considered it in danger of neglecting the value of traditional tools. It is well known that Einstein had to go beyond the classroom curriculum of the Zurich Polytechnic for the most recent developments in physics, privately studying Maxwell in the treatment published by August Föppl and reading up on electron theory through the texts of Paul Drude, for example. Drawing on Mach, Boltzmann, and Poincaré as he did in the years before 1905, Einstein was well aware of long-standing critical approaches to foundational issues, and in a period of rapid change himself joined in a search for reliable principles that he knew had been initiated by others. When he looked back on this period much later, however, Einstein painted a different picture in broad brushstrokes, one that was very likely as strongly shaped by Planck's description of the past as classical as it was by his own experience. Referring to the physics he met in his student days in his 1949 "Autobiographical Notes," Einstein wrote that in spite of the fruitfulness of particulars, "dogmatic rigidity prevailed in matters of principles: In the beginning (if there was such a thing) God created Newton's laws of motion together with the necessary masses and forces. This is all; everything beyond this follows from the development of appropriate mathematical methods by means of deduction." The success of mechanics in different realms meant, "We must not be surprised, therefore, that, so to speak, all physicists of the last century saw in classical mechanics a firm and final foundation for all physics, yes indeed, for all natural science."[43]

By subsuming all physics before 1900 within the label of *classical physics*, Planck had helped his discipline imagine that Einstein's generation was singularly unlike all others, that they alone had been able break through the dogmatic rigidity of Newtonian principles. And this too became a memory. Physicists now look back on Einstein's generation as uniquely free of the classical past. In spite of the achievements of Einstein's era, this book has argued that we must recognize this view as a myth, originally built as it was on a false image of the past that was created in the service of a new future.

43. See Einstein, "Autobiographical Notes" (1949), 19 and 21. The broad brushstrokes Einstein uses in these passages are in fact contradicted by the more detailed description that follows.

Bibliography

Archival Sources Consulted

Archives de l'Académie des Sciences de Institut de France.

Archive for the History of Quantum Physics. Deutsches Museum (Munich), National Museum of Science and Industry (London) and other locations (AHQP).

Bundesarchiv Deutschland (Berlin-Lichterfelde). Collections on the Physikalisch-Technische Reichsanstalt 1869–1950 and Ausländisches Maß- und Gewichtswesen 1871–1940 and Abt. I Ausländisches Maß- und Gewichtswesen 1871–1950.

Bureau International des Poids et Mesures.

Clark University Archives (CUA).

Collège de France, Paris.

Deutsches Museum (Munich) Archiv (DMA).

George Ellery Hale Collection. California Institute of Technology Archives.

Hale Observatory Archives, Pasadena (HOA).

Harvard College Observatory Papers. Harvard University Archives (HCO-HUA).

Hull, Gordon Ferrie, Papers. Dartmouth College Library.

Langevin, Paul, Papers. L'Ecole Supérieure de Physique et de Chimie Industrielles de la Ville de Paris.

Leroy Roberts Pratt & Whitney Collection. University of Connecticut Libraries, Archives & Special Collections at the Thomas J. Dodd Research Center (LRP&W-UC).

Mary Lea Shane Archives of the Lick Observatory. University Library, University of California–Santa Cruz.

Michelson Collection. U.S. Naval Academy, Nimitz Library (this now holds material initially gathered for the Michelson Museum in China Lake) (MC-USNA).

Records of the National Institute of Standards and Technology. Records of the Office of Standard Weights and Measures, Correspondence of the Office, 1859–93 (NIST-RG167-NARA).

National Museum of Science and Industry Library (London).

Naval Historical Foundation Collection. Library of Congress (NHFC-LC).
Newcomb, Simon, Papers. Library of Congress (SNP-LC).
Niedersächsische Staats- und Universitätsbibliothek Göttingen, Handschriften und seltene Drucke.
Pickering, Edward C., Papers. Harvard University Archives (ECP-HUA).
Rayleigh Archives. Hanscom Air Force Base, Research Library, Bedford, MA and other locations (RA-HAFB).
Records of the U.S. Naval Observatory. Record Group 78, National Archives and Records Administration (USNO-RG78-NARA).
Rheinische Friedrich-Wilhelms-Universität Bonn Archiv.
Rowland, Henry A., Papers. Johns Hopkins University, Milton S. Eisenhower Library, Special Collections (HAR-JHU).
Royal Swedish Academy of Sciences, Nobel Archive.
Staatsbibliothek Preußischer Kulturbesitz Berlin, Haus II, Handschriftenabteilung.
Universitätsarchiv Stuttgart.
Universites de Paris, Bibliotheque de la Sorbonne, Service du Livre Ancien.
University of Chicago Library, Special Collections (UCL).
Warner & Swasey Collection. Case Western Reserve University Archives and Special Collections (W&SC-CWRUA).
Yerkes Observatory Archives, Chicago (YOA).

Books and Articles

"1902. Hendrik Antoon Lorentz and Pieter Zeeman. Presentation by Hj. Théel." 1967. In *Nobel Lectures: Including Presentation Speeches and Laureates' Biographies: Physics, 1901–1921*, 11–13. Amsterdam, London, New York: Elsevier Publishing Co. for the Nobel Foundation.
"1903. Antoine Henri Becquerel, Pierre Curie and Marie Sklodowska-Curie. Presentation by H. R. Törnebladh." 1967. In *Nobel Lectures: Including Presentation Speeches and Laureates' Biographies: Physics, 1901–1921*, 47–51. Amsterdam, London, New York: Elsevier Publishing Co. for the Nobel Foundation.
"1905. Philipp Eduard Anton von Lenard. Presentation by A. Lindstedt." 1967. In *Nobel Lectures: Including Presentation Speeches and Laureates' Biographies: Physics, 1901–1921*, 101–4. Amsterdam, London, New York: Elsevier Publishing Co. for the Nobel Foundation.
"1906. Joseph John Thomson. Presentation by J. P. Klason." 1967. In *Nobel Lectures: Including Presentation Speeches and Laureates' Biographies: Physics, 1901–1921*, 141–44. Amsterdam, London, New York: Elsevier Publishing Co. for the Nobel Foundation.
"1907. Albert Abraham Michelson. Presentation by K. B. Hasselberg." 1967. In *Nobel Lectures: Including Presentation Speeches and Laureates' Biographies: Physics, 1901–1921*, 159–65. Amsterdam, London, New York: Elsevier Publishing Co. for the Nobel Foundation.

"1913. Heike Kamerlingh Onnes. Presentation by Th. Nordström." 1967. In *Nobel Lectures: Including Presentation Speeches and Laureates' Biographies: Physics, 1901–1921*, 303–5. Amsterdam, London, New York: Elsevier Publishing Co. for the Nobel Foundation.

A. W. R. 1874. "Pickering's 'Physical Manipulation.'" *Nature* 10, 2 July:160–61.

Abraham, Max. 1900. "H. A. Lorentz, Über die Theorie der neuentdeckten magnetooptischen Erscheinungen." *Physikalische Zeitschrift* 2:88.

———. 1902. "Dynamik des Elektrons." *Königliche Gesellschaft der Wissenschaften zu Göttingen. Mathematisch-Physikalische Klasse. Nachrichten*:22–41.

———. 1903. "Prinzipien der Dynamik des Elektrons." *Annalen der Physik* 10:105–79.

———. 1903. "Prinzipien der Dynamik des Elektrons." *Physikalische Zeitschrift* 4:57–63.

———. 1904. "Die Grundhypothesen der Elektronentheorie." *Physikalische Zeitschrift* 5:576–79.

———. 1905. *Theorie der Elektrizität*. Vol. 2: *Elektromagnetische Theorie der Strahlung*. 2 vols. Leipzip: B. G. Teubner.

Adams, Henry. 1918. *The Education of Henry Adams: An Autobiography*, edited by Henry Cabot Lodge. Boston: Houghton Mifflin.

Albisetti, James C. 1983. *Secondary School Reform in Imperial Germany*. Princeton, NJ: Princeton Univ. Press.

Ames, J.-S. 1900. "L'équivalent mécanique de las chaleur." In *Rapports présentés au Congrès international de physique réuni à Paris en 1900 sous les auspices de la Société française de physique*. Tome I. *Questions générales. Métrologie. Physique mécanique. Physique moléculaire*, edited by Lucien Poincaré and Charles-Édouard Guillaume, 178–213. Paris: Gauthier-Villars.

Andrade, Jules Frédéric Charles. 1898. *Leçons de mécanique physique*. Paris: Société d'éditions scientifiques.

Arabatzis, Theodore. 1992. "The Discovery of the Zeeman Effect: A Case Study of the Interplay between Theory and Experiment." *Studies in the History and Philosophy of Science* 23:365–88.

———. 1996. "Rethinking the 'Discovery' of the Electron." *Studies in the History and Philosophy of Modern Physics* 27:405–35.

———. 2006. *Representing Electrons: A Biographical Approach to Theoretical Entities*. Chicago: Univ. of Chicago Press.

Barkan, Diana Kormos. 1992. "A Usable Past: Creating Disciplinary Space for Physical Chemistry." In *The Invention of Physical Science: Intersections of Mathematics, Theology, and Natural Philosophy since the Seventeenth Century; Essays in Honor of Erwin N. Hiebert*, edited by Mary Jo Nye, Joan L. Richards, and Roger Stuewer, 175–202. Dordrecht: Kluwer.

———. 1993. "The Witches Sabbath: The First International Solvay Congress in Physics." In *Einstein in Context*, edited by Mara Beller and Jürgen Renn, 59–82. Cambridge: University of Cambridge.

———. 1999. *Walther Nernst and the Transition to Modern Physical Science*. Cambridge: Cambridge Univ. Press.

Bartky, Ian R. 1989. "The Adoption of Standard Time." *Technology and Culture* 30:25–56.

———. 2000. *Selling the True Time: Nineteenth-Century Timekeeping in America*. Stanford, CA: Stanford Univ. Press.

Barus, Carl. 1900. "Les progrès de la pyrométrie." In *Rapports présentés au Congrès international de physique réuni à Paris en 1900 sous les auspices de la Société française de physique*. Tome I. *Questions générales. Métrologie. Physique mécanique. Physique moléculaire*, edited by Lucien Poincaré and Charles-Édouard Guillaume, 148–77. Paris: Gauthier-Villars.

———. 1905. "The Progress of Physics in the Nineteenth Century." *Science* 22, no. 560:353–69, 385–97.

Battimelli, Giovanni, Maria Grazia Ianniello, and Otto Kresten, eds. 1993. *Proceedings of the International Symposium on Ludwig Boltzmann (Rome, February 9–11, 1989)*. Vienna: Verlag der Österreichischen Akademie der Wissenschaften.

Becquerel, Henri. 1900. "Sur le rayonnement de l'uranium et sur diverses propriétés physiques du rayonnement des corps radio-actifs." In *Rapports présentés au Congrès international de physique réuni à Paris en 1900 sous les auspices de la Société française de physique*. Tome III. *Électro-optique et ionisation. Applications. Physique cosmique. Physique biologique*, edited by Lucien Poincaré and Charles-Édouard Guillaume, 47–78. Paris: Gauthier-Villars.

Bell, Louis. 1887. "On the Absolute Wave-Length of Light." *American Journal of Science* 33:167–82.

———. 1902. "On the Discrepancy between Grating and Interference Measurements." *Astrophysical Journal* 15:157–71.

———. 1903. "The Perot-Fabry Corrections of Rowland's Wave-Lengths." *Astrophysical Journal* 18:191–97.

Bennett, Jim, Robert Brain, Kate Bycroft, Simon Schaffer, Heinz Otto Sibum, and Richard Staley. 1993. *Empires of Physics: A Guide to the Exhibition*. Cambridge: Whipple Museum of the History of Science.

Bennett, Jim, Robert Brain, Simon Schaffer, Heinz Otto Sibum, and Richard Staley. 1994. *1900: The New Age; A Guide to the Exhibition*. Cambridge: Whipple Museum of the History of Science.

Benoît, J.-René. 1900. "De la précision dans la détermination des longueurs en métrologie." In *Rapports présentés au Congrès international de physique réuni à Paris en 1900 sous les auspices de la Société française de physique*. Tome I. *Questions générales. Métrologie. Physique mécanique. Physique moléculaire*, edited by Lucien Poincaré and Charles-Éduoard Guillaume, 30–77. Paris: Gauthier-Villars.

Bestelmeyer, Adolf. 1907. "Spezifische Ladung und Geschwindigkeit der durch Röntgenstrahlen erzeugten Kathodenstrahlen." *Annalen der Physik* 22:429–47.

———. 1909. "Bemerkungen zu der Abhandlung Hrn. A. H. Bucherers, 'Die experimentelle Bestätigung des Relativitätsprinzips.'" *Annalen der Physik* 30:166–74.

Bigg, Charlotte. 2002. "Behind the Lines: Spectroscopic Enterprises in Early Twentieth Century Europe." PhD dissertation, University of Cambridge.

———. 2003. "Spectroscopic Metrologies." *Nuncius* 18, no. 2:765–77.

Blackmore, John. 1992. *Ernst Mach—A Deeper Look: Documents and New Perspectives*. Dordrecht and Boston: Kluwer.

———. 1995. *Ludwig Boltzmann: His Later Life and Philosophy, 1900–1906. Book One: A Documentary History*. Vol. 168, Boston Studies in the Philosophy of Science. Dordrecht: Kluwer.

———. 1995. *Ludwig Boltzmann: His Later Life and Philosophy, 1900–1906. Book Two: The Philosopher*. Vol. 174, Boston Studies in the Philosophy of Science. Dordrecht: Kluwer.

Boltzmann, Ludwig. 1897. *Vorlesungen über die Principe der Mechanik*. Leipzig: J. A. Barth.

———. 1905 [1899]. "Über die Entwickelung der Methoden der theoretischen Physik in neuerer Zeit." In *Populäre Schriften*, 198–227. Leipzig: J. A. Barth.

———. 1974 [1899]. "On the Development of the Methods of Theoretical Physics in Recent Times." In *Theoretical Physics and Philosophical Problems: Selected Writings*, edited by Brian McGuinness, 77–100. Dordrecht and Boston: D. Reidel.

———. 1974 [1900, 1902]. "On the Principles of Mechanics, I, II." In *Theoretical Physics and Philosophical Problems: Selected Writings*, edited by Brian McGuinness, 129–52. Dordrecht and Boston: D. Reidel.

Bond, George M., ed. 1887. *Standards of Length and Their Practical Application*. Hartford, CT: Pratt & Whitney.

———. 1887. "Standards of Length and Their Subdivision." In *Standards of Length and Their Practical Application*, edited by George M. Bond, 119–38. Hartford, CT: Pratt & Whitney.

———. 1887. "Standards of Length as Applied to Gauge Dimensions." In *Standards of Length and Their Practical Application*, edited by George M. Bond, 139–72. Hartford, CT: Pratt & Whitney.

Boorse, Henry A., and Lloyd Motz, eds. 1966. *The World of the Atom*. 2 vols. Vol. 1. New York and London: Basic Books.

Bork, Alfred M. 1966. "The 'FitzGerald' Contraction." *Isis* 57:199–207.

Born, Max. 1909. "Die Theorie des starren Elektrons in der Kinematik des Relativitätsprinzip." *Annalen der Physik* 30:1–56.

———. 1909. "Die träge Masse und das Relativitätsprinzip." *Annalen der Physik* 28:571–84.

———. 1909. "Über die Dynamik des Elektron in der Kinematik des Relativitätsprinzip." *Physikalische Zeitschrift* 10:814–17.

———. 1910. "Über die Definition des starren Körpers in der Kinematik des Relativitätsprinzip." *Physikalische Zeitschrift* 11:233–34.

———. 1910. "Zur Kinematik des starren Körpers im System des Relativitätsprinzips." *Königliche Gesellschaft der Wissenschaften zu Göttingen. Mathematisch-Physikalische Klasse. Nachrichten*:161–79.

———. 1914. "P. Langevin u. M. de Broglie, La théorie du rayonnement et les quanta." *Physikalische Zeitschrift* 15:166–67.

Born, Max, and Emil Wolf. 1999. *Principles of Optics: Electromagnetic Theory of Propagation, Interference, and Diffraction of Light*. 7th expanded ed. London: Pergamon.

Brachner, Alto. 1985. "German Nineteenth-Century Scientific Instruments." In *Nineteenth-Century Instruments and Their Makers*, edited by P. R. de Clercq, 117–57. Amsterdam: Rodopi; Leiden: Museum Boerhaave.

Brain, Robert M. 1993. *Going to the Fair: Readings in the Culture of Nineteenth-Century Exhibitions*. Cambridge: Whipple Museum of the History of Science.

Brannigan, Augustine. 1981. *The Social Basis of Scientific Discoveries*. Cambridge: Cambridge Univ. Press.

Brashear, John A. 1988 [1924]. *A Man Who Loved the Stars: The Autobiography of John A. Brashear*. First published in 1924 by the American Society of Mechanical Engineers. Pittsburgh: Univ. of Pittsburgh Press.

Breene, Robert G., Jr. 1961. *The Shift and Shape of Spectral Lines*. New York: Pergamon.

Broda, Engelbert, and Hans (Mitarb.) Thirring. 1957. *Ludwig Boltzmann: Mensch, Physiker, Philosoph*. Berlin: Dt. Verlag der Wissenschaften.

Brooks, Randall C. 1988. "Standard Screw Threads for Scientific Instruments. Part II: The British Association Screw Gauge." *History and Technology* 6:45–59.

———. 1993. "Towards the Perfect Screw Thread: The Making of Precision Screws in the 17th–19th Centuries." *Transactions of the Newcomen Society for the Study of the History of Engineering and Technology* 64:101–20.

Brush, Stephen G. 1967. "Note on the History of the FitzGerald-Lorentz Contraction." *Isis* 58:230–32.

———. 1976. *The Kind of Motion We Call Heat: A History of the Kinetic Theory of Gases in the 19th Century*. Vol. 2: *Statistical Physics and Irreversible Processes*. Amsterdam and New York: North-Holland.

———. 1999. "Why Was Relativity Accepted?" *Physics in Perspective* 1:184–214.

Bryan, G. H. 1903. "Allgemeine Grundlegung der Thermodynamik." In *Encyclopädie der mathematischen Wissenschaft*. Vol. 5, *Physik*, 1. Teil, edited by Arnold Sommerfeld, 71–160. Leipzig: B. G. Teubner.

———. 1905. "Three Cambridge Mathematical Works (Book Review)." *Nature* 71:601–3.

Bucherer, A. H. 1908. "Messungen an Becquerelstrahlen. Die experimentelle Bestätigung der Lorentz-Einsteinschen Theorie." *Physikalische Zeitschrift* 9: 755–62.

———. 1909. "Die experimentelle Bestätigung des Relativitätsprinzips." *Annalen der Physik* 28:513–36.

Buchwald, Jed Z., and Andrew Warwick, eds. 2001. *Histories of the Electron: The Birth of Microphysics*. Dibner Institute Studies in the History of Science and Technology. Cambridge, MA: MIT Press.

Budde, E. 1911. "Zur Theorie des Michelsons Versuches." *Physikalische Zeitschrift* 12:979–91.

———. 1912. "Zur Theorie des Michelsons Versuches. II." *Physikalisches Zeitschrift* 13:825.

Büttner, Jochen, Jürgen Renn, and Matthias Schemmel. 2003. "Exploring the Limits of Classical Physics: Planck, Einstein, and the Structure of a Scientific Revolution." *Studies in the History and Philosophy of Modern Physics* 34:37–59.

Cahan, David. 1989. *An Institute for an Empire: The Physikalisch-Technische Reichsanstalt, 1871–1918*. Cambridge: Cambridge Univ. Press.

Cahan, David, and M. Eugene Rudd. 2000. *Science at the American Frontier: A Biography of DeWitt Bristol Brace*. Lincoln: Univ. of Nebraska Press.

Cajori, Florian. 1899. *A History of Physics in Its Elementary Branches: Including the Evolution of Physical Laboratories*. New York: Macmillan.

Campbell, John T. 1973. "Kaufmann, Walter." In *Dictionary of Scientific Biography*, edited by Charles S. Gillespie, 263–65. New York: Scribner.

Campbell, Norman Robert. 1907. *Modern Electrical Theory*. Cambridge: Cambridge Univ. Press.

Canales, Jimena. 2002. "Photogenic Venus." *Isis* 93, no. 4:585–613.

Capehart [Director], Alexander S. 1901. "Report of the Department of Liberal Arts and Chemical Industries." In *Report of the Commissioner-General for the United States to the International Universal Exposition, Paris, 1900*, 5–98. Washington, DC: Government Printing Office.

Cassidy, David C. 1986. "Understanding the History of Special Relativity." *Historical Studies in the Physical Sciences* 16:177–95.

———. 2004. *Einstein and Our World*. 2nd ed. Amherst, NY: Humanity Books.

Cattermole, M. J. G., and A. F. Wolfe. 1987. *Horace Darwin's Shop: A History of the Cambridge Scientific Instrument Company, 1878 to 1968*. Bristol: Hilger.

Ceranski, Beate. 2004. "Die Arbeit mit dem Radium. Radioaktivitätsforschung 1896–1914." Habilitation, Stuttgart Universität.

Cercignani, Carlo. 1998. *Ludwig Boltzmann: The Man Who Trusted Atoms*. Oxford: Oxford Univ. Press.

Chandler, Arthur. 1987. "Culmination: The Paris Exposition Universelle." *World's Fair* 7, no. 3:8–14.

Chappuis, P. 1900. "L'échelle thermométrique normale et les échelles pratiques pour la mesure des températures." In *Rapports présentés au Congrès international de physique réuni à Paris en 1900 sous les auspices de la Société française de physique*. Tome I. *Questions générales. Métrologie. Physique mécanique. Physique moléculaire*, edited by Lucien Poincaré and Charles-Édouard Guillaume, 131–47. Paris: Gauthier-Villars.

Clerke, Agnes. 1887. *A Popular History of Astronomy during the Nineteenth Century*. 2nd ed. London: Adam and Charles Black.

Clow, Nani. 1999. "The Laboratory of Victorian Culture: Experimental Physics, Industry, and Pedagogy in the Liverpool Laboratory of Oliver Lodge, 1881–1900." PhD dissertation, Harvard University.

Coen, Deborah R. 2004. "A Scientific Dynasty: Probability, Liberalism, and the Exner Family in Imperial Austria." PhD dissertation, Harvard University.

Collins, Harry. 1985. *Changing Order: Replication and Induction in Scientific Practice*. London: Sage.
———. 2001. "Tacit Knowledge, Trust, and the Q of Sapphire." *Social Studies of Science* 31:71–85.
Collins, Harry, and Trevor Pinch. 1993. *The Golem: What Everyone Should Know about Science*. Cambridge: Cambridge Univ. Press.
———. 1998. *The Golem: What You Should Know about Science*. 2nd ed. Cambridge: Canto.
Comité international des poids et mesures. 1897. *Procès-verbaux des séances de 1897*.
———. 1902 [1901]. *Procès-verbaux des séances. Deuxième série*. Tome I. *Session de 1901*. Paris: Gauthier-Villars.
Cooper, Herbert J., ed. 1952. *Scientific Instruments: Described by Specialists under the Editorship of Herbert J. Cooper*. 2nd ed. 2 vols. Vol. 1. London: Hutchinson's Scientific and Technical Publications.
Cope, Kenneth L., ed. 1997. *A Brown & Sharpe Catalogue Collection: 1868, 1887, 1899*. Mendham, NJ: Astragal Press.
Cornu, Alfred. 1873. "Détermination nouvelle de la vitesse de la lumière." *Académie des sciences, Paris, Comptes rendus* 76:338–42.
———. 1874. "Détermination nouvelle de la vitesse de la lumière." *Journal de l'École Polytechnique* 27:133–80.
———. 1900. "Sur la vitesse de la lumière." In *Rapports présentés au Congrès international de physique réuni à Paris en 1900 sous les auspices de la Société française de physique*. Tome II. *Optique. Électricité. Magnétisme*, edited by Lucien Poincaré and Charles-Édouard Guillaume, 225–46. Paris: Gauthier-Villars.
Corry, Leo. 1997. "Hermann Minkowski and the Postulate of Relativity." *Archive for History of Exact Sciences* 51:273–314.
Crawford, Elisabeth. 1984. *The Beginnings of the Nobel Institution: The Science Prizes, 1901–1915*. Cambridge: Cambridge Univ. Press; Paris: Éditions de la Maison des Sciences de l'Homme.
Crew, Henry. 1904. "Remarks on Standard Wave-Lengths." *Astrophysical Journal* 20:313–17.
Cunningham, Ebenezer. 1907. "On the Electromagnetic Mass of a Moving Electron." *Philosophical Magazine* 14:538–47.
Curie, Marie. 1904. *Untersuchungen über die radioaktiven Substanzen*. Translated by Walter Kaufmann. Braunschweig: F. Vieweg und Sohn.
Curie, P., and Mme. Curie. 1900. "Les nouvelles substances radioactives et les rayons qu'elles émettent." In *Rapports présentés au Congrès international de physique réuni à Paris en 1900 sous les auspices de la Société française de physique*. Tome III. *Électro-optique et ionisation. Applications. Physique cosmique. Physique biologique*, edited by Lucien Poincaré and Charles-Édouard Guillaume, 79–114. Paris: Gauthier-Villars.
D'Agostino, Salvo. 1990. "Boltzmann and Hertz on the *Bild*-Conception of Physical Theory." *History of Science* 28:380–98.

Darrigol, Olivier. 1988. "Statistics and Combinatorics in Early Quantum Theory." *Historical Studies in the Physical and Biological Sciences* 19:17–80.

———. 1991. "Statistics and Combinatorics in Early Quantum Theory, II: Early Symptoms of Indistinguishability and Holism." *Historical Studies in the Physical and Biological Sciences* 21, no. 2:237–98.

———. 1992. *From c-Numbers to q-Numbers: The Classical Analogy in the History of Quantum Theory.* California Studies in the History of Science. Berkeley: Univ. of California Press.

———. 1995. "Henri Poincaré's Criticism of *Fin de Siècle* Electrodynamics." *Studies in History and Philosophy of Modern Physics* 26:1–44.

———. 1996. "The Electrodynamic Origins of Relativity Theory." *Historical Studies in the Physical and Biological Sciences* 26:241–312.

———. 2000. *Electrodynamics from Ampère to Einstein.* Oxford: Oxford Univ. Press.

———. 2001. "The Historians' Disagreements over the Meaning of Planck's Quantum." *Centaurus* 43:219–39.

———. 2004. "The Mystery of the Einstein-Poincaré Connection." *Isis* 95:614–26.

Deltete, Robert J. 2000. "Helm's History of Energetics: A Reading Guide." In *The Historical Development of Energetics*, by Georg Helm, 4–45. Translated by Robert Deltete. Dordrecht, Boston, London: Kluwer.

Dennis, Michael Aaron. 1987. "Accounting for Research: New Histories of Corporate Laboratories and the Social History of American Science." *Social Studies of Science* 17:479–518.

Deutschland. Reichskommissar für die Weltausstellung in Paris 1900 [ed. Otto N. Witt]. 1900. *Exposition universelle de 1900, Paris: Catalogue officiel de la section Allemande.* Berlin: Impr. Imperiale.

———. 1900. *International Exposition Paris 1900: Official Catalogue, Exhibition of the German Empire.* Berlin: J. A. Stargardt.

———. 1900. *Weltausstellung in Paris 1900: Amtlicher Katalog der Ausstellung des Deutschen Reichs.* Berlin: J. A. Stargardt.

DeVorkin, David. 1975. "Michelson and the Problem of Stellar Diameters." *Journal for the History of Astronomy* 6:1–18.

"Discussion du rapport de M. Einstein." 1912. In *La Théorie du Rayonnement et les Quanta: Rapports et discussions de la réunion tenue à Bruxelles, du 30 Octobre au 3 Novembre 1911*, ed. Paul Langevin and M. de Broglie, 436–50. Paris: Gauthier-Villars.

"Discussion on Radiation." 1913. *Reports of the British Association for the Advancement of Science*:376–386.

Drake, Francis E. 1901. "Report of the Director of the Department of Machinery and Electricity." In *Report of the Commissioner-General for the United States to the International Universal Exposition, Paris, 1900.* Vol. 3, 99–173. Washington, DC: Government Printing Office.

Drude, Paul. 1900. *Lehrbuch der Optik.* Leipzig: S. Hirzel.

———. 1900. "Théorie de la dispersion dans les métaux fondée sur la considération des électrons." In *Rapports présentés au Congrès international de physique réuni à Paris en 1900 sous les auspices de la Société française de physique*. Tome III. *Électro-optique et ionisation. Applications. Physique cosmique. Physique biologique*, edited by Lucien Poincaré and Charles-Édouard Guillaume, 34–46. Paris: Gauthier-Villars.

Dugas, René. 1955. *A History of Mechanics*. Translated by J. R. Maddox. Neuchâtel: Éditions du Griffon; New York: Central Book Company.

Dunsch, Lothar, and Hella Müller. 1989. *Ein Fundament zum Gebäude der Wissenschaften: Einhundert Jahre Ostwalds Klassiker der exakten Wissenschaften (1889–1989)*. Ostwald's Klassiker der exakten Wissenschaften. Leipzig: Akademische Verlagsgesellschaft.

Earman, John, and Clark Glymour. 1980. "Relativity and Eclipses: The British Expeditions of 1919 and Their Predecessors." *Historical Studies in the Physical Sciences* 11:49–85.

Ebert, Hermann. 1887. "Über die Abhängigkeit der Wellenlänge des Lichtes von seiner Intensität." *Annalen der Physik* 32:337–84.

———. 1888. "Die Methode der hohen Interferenzen in ihrer Verwendbarkeit für Zwecke der quantitativen Spectralanalyse." *Annalen der Physik* 34:39–90.

———. 1891. "Einfluss der Helligkeitsvertheilung in den Spectrallinien auf die Interferenzerscheinungen." *Annalen der Physik* 43:790–807.

Ehrenfest, Paul. 1905. "Über die physikalische Voraussetzungen der Planck'schen Theorie der irreversiblen Strahlungsvorgänge." *Kaiserliche Akademie der Wissenschaften (Vienna). Mathematisch-naturwissenschaftliche Classe. Zweite Abtheilung. Sitzungsberichte* 114:1301–14.

———. 1906. "Ludwig Boltzmann." *Mathematisch-Naturwissenschaftliche Blätter* 3.

———. 1906. "Zur Planckschen Strahlungstheorie." *Physikalische Zeitschrift* 7: 528–32.

———. 1907. "Die Translation deformierbarer Elektronen und der Flächensatz." *Annalen der Physik* 23:204–5.

———. 1909. "Gleichförmige Rotation starrer Körper und Relativitätstheorie." *Physikalische Zeitschrift* 10:918.

Einstein, Albert. 1905. "Über die von der molekularkinetischen Theorie der Wärme geforderte Bewegung von in ruhenden Flüssigkeiten suspendierten Teilchen." *Annalen der Physik* 17:549–60.

———. 1905. "Über einen die Erzeugung und Verwandlung des Lichtes betreffenden heuristischen Gesichtspunkt." *Annalen der Physik* 17:132–48.

———. 1905. "Zur Elektrodynamik bewegter Körper." *Annalen der Physik* 17:891–921.

———. 1906. "Über eine Methode zur Bestimmung des Verhältnisses der transversalen und longitudinalen Masse des Elektrons." *Annalen der Physik* 21:583–86.

———. 1906. "Zur Theorie der Brownschen Bewegung." *Annalen der Physik* 19:371–81.

———. 1906. "Zur Theorie der Lichterzeugung und Lichtabsorption." *Annalen der Physik* 20:199–206.

———. 1907. "Bemerkungen zu der Notiz von Hrn. Paul Ehrenfest: 'Die translation deformierbarer Elektronen und der Flächensatz.'" *Annalen der Physik* 23:206–8.

———. 1907. "Die Plancksche Theorie der Strahlung und die Theorie der spezifischen Wärme." *Annalen der Physik* 22:180–90.

———. 1907. "Über das Relativitätsprinzip und die aus demselben gezogenen Folgerungen." *Jahrbuch der Radioaktivität und Elektronik* 4:411–62.

———. 1907. "Über die Möglichkeit einer neuen Prüfung des Relativitätsprinzips." *Annalen der Physik* 23:197–98.

———. 1907. "Über die vom Relativitätsprinzip geforderte Trägheit der Energie." *Annalen der Physik* 23:371–84.

———. 1908. "Bemerkung zu der Arbeit von D. Mirimanoff 'Über die Grundgleichungen der Elektrodynamik bewegter Körper von Lorentz und das Prinzip der Relativität.'" *Annalen der Physik* 28:885–88.

———. 1909. "Über die Entwicklung unserer Anschauungen über das Wesen und die Konstitution der Strahlung." *Physikalische Zeitschrift* 10:817–26.

———. 1909. "Zum gegenwärtigen Stand des Strahlungsproblems." *Physikalische Zeitschrift* 10:185–93.

———. 1917. *Über die spezielle und die allgemeine Relativitätstheorie: Gemeinverständlich.* Braunschweig: Vieweg.

———. 1920. *Relativity: The Special and General Theory.* Translated by Robert W. Lawson. New York: Henry Holt.

———. 1949. "Autobiographical Notes." In *Albert Einstein: Philosopher-Scientist,* edited by P. A. Schilpp, 2–94. Evanston, IL: Library of Living Philosophers.

———. 1974 [1916]. "Die Grundlage der allgemeinen Relativitätstheorie." In *Das Relativitätsprinzip: Eine Sammlung von Abhandlungen,* by H. A. Lorentz, Albert Einstein, Hermann Minkowski, and Hermann Weyl, 81–124. Leipzig and Berlin: Teubner.

———. 1989. *The Collected Papers of Albert Einstein.* Vol. 2, *The Swiss Years: Writings 1900–1909,* edited by John Stachel. Princeton, NJ: Princeton Univ. Press.

———. 1993. *The Collected Papers of Albert Einstein.* Vol. 5, *The Swiss Years: Correspondence, 1902–1914,* edited by Martin J. Klein, A. J. Kox, and Robert Schulmann. Princeton, NJ: Princeton Univ. Press.

Einstein, Albert, and Jakob Laub. 1908. "Über die elektromagnetischen Grundgleichungen fur bewegte Körpern." *Annalen der Physik* 26:532–40.

———. 1908. "Über die im elektromagnetischen Felde auf ruhende Körper ausgeubten ponderomotorischen Kräfte." *Annalen der Physik* 26:541–50.

Encyclopédie du siècle. L'Exposition de Paris de 1900. 1900. Paris.

Encyklopädie der mathematischen Wissenschaften mit Einschluss ihrer Anwendungen. 1898–1935. Various editors. 6 vols. Leipzig: B. G. Teubner.

Eucken, Arnold, ed. 1913. *Die Theorie der Strahlung und der Quanten. Verhandlungen auf einer von E. Solvay einberufenen Zusammenkunft (30. Oktober bis 3. November 1911).* Berlin: Verlag Chemie GMBH.

Evans, Chris. 1989. *Precision Engineering: An Evolutionary View.* Bedford: Cranfield Press.

Evans, Chris J., and Deborah Jean Warner. 1988. "Precision Engineering and Experimental Physics: William A. Rogers, the First Academic Mechanician in the U.S." In *The Michelson Era in American Science: 1870–1930*, edited by Stanley Goldberg and Roger H. Stuewer, 2–12. New York: American Institute of Physics.

Fabry, Charles, J. Macé de Lépinay, and Alfred Pérot. 1899. "Sur la masse du décimètre cube d'eau." *Académie des sciences, Paris, Comptes rendus* 128:1317–19.

———. 1899. "Sur la mesure en logueurs d'onde des dimensions d'un cube de quartz de 4cm de côte." *Académie des sciences, Paris, Comptes rendus* 128:1317–19.

Fabry, Charles, and Alfred Pérot. 1896. "Mesure de petites èpaisseures en valeur absolue." *Académie des sciences, Paris, Comptes rendus* 123:802–5.

———. 1898. "Sur une nouvelle méthode de spectroscopie intéferentielle." *Académie des sciences, Paris, Comptes rendus* 126:34–36.

———. 1899. "Méthode interférentielles pour la mesure des grandes épaisseurs et la comparison des longueurs d'onde." *Annales de Chimie et de Physique* 16:289–338.

———. 1899. "Sur une source intense de lumière monochromatique." *Académie des sciences, Paris, Comptes rendus* 128:1156–58.

———. 1902. "Measures of Absolute Wave-Lengths in the Solar Spectrum and in the Spectrum of Iron." *Astrophysical Journal* 15:73–96, 261–73.

———. 1903. "A Reply to the Recent Article by Louis Bell." *Astrophysical Journal* 16:37.

———. 1904. "Rapport sur la nécessité d'établir un nouveau système de longueurs dónde étalons." *Astrophysical Journal* 20:318–26.

Falconer, Isobel. 1987. "Corpuscles, Electrons, and Cathode Rays: J. J. Thomson and the 'Discovery of the Electron.'" *British Journal for the History of Science* 20:241–76.

———. 2001. "Corpuscles to Electrons." In *Histories of the Electron: The Birth of Microphysics*, edited by Jed Z. Buchwald and Andrew Warwick, 77–100. Cambridge, MA: MIT Press.

Faragó, P. S., and L. Jánossy. 1957. "Review of the Experimental Evidence for the Law of Variation of the Electron Mass with Velocity." *Il Nouvo Cimento* 5:1411–36.

Feffer, Stuart M. 1989. "Arthur Schuster, J. J. Thomson, and the Discovery of the Electron." *Historical Studies in the Physical and Biological Sciences* 20:33–61.

FitzGerald, G. F. 1889. "The Ether and the Earth's Atmosphere." *Science* 13:390.

Fölsing, Albrecht. 1997. *Albert Einstein: A Biography*. New York: Viking.

Forman, Paul. 1975. "Ritz, Walter." In *Dictionary of Scientific Biography*, edited by Charles C. Gillespie, 475–81. New York: Scribner.

Forman, Paul, John L. Heilbron, and Spencer R. Weart. 1975. "Physics *circa* 1900: Personnel, Funding, and Productivity of the Academic Establishments." *Historical Studies in the Physical Sciences* 5:1–185.

Foucault, Léon. 1878 [1862]. "Détermination expérimentale de la vitesse de la lumière." In *Recueil des travaux scientifiques de Léon Foucault*. Paris: Gauthier-Villars.

Frank, Philipp. 1908. "Das Relativitätsprinzip der Mechanik und die Gleichungen für die elektromagnetischen Vorgänge in bewegten Körpern." *Annalen der Physik* 27:897–902.

———. 1908. "Relativitätstheorie und Elektronentheorie in ihrer Anwendung zur Ableitung der Grundgleichungen für die elektromagnetischen Vorgänge in bewegten ponderablen Körpern." *Annalen der Physik* 27:1059–65.

———. 1909. "Die Stellung des Relativitätsprinzips im System der Mechanik und der Elektrodynamik." *Kaiserliche Akademie der Wissenschaften (Vienna). Mathematisch-naturwissenschaftliche Classe. Zweite Abtheilung. Sitzungsberichte* 118:373–446.

Frank, Philipp, and Hermann Rothe. 1911. "Über die Transformation der Raum-Zeitkoordinaten von ruhenden auf bewegte Systeme." *Annalen der Physik* 34:825–55.

Franklin, Allan. 1986. *The Neglect of Experiment*. Cambridge: Cambridge Univ. Press.

Frercks, Jan. 2000. "Creativity and Technology in Experimentation: Fizeau's Terrestrial Determination of the Speed of Light." *Centaurus* 42:249–87.

———. 2005. "Fizeau's Research Program on Ether Drag: A Long Quest for a Publishable Paper." *Physics in Perspective* 7:35–65.

Frewer, Magdalene. 1979. "Das mathematische Lesezimmer der Universität Göttingen unter der Leitung von Felix Klein (1886–1922)." Bibliothekar-Lehrinstitut das Landes Nordrhein-Westfalen.

Fried, Bart. 1991. "The Masterful Techniques of John A. Brashear." *Sky and Telescope* 81, April:432–38.

Friedman, Robert Marc. 2001. *The Politics of Excellence: Behind the Nobel Prize in Science*. New York: Freeman.

Fuchs, Eckhardt. 1999. "Das Deutsche Reich auf den Weltausstellungen vor dem Ersten Weltkrieg." In *Weltausstellungen im 19. Jahrhundert*, edited by Eckhardt Fuchs, 61–88. Leipzig: Leipziger Universitätsverlag.

Gale, Henry G. 1902. "On the Relation between Density and Index of Refraction of Air." *Physical Review* 14:1–16.

Gale, Henry G., and Harvey B. Lemon. 1910. "The Analysis of the Principal Mercury Lines by a Diffraction Grating and a Comparison with the Results Obtained by Other Methods." *Astrophysical Journal* 31:78–87.

Galison, Peter. 1979. "Minkowski's Space-Time: From Visual Thinking to the Absolute World." *Historical Studies in the Physical Sciences* 10:85–121.

———. 1983. "Re-Reading the Past from the End of Physics: Maxwell's Equations in Retropsect." In *Functions and Uses of Disciplinary Histories*, ed. Loren R. Graham, Wolf Lepenies and Peter Weingart, 35–51. Dordrecht: Reidel.

———. 1987. *How Experiments End*. Chicago: Univ. of Chicago Press.

———. 1997. *Image and Logic: The Material Culture of Microphysics*. Chicago: Univ. of Chicago Press.

———. 2003. *Einstein's Clocks, Poincaré's Maps: Empires of Time*. New York: W. W. Norton.

Gearhart, Clayton A. 2002. "Planck, the Quantum, and the Historians." *Physics in Perspective* 4:170–215.

Gehrcke, Ernst. 1906. *Die Anwendung Interferenzen in der Spektroskopie und Metrologie*. Braunschweig: Friedrich Vieweg und Sohn.

Geppert, Alexander C. T. 2002. "Welttheater: Die geschichte des europäischen Austellungswesens im 19. und 20. Jahrhundert. Ein Forschungsbericht." *Neue politische Literatur* 47, no. 1:10–61.

Gibbs, J. W. 1886. "[Review of] Astronomical Papers Prepared for the Use of the American Ephemeris and Nautical Almanac. Vol. II, parts 3 and 4. Velocity of Light in Air and Refracting Media." *American Journal of Science* 31:62–64.

Gilbert, G. Nigel, and Michael Mulkay. 1984. "Experiments Are the Key: Participants' Histories and Historians' Histories of Science." *Isis* 75:105–25.

Glazebrook, R. T. 1901. "The Aims of the National Physical Laboratory of Great Britain." *Smithsonian Institution Annual Report*:341–57.

Goldberg, Stanley. 1970. "Drude, Paul Karl Ludwig." In *Dictionary of Scientific Biography*, edited by Charles C. Gillespie, 189–93. New York: Scribner.

———. 1969. "The Early Response to Einstein's Special Theory of Relativity, 1905–1911: A Case Study in National Differences." PhD dissertation, Harvard University.

———. 1969. "The Lorentz Theory of Electrons and Einstein's Theory of Relativity." *American Journal of Physics* 35:982–94.

———. 1970. "Poincaré's Silence and Einstein's Relativity: The Role of Theory and Experiment in Poincaré's Physics." *British Journal for the History of Science* 5:73–84.

———. 1970. "The Abraham Theory of the Electron: The Symbiosis between Theory and Experiment." *Archive for History of Exact Sciences* 7:9–25.

———. 1976. "Max Planck's Philosophy of Nature and His Elaboration of the Special Theory of Relativity." *Historical Studies in the Physical Sciences* 7:125–60.

———. 1979. "Henri Poincaré and Einstein's Theory of Relativity." *Historical Studies in the Physical Sciences* 10:85–121.

———. 1984. *Understanding Relativity: Origin and Impact of a Scientific Revolution*. Boston: Birkhäuser.

Goldberg, Stanley, and Roger H. Stuewer, eds. 1988. *The Michelson Era in American Science: 1870–1930*. New York: American Institute of Physics.

Golden, Frederic. 1999. "Albert Einstein (Person of the Century)." *Time*, 31 December.

Gooday, Graeme. 1990. "Precision Measurement and the Genesis of Physics Teaching Laboratories in Victorian Britain." *British Journal for the History of Science* 23:25–51.

———. 2001. "The Questionable Matter of Electricity: The Reception of J. J. Thomson's 'Corpuscle' among Electrical Theorists and Technologists." In

Histories of the Electron: The Birth of Microphysics, edited by Jed Z. Buchwald and Andrew Warwick, 101–34. Cambridge, MA: MIT Press.

———. 2004. *The Morals of Measurement: Accuracy, Irony, and Trust in Late Victorian Electrical Practice*. Cambridge: Cambridge Univ. Press.

Gooding, David, Trevor Pinch, and Simon Schaffer, eds. 1989. *The Uses of Experiment: Studies in the Natural Sciences*. Cambridge: Cambridge Univ. Press.

Gorham, Geoffrey. 1991. "Planck's Principle and Jeans's Conversion." *Studies in History and Philosophy of Science* 22:471–97.

Gottfried, Kurt, and Kenneth G. Wilson. 1997. "Science as a Cultural Construct." *Nature* 386:545–47.

Graham, Loren R., Wolf Lepenies, and Peter Weingart, eds. 1983. *Functions and Uses of Disciplinary Histories*. Vol. 7, Sociology of the Sciences. Dordrecht, Boston, Lancaster: D. Reidel.

Great Britain. Royal Commission for the Paris Exposition 1900. 1900. *Paris Exhibition, 1900: British Official Catalogue*. London: Royal Commission.

Greenspan, Nancy Thorndike. 2005. *The End of the Certain World: The Life and Science of Max Born, the Nobel Physicist Who Ignited the Quantum Revolution*. New York: Basic Books.

Griffiths, E.-H. 1900. "La chaleur spécifique de l'eau (Appendice)." In *Rapports présentés au Congrès international de physique réuni à Paris en 1900 sous les auspices de la Société française de physique*. Tome I. *Questions générales. Métrologie. Physique mécanique. Physique moléculaire*, edited by Lucien Poincaré and Charles-Édouard Guillaume, 214–27. Paris: Gauthier-Villars.

Guggenheimer, S. 1901. "J. J. Thomson, Über Andeutungen hinsichtlich der Konstitution der Materie, welche durch die neueren Untersuchungen über die Entladung der Elektrizität in Gasen geliefert werden." *Physikalische Zeitschrift* 2:564–65.

Guillaume, Charles-Édouard. 1899. "International Congress of Physics." *Science* 10, no. 248:459–61.

———. 1900. "The International Physical Congress." *Nature* 62:425–28.

Gumlich, Ernst. 1902. *Präcisionsmessungen mit Hülfe der Wellenlänge des Lichts*. Berlin: C. A. Schwetscke und Sohn.

Hale, George Ellery. 1902. "Spectroscopy." In *Encyclopædia Britannica*, 775–85.

———. 1904. "Co-Operation in Solar Research." *Astrophysical Journal* 20:306–12.

———. 1931. "Some of Michelson's Researches." *Publications of the Astronomical Society of the Pacific* 43:174–85.

Hallock, William. 1901. "The American Physical Society." *Science* 13, no. 316:101–2.

Hamerla, Ralph R. 2006. *An American Scientist on the Research Frontier: Edward Morley, Community, and Radical Ideas in Nineteenth-Century American Science*. Dordrecht: Springer Academic Publishing.

Harrison, George R. 1949. "The Production of Diffraction Gratings: I. Development of the Ruling Art." *Journal of the Optical Society of America* 39:413–26.

Haubold, Barbara, Hans Joachim Haubold, and Lewis Pyenson. 1988. "Michelson's First Ether-Drift Experiment in Berlin and Potsdam." In *The Michelson Era in American Science: 1870–1930*, edited by Stanley Goldberg and Roger H. Stuewer, 42–54. New York: American Institute of Physics.

Heil, W. 1910. "Diskussion der Versuche über die träge Masse bewegter Elektronen." *Annalen der Physik* 31:519–46.

———. 1910. "Zur Diskussion der Hupkaschen Versuche über die träge Masse bewegter Elektronen." *Annalen der Physik* 33:403–13.

Heilbron, J. L. 1981. *Historical Studies in the Theory of Atomic Structure*. New York: Arno Press.

———. 1982. "*Fin-de-Siècle* Physics." In *Science, Technology, and Society in the Time of Alfred Nobel*, edited by Carl Gustaf Bernhard, Elisabeth Crawford, and Per Sörbom, 51–73. Oxford: Nobel Foundation.

———. 1986. *The Dilemmas of an Upright Man: Max Planck as Spokesman for German Science*. Berkeley: Univ. of California Press.

———. 2000. *The Dilemmas of an Upright Man: Max Planck and the Fortunes of German Science*. With new afterword edition. Cambridge, MA: Harvard Univ. Press.

Helm, Georg. 2000. *The Historical Development of Energetics*. Translated by Robert J. Deltete. Vol. 209, Boston Studies in the Philosophy of Science. Dordrecht, Boston, London: Kluwer.

Hentschel, Klaus. 1992. "Einstein's Attitude towards Experiments: Testing Relativity Theory, 1907–1927." *Studies in History and Philosophy of Science* 23:593–624.

———. 1993. "The Discovery of the Redshift of Solar Fraunhaufer Lines by Rowland and Jewell in Baltimore around 1890." *Historical Studies in the Physical and Biological Sciences* 23, no. 2:219–77.

———. 1996. *Physics and National Socialism: An Anthology of Primary Sources*. Translated by Ann M. Hentschel. Vol. 18, Science Networks Historical Studies. Basel and Boston: Birkhäuser Verlag.

———. 1999. "Photographic Mapping of the Solar Spectrum, 1864–1900, Part 1." *Journal for the History of Astronomy* 30:93–119.

———. 1999. "Photographic Mapping of the Solar Spectrum, 1864–1900, Part 2." *Journal for the History of Astronomy* 30:201–24.

———. 2002. *Mapping the Spectrum: Techniques of Visual Representation in Research and Teaching*. Oxford and New York: Oxford Univ. Press.

Herbert, Christopher. 2001. *Victorian Relativity: Radical Thought and Scientific Discovery*. Chicago: Univ. of Chicago Press.

Herglotz, Gustav. 1909. "Über den vom Standpunkt des Relativitätsprinzips als 'starr' zu bezeichnenden Körpe." *Annalen der Physik* 31:393–415.

Hermann, Armin. 1971. *The Genesis of Quantum Theory, 1899–1913*. Translated by Claude W. Nash. Cambridge, MA: MIT Press.

Hertz, Heinrich. 1894. *Die Prinzipien der Mechanik in neuem Zusammenhange dargestellt*. Leipzig: J. A. Barth.

Hicks, W. M. 1902. "On the Michelson-Morley Experiment Relating to the Drift of the Aether." *Philosphical Magazine* 3:9–42.

———. 1902. "The FitzGerald-Lorentz Effect (Letter to the Editor)." *Nature* 65:343.

———. 1902. "The Michelson-Morley Experiment. To the Editors of the Philosophical Magazine." *Philosphical Magazine* 3:556.

———. 1902. "The Michelson-Morley Experiment. To the Editors of the Philosophical Magazine." *Philosphical Magazine* 3:256.

Hiebert, Erwin N. 1979. "*Fin-de-Siècle* Physics." In *Rutherford and Physics at the Turn of the Century*, edited by Mario Bunge and William R. Shea, 3–22. New York: Dawson and Science History Publications.

———. 1980. "Nernst, Hermann Walther." In *Dictionary of Scientific Biography*, edited by Charles C. Gillespie, 432–53. New York: Scribner.

Hirosige, Tetu. 1969. "Origins of Lorentz' Theory of Electrons and the Concept of the Electromagnetic Field." *Historical Studies in the Physical Sciences* 1:151–209.

———. 1976. "The Ether Problem, the Mechanistic Worldview, and the Origins of the Theory of Relativity." *Historical Studies in the Physical Sciences* 7:3–82.

Hoffmann, Dieter, and Jost Lemmerich. 2000. *Quantum Theory Centenary*. Translated by Ann Hentschel. Berlin: Deutsche Physikalische Gesellschaft, Bad Honnef.

Holton, Gerald. 1988. "Einstein, Michelson, and the 'Crucial' Experiment." In *Thematic Origins of Scientific Thought: Kepler to Einstein*, rev. ed., 279–370. Cambridge, MA: Harvard Univ. Press.

———. 1988. "Mach, Einstein, and the Search for Reality." In *Thematic Origins of Scientific Thought: Kepler to Einstein*, rev. ed., 237–77. Cambridge, MA: Harvard Univ. Press.

———. 1988. "On the Origins of the Special Theory of Relativity." In *Thematic Origins of Scientific Thought: Kepler to Einstein*, rev. ed., 191–236. Cambridge, MA: Harvard Univ. Press.

———. 1988. *Thematic Origins of Scientific Thought: Kepler to Einstein*. Rev. ed. Cambridge, MA: Harvard Univ. Press.

———. 1993. "More on Einstein, Michelson, and the 'Crucial' Experiment." *Methodology and Science: International Journal for the Empirical Study of the Foundations of Science and their Methodology* 26, no. 1:6–7.

———. 1993. "Quanta, Relativity, and Rhetoric." In *Science and Anti-Science*, 74–108. Cambridge, MA: Harvard Univ. Press.

Hon, Giora. 1995. "Is the Identification of Experimental Error Contextually Dependent? The Case of Kaufmann's Experiment and Its Varied Reception." In *Scientific Practice: Theories and Stories of Doing Physics*, edited by Jed Z. Buchwald, 170–223. Chicago: Univ. of Chicago Press.

Hounshell, David A. 1980. "Edison and the Pure Science Ideal in 19th-Century America." *Science* 207, no. 4431:612–17.

———. 1981. "The *System*: Theory and Practice." In *Yankee Enterprise: The Rise of the American System of Manufactures*, edited by Otto Mayr and Robert C. Post, 127–52. Washington, DC: Smithsonian Institution Press.

Howard, Don, and John Stachel, eds. 2000. *Einstein: The Formative Years, 1879–1909*. Vol. 8, Einstein Studies. Boston, Basel, Berlin: Birkhäuser.

Howard, John N. 1967. "The Michelson-Rayleigh Correspondence of the AFCRL Rayleigh Archives." *Isis* 58:88–89.

Hudson, Robert G. 1989. "James Jeans and Radiation Theory." *Studies in History and Philosophy of Science* 20:57–76.

———. 1997. "Classical Physics and Early Quantum Theory: A Legitimate Case of Theoretical Underdetermination." *Synthese: International Journal for Epistemology, Methodology and Philosophy of Science* 110:217–56.

Hughes, Thomas Parke. 1983. *Networks of Power: Electrification in Western Society, 1880–1930*. Baltimore: Johns Hopkins Univ. Press.

Huijnen, Pim, and A. J. Kox. 2007. "Paul Ehrenfest's Rough Road to Leiden: A Physicist's Search for a Position, 1904–1912." *Physics in Perspective* 9:186–211.

Humphreys, W. J., and J. F. Mohler. 1896. "Effect of Pressure on the Wave-Lengths of Lines in the Arc-Spectra of Certain Elements." *Astrophysical Journal* 3:114–37.

Hunt, Bruce J. 1986. "Experimenting on the Ether: Oliver J. Lodge and the Great Whirling Machine." *Historical Studies in the Physical and Biological Sciences* 16:111–34.

———. 1997. "Doing Science in a Global Empire: Cable Telegraphy and Electrical Physics in Victorian Britain." In *Victorian Science in Context*, edited by Bernard Lightman, 312–33. Chicago: Univ. of Chicago Press.

Hupka, E. 1910. "Beitrag zur Kenntnis der trägen Masse bewegter Elektronen." *Annalen der Physik* 31:169–204.

———. 1910. "Zur Frage der trägen Masse bewegter Elektronen." *Annalen der Physik* 33:400–402.

Ignatowski, Woldemar von. 1910. "Einige allgemeine Bemerkungen zum Relativitätsprinzip." *Physikalische Zeitschrift* 11:972–76.

Im, Gyeong Soon. 1991. "Max Born und die Quantentheorie." PhD dissertation, Universität Hamburg.

"Instruments of Precision at the Paris Exhibition." 1900. *Nature* 63:61–62.

"International Catalogue of Scientific Literature." 1899. *Nature* 61:139.

"International Co-Operation in Solar Research. Minutes of the Meeting of Delegates to the Conference on Solar Research, Held in the Hall of Congresses, St. Louis, September 23, 1904." 1904. *Astrophysical Journal* 20:301–5.

"Internationaler Kongress für Physik zu Paris vom 6. bis 12. August 1900." 1899. *Physikalische Zeitschrift* 1:35.

Jackson, Myles. 2000. *Spectrum of Belief: Joseph von Fraunhofer and the Craft of Precision Optics*. Cambridge, MA: MIT Press.

Jaffe, Bernard. 1960. *Michelson and the Speed of Light*. Science Study Series. Garden City, NY: Anchor.

Jamin, J. 1858. "Mémoire sur les variations de l'indice de réfraction de l'eau a diverses pressions." *Annales de chimie et de physique* 52:163–71.

Janssen, Michel. 1995. "A Comparison between Lorentz's Ether Theory and Special Relativity in the Light of the Experiments of Trouton and Noble." PhD dissertation, University of Pittsburgh.

———. 2002. "Reconsidering a Scientific Revolution: The Case of Einstein versus Lorentz." *Physics in Perspective* 4:421–46.

Janssen, Michel, and Matthew Mecklenburg. 2004. "Electromagnetic Models of the Electron and the Transition from Classical to Relativistic Mechanics." *Max Planck Institute for the History of Science Preprint* 277.

Jeans, J. H. 1901. "The Distribution of Molecular Energy." *Philosophical Transactions of the Royal Society of London. Series A, Containing Papers of a Mathematical or Physical Character* 196:397–430.

———. 1904. *The Dynamical Theory of Gases*. Cambridge: Cambridge Univ. Press.

———. 1905. "A Comparison between Two Theories of Radiation." *Nature* 72:293–94.

———. 1905. "On the Partition of Energy between Matter and Aether." *Philosophical Magazine* 10, July:91–98.

———. 1905. "The Dynamical Theory of Gases." *Nature* 71:607.

———. 1910. "On Non-Newtonian Mechanical Systems, and Planck's Theory of Radiation." *Philosophical Magazine* 20:943–54.

———. 1913. "Die kinetische Theorie der spezifischen Wärme nach Maxwell und Boltzmann." In *Die Theorie der Strahlung und der Quanten. Verhandlungen auf einer von E. Solvay einberufenen Zusammenkunft (30. Oktober bis 3. November 1911)*, edited by Arnold Eucken, 45–64. Berlin: Verlag Chemie GMBH.

———. 1914. *Report on Radiation and Quantum-Theory*. London: The Electrician.

Jewell, L. E. 1905. "The Revision of Rowland's System of Standard Wave-Lengths." *Astrophysical Journal* 21:23–34.

Joerges, Bernward, and Terry Shinn, eds. 2001. *Instrumentation: Between Science, State, and Industry*. Vol. 22, *Sociology of the Sciences*. Dordrecht and Boston: Kluwer.

Johnston, Sean F. 2001. *A History of Light and Colour Measurement: Science in The Shadows*. Bristol and Philadelphia: Institute of Physics.

———. 2003. "An Unconvincing Transformation? Michelson's Interferential Spectroscopy." *Nuncius* 18, no. 2:803–23.

Johonnott, Edwin S. 1899. "Thickness of the Black Spot in Liquid Films." *Philosophical Magazine* 47:501–22.

Josephson, Matthew. 1959. *Edison: A Biography*. New York: McGraw Hill.

Jungnickel, Christa, and Russell McCormmach. 1986. *Intellectual Mastery of Nature: Theoretical Physics from Ohm to Einstein*, 2 vols. Vol. 1: *The Torch of Mathematics, 1800–1870*. Vol. 2: *The Now Mighty Theoretical Physics, 1870–1925*. Chicago: Univ. of Chicago Press.

Kanigel, Robert. 1997. *One Best Way: Frederick Winslow Taylor and the Enigma of Efficiency*. New York: Viking.

Kargon, Robert. 1986. "Henry Rowland and the Physics Discipline in America." *Vistas in Astronomy* 29:131–36.

Katzir, Shaul. 2005. "On 'The Electromagnetic World-View': A Comment on an Article by Suman Seth." *Historical Studies in the Physical and Biological Sciences* 36, no. 1:189–92.

———. 2005. "Poincaré's Relativistic Physics: Its Origins and Nature." *Physics in Perspective* 7:268–92.

Kaufmann, Walter. 1897. "Die magnetische Ablenkbarkeit der Kathodenstrahlen und ihre Abhängigkeit vom Entladungspotential." *Annalen der Physik und Chemie* 61:544–52.

———. 1897. "Nachtrag zu der Abhandlung 'Die magnetische Ablenkbarkeit der Kathodenstrahlen etc.'" *Annalen der Physik und Chemie* 62:596–98.

———. 1899. "Ein mechanisches Modell zur Darstellung des Verhaltens Geisslersche Röhren." *Physikalische Zeitschrift* 1:59–60.

———. 1899. "Grundzüge einer elektrodynamischen Theorie der Gasentladungen. (I. Mitteilung)." *Königliche Gesellschaft der Wissenschaften zu Göttingen. Mathematisch-Physikalische Klasse. Nachrichten*:243–50.

———. 1899. "Grundzüge einer elektrodynamischen Theorie der Gasentladungen. (II. Mitteilung: Versuchsergebnisse.)." *Königliche Gesellschaft der Wissenschaften zu Göttingen. Mathematisch-Physikalische Klasse. Nachrichten*: 251–59.

———. 1899. "Über den sogenannten 'Widerstand' leitender Gase." *Physikalische Zeitschrift* 1:348–49.

———. 1899. "Über Ionenwanderung in Gasen." *Physikalische Zeitschrift* 1:22–25.

———. 1900. "Elektrodynamische Eigentümlichkeiten leitender Gase." *Annalen der Physik* 2:158–78.

———. 1900. "H. Pellat, Die Physikalisch Technischen Staatslaboratorien. 6 Seiten." *Physikalische Zeitschrift* 1:486.

———. 1900. "P. Villard, Die Kathodenstrahlen." *Physikalische Zeitschrift* 2:137–40.

———. 1901. "Die magnetische und electrische Ablenkbarkeit der Becquerelstrahlen und die scheinbare Masse der Elektronen." *Königliche Gesellschaft der Wissenschaften zu Göttingen. Mathematisch-Physikalische Klasse. Nachrichten*:143–55.

———. 1901. "H. Becquerel, Über die Uranstrahlen und die verschiedenen physikalischen Eigenschaften der von radioaktiven Körpern herrührenden Strahlung. 31 S.; P. Curie und Frau Curie, Die neuen radioaktiven Substanzen und die von ihnen emittierten Strahlen. 35 S." *Physikalische Zeitschrift* 2:563–64.

———. 1901. "J. J. Thomson, Die Entladung der Elektrizität durch Gase (Besprechung)." *Physikalische Zeitschrift* 2:610.

———. 1901. "Methode zur exakten Bestimmung von Ladung und Geschwindigkeit der Becquerelstrahlen." *Physikalische Zeitschrift* 2:602–3.

———. 1901. "The Development of the Electron Idea." *Electrician* 8 November:95–97.

———. 1902. "Die Entwicklung des Elektronenbegriffs." *Physikalische Zeitschrift* 3:9–15.

———. 1902. "Über die elektromagnetische Masse des Elektrons." *Königliche Gesellschaft der Wissenschaften zu Göttingen. Mathematisch-Physikalische Klasse. Nachrichten*:291–303.

———. 1903. "Die elektromagnetische Masse des Elektrons." *Physikalische Zeitschrift* 4:54–57.

———. 1903. "Über die 'Elektromagnetische Masse' der Elektronen." *Königliche Gesellschaft der Wissenschaften zu Göttingen. Mathematisch-Physikalische Klasse. Nachrichten*:90–103, 148-Berichtigung.

———. 1905. "Über die Konstitution des Elektrons." *Königlich preussische Akademie der Wissenschaften (Berlin). Sitzungsberichte*:949–56.

———. 1906. "Nachtrag zu der Abhandlungen: Über die Konstitution des Elektrons." *Annalen der Physik* 20:639–40.

———. 1906. "Über die Konstitution des Elektrons." *Annalen der Physik* 19:487–553.

———. 1907. "Bemerkungen zu Herrn Plancks: 'Nachtrag zu der Besprechung der Kaufmannschen Ablenkungsmessungen.'" *Deutsche Physikalische Gesellschaft. Verhandlungen* 9:667–73.

———. 1907. "Neue Hilfsmittel für Laboratorium und Hörsaal (mit Demonstrationen)." *Physikalische Zeitschrift* 8:748–52.

———. 1908. "Erwiderung an Herrn Stark." *Deutsche Physikalische Gesellschaft. Verhandlungen* 10:91–95.

———. 1911. "Eine einfacher Vorlesungsapparat." *Physikalische Zeitschrift* 12:29.

Kaufmann, Walter, and E. Aschkinass. 1897. "Über die Deflexion der Kathodenstrahlen." *Annalen der Physik und Chemie* 62:588–95.

Kaufmann, Walter, and Alfred Coehn. 1909. *Magnetismus und Elektrizität*, edited by Leopold Pfaundler. Vol. 4, Buch 5, Abteilung 1, Müller-Poillet's *Lehrbuch der Physik und Meteorologie*. Braunschweig: Vieweg.

Kaufmann, Walter, Alfred Coehn, and Alfred Nippoldt. 1914. *Magnetismus und Elektrizität*, edited by Leopold Pfaundler. 10., umgearb. u. verm. Aufl. ed. Vol. 4, Buch 5, Abteilung 2 u. 3, Müller-Pouillet's *Lehrbuch der Physik und Meteorologie*. Braunschweig: Vieweg.

Kelvin, Lord. 1900. "On the Motion Produced in an Infinite Elastic Solid by the Motion through the Space Occupied by It of a Body Acting on It Only by Attraction or Repulsion." *Philosophical Magazine* 50:181–98, 305–7.

———. 1900. "Sur le mouvement d'un solide élastique traversé par un corps agissant sur lui par attraction ou répulsion," and "Addition au Rapport précédent." In *Rapports présentés au Congrès international de physique réuni à Paris en 1900 sous les auspices de la Société française de physique*. Tome II. *Optique. Électricité. Magnétisme*, edited by Lucien Poincaré and Charles-Édouard Guillaume, 1–19 and 19–22. Paris: Gauthier-Villars.

———. 1904. *The Baltimore Lectures*. London: C. J. Clay and Sons.

———. 1970 [1900]. "Nineteenth Century Clouds over the Dynamical Theory of Heat and Light." In *The Royal Institution Library of Science: Physical Sciences*, edited by W. L. Bragg and G. Porter, 324–58. Barking, Essex, Amsterdam, New York: Elsevier.

Kevles, Daniel J. 1988. "Physics and National Power, 1870–1930." In *The Michelson Era in American Science: 1870–1930*, edited by Stanley Goldberg and Roger H. Stuewer, 248–57. New York: American Institute of Physics.

———. 1995. *The Physicists: The History of a Scientific Community in Modern America*. Cambridge, MA: Harvard Univ. Press.

Kirchhoff, Gustav. 1876. *Vorlesungen über mathematische Physik: Mechanik*. Leipzig: B. G. Teubner.

Klein, Martin J. 1962. "Max Planck and the Beginning of the Quantum Theory." *Archive for History of Exact Sciences* 1:459–49.

———. 1965. "Einstein, Specific Heats, and the Early Quantum Theory." *Science* 148:173–80.

———. 1970. *Paul Ehrenfest*. Vol. 1, *The Making of a Theoretical Physicist*. Amsterdam: North-Holland.

Knott, C. G. 1902. "Light." In *Encyclopædia Britannica*, 235–51.

Kohl, Emil. 1909. "Über den Michelsonschen Versuch." *Annalen der Physik* 28:259–307.

Kraemer, Hans. 1900. *Das XIX. Jahrhundert in Wort und Bild. Politische und Kultur-Geschichte*. Vol. 4. Berlin: Bong.

Kragh, Helge. 1999. *Quantum Generations: A History of Physics in the Twentieth Century*. Princeton, NJ: Princeton Univ. Press.

Kretschmer, Winfried. 1999. *Geschichte der Weltausstellungen*. Frankfurt am Main and New York: Campus.

Kuhn, Thomas S. 1983. "Revisiting Planck." *Historical Studies in the Physical and Biological Sciences* 14:231–51.

———. 1987. *Black-Body Theory and the Quantum Discontinuity 1894–1912*. Reprint of the 1978 edition. Chicago: Univ. of Chicago Press.

L'industrie française des instruments de précision. 1901. Reprinted, Paris, Éditions Alain Brieux, 1980 edition. Paris: Syndicat des constructeurs en instruments d'optique et de prècision.

Labinger, Jay A., and Harry Collins, eds. 2001. *The One Culture?: A Conversation about Science*. Chicago: Univ. of Chicago Press.

Langevin, Paul, and M. de Broglie, eds. 1912. *La théorie du rayonnement et les quanta: Rapports et discussions de la réunion tenue à Bruxelles, du 30 Octobre au 3 Novembre 1911*. Paris: Gauthier-Villars.

Larmor, Joseph. 1900. "Address of the President of the Mathematical and Physical Section of the British Association for the Advancement of Science." *Science* 12, no. 299:417–36.

———. 1905. "Introduction." In *Science and Hypothesis*, by Henri Poincaré, xi–xxvii. London: Walter Scott.

———. 1909. "Bakerian Lecture: On the Statistical and Thermodynamical Relations of Radiant Energy." *Proceedings of the Royal Society of London. Series A, Containing Papers of a Mathematical and Physical Character* 83, no. 560:82–95.

Laub, Jakob. 1907. "Zur Optik der bewegten Körper." *Annalen der Physik* 23:738–44.

———. 1910. "Über die experimentellen Grundlagen des Relativitätsprinzip." *Jahrbuch der Radioaktivität und Elektronik* 7:405–63.

Laudan, Rachel. 1993. "Histories of the Sciences and Their Uses: A Review to 1913." *Social Studies of Science* 31:1–34.

Laue, Max. 1907. "Die Mitführung des Lichtes durch bewegte Körper nach dem relativitätsprinzip." *Annalen der Physik* 23:989–90.

———. 1909. "Die Wellenstrahlung einer bewegten Punktladung nach dem Relativitätsprinzip." *Annalen der Physik* 28:436–42.

———. 1910. "Ist der Michelsonversuch beweisend?" *Annalen der Physik* 33:186–91.

———. 1911. *Das Relativitätsprinzip*. Braunschweig: Friedrich Vieweg und Sohn.

———. 1911. "Zur Diskussion über den Starrer Korper in der Relativitätstheorie." *Physikalische Zeitschrift* 12:85–87.

———. 1912. "Zur Theorie des Michelsonsversuches." *Physikalische Zeitschrift* 13:501–6.

———. 1913. *Das Relativitätsprinzip*. 2nd expanded ed. Braunschweig: Friedrich Vieweg und Sohn.

Lelong, Benoit. 2001. "Paul Villard, J. J. Thomson, and the Composition of Cathode Rays." In *Histories of the Electron: The Birth of Microphysics*, edited by Jed Z. Buchwald and Andrew Warwick, 135–67. Cambridge, MA: MIT Press.

Lenzen, Victor F. 1965. "The Contributions of Charles S. Peirce to Metrology." *Proceedings of the American Philosophical Society* 109, no. 1:29–46.

Lessing, Julius. 1900. "Das halbe Jahrhundert der Weltaustellungen: Vortrag gehalten in der Volkswirthschaftlichen Gesellschaft zu Berlin, März 1900." *Volkswirtschaftliche Zeitfragen* 22, no. 6:1–30.

Lindemann, F. A., and Francis Simon. 1942. "Walther Nernst, 1864–1941." *Obituary Notices of Fellows of the Royal Society* 4:101–12.

Livingston, Dorothy Michelson. 1973. *The Master of Light: A Biography of Albert A. Michelson*. New York: Scribner.

Lodge, Oliver. 1888. "Sketch of the Electrical Papers in Section A at the Recent Bath Meeting of the British Association." *The Electrician* 21:622–25.

———. 1893. "Aberration Problems: a Discussion concerning the Motion of the Ether near the Earth, and concerning the Connexion between Ether and Gross Matter; with Some New Experiments." *Philosophical Transactions A* 184:727–804.

———. 1897. "Experiments on the Absence of Mechanical Connection between Ether and Matter." *Philosophical Transactions* 189:149–66.

———. 1898. "Note on Mr. Sutherland's Objection to the Conclusiveness of the Michelson-Morley Experiment." *Philosophical Magazine* 46:343–44.

Lorentz, H. A. 1900. "Théorie des phénomènes magnéto-optiques récemment découverts." In *Rapports présentés au Congrès international de physique réuni à Paris en 1900 sous les auspices de la Société française de physique*. Tome III. *Électro-optique et ionisation. Applications. Physique cosmique. Physique biologique*, edited by Lucien Poincaré and Charles-Édouard Guillaume, 1–33. Paris: Gauthier-Villars.

———. 1900. "Über die scheinbare Masse der Ionen." *Physikalische Zeitschrift* 2:78–80.

———. 1901. "Sur la méthode du miroir tournant pour la détermination de la vitesse de la lumière." *Archives Néerlandaises des Sciences Exactes et Naturelles* 6:303–18.

———. 1906 [1895]. *Versuch einer Theorie der elektrischen und optischen Erscheinungen in bewegten Körpern*. Reprint of 1895 edition. Leipzig.

———. 1909. *The Theory of Electrons*. Leipzig: Teubner.

———. 1910. "Alte und neue Fragen der Physik." *Physikalische Zeitschrift* 11:1234–57.

———. 1934 [1908]. "Le partage de l'énergie entre la matière pondérable et l'éther." In *Collected Papers*, 317–43. The Hague: Martinus Nijhoff.

———. 1934 [1908]. "Zur Strahlungstheorie." In *Collected Papers*, 344–46. The Hague: Martinus Nijhoff.

———. 1937 [1886]. "De l'Influence du mouvement de la terre sur les phénomènes lumineux." In *Collected Papers*, 153–214. The Hague: Martinus Nijhoff.

———. 1937 [1904]. "On Electromagnetic Phenomena in a System Moving with Any Velocity Less Than That of Light." In *Collected Papers*, 172–97. The Hague: Martinus Nijhoff.

Lorentz, H. A., Albert Einstein, Hermann Minkowski, and Hermann Weyl. 1974 [1913]. *Das Relativitätsprinzip: Eine Sammlung von Abhandlungen*. 7th ed. Leipzig and Berlin: Teubner.

———. n.d. *The Principle of Relativity*. Translated by W. Perret and G. B. Jeffrey. New York: Dover.

Lovering, Joseph. 1889. "Michelson's Recent Researches on Light." *Smithsonian Institution Annual Report*:449–68.

Lummer, Otto. 1900. "Les rayonnement des corps noirs." In *Rapports présentés au Congrès international de physique réuni à Paris en 1900 sous les auspices de la Société française de physique*. Tome II. *Optique. Électricité. Magnétisme*, edited by Lucien Poincaré and Charles-Édouard Guillaume, 41–99. Paris: Gauthier-Villars.

———. 1902. "Die planparallelen Platten als Interferenzspektroskop." *Physikalische Zeitschrift* 3:172–75.

Lummer, Otto, and Ernst Pringsheim. 1908. "Über die Jeans-Lorentzsche Strahlungsformel." *Physikalische Zeitschrift* 9:449–50.

Lynch, Michael. 1985. *Art and Artifact in Laboratory Science: A Study of Shop Work and Shop Talk in a Research Laboratory*. London: Routledge and Kegan Paul.

Macé de Lépinay, J. 1886. "Détermination de la valeur absolue de la longueur d'onde de la raie D2." *Journal de physique theorique et appliquee* 5:411–16.

———. 1900. "Déterminations métrologiques par les méthodes interférentielles." In *Rapports présentés au Congrès international de physique réuni à Paris en 1900 sous les auspices de la Société française de physique*. Tome I. *Questions générales. Métrologie. Physique mécanique. Physique moléculaire*, edited by Lucien Poincaré and Charles-Éduoard Guillaume, 108–30. Paris: Gauthier-Villars.

Mach, Ernst. 1883. *Die Mechanik in ihrer Entwickelung: Historisch-kritisch dargestellt*. Leipzig: F. A. Brockhaus.

Mach, Ludwig. 1892. "Über einen Interferenzrefraktor." *Zeitschrift für Instrumentenkunde* 12:89–93.

Maltese, Giulio, and Lucia Orlando. 1996. "The Definition of Rigidity in the Special Theory of Relativity and the Genesis of the General Theory of Relativity." *Studies in the History and Philosophy of Modern Physics* 26, no. 3:263–306.

Mandell, Richard D. 1967. *Paris 1900: The Great World's Fair*. Toronto: Univ. of Toronto Press.

Manegold, Karl-Heinz. 1970. *Universität, Technische Hochschule und Industrie: Ein Beitrag zur Emanzipation der Technik im 19. Jahrhundert unter besonderer Berücksichtigung der Bestrebungen Felix Kleins*, Schriften zur Wirtschafts- und Sozialgeschichte, Bd. 16. Berlin: Duncker and Humblot.

Marage, Pierre, and Grégoire Wallenborn, eds. 1999. *The Solvay Councils and the Birth of Modern Physics*. Basel and Boston: Birkhäuser Verlag.

Martínez, Alberto A. 2004. "Ritz, Einstein, and the Emission Hypothesis." *Physics in Perspective* 6:4–28.

Mascart, E. E. 1874. "Sur la dispersion des gaz." *Académie des sciences, Paris, Comptes rendus* 78:679–82.

———. 1874. "Sur la réfraction de le'eau comprimée." *Académie des sciences, Paris, Comptes rendus* 78:801–7.

———. 1874. "Sur la réfraction des gaz." *Académie des sciences, Paris, Comptes rendus* 78:617–21.

———. 1889. "On the Achromatism of Interference." *Philosophical Magazine* 27:519–24.

Maxwell, James Clerk. 1880. "Letter to David Peck Todd, 19 March 1879." *Nature* 21:314–15.

McCormmach, Russell. 1967. "Henri Poincaré and the Quantum Theory." *Isis* 58:37–55.

———. 1970. "Einstein, Lorentz, and the Electron Theory." *Historical Studies in the Physical Sciences* 2:41–87.

———. 1970. "H. A. Lorentz and the Electromagnetic View of Nature." *Isis* 61:459–97.

———. 1973. "Lorentz, Hendrik Antoon." In *Dictionary of Scientific Biography*, edited by Charles S. Gillespie, 487–500. New York: Scribner.

———. 1982. *Night Thoughts of a Classical Physicist*. New York: Avon.

McMahon, A. Michal. 1984. *The Making of a Profession: A Century of Electrical Engineering in America*. New York: Institute of Electrical and Electronics Engineers.

Mehra, Jagdish. 1975. *The Solvay Conferences on Physics: Aspects of the Development of Physics since 1911*. Dordrecht: Reidel.

Mehrtens, Herbert. 1990. *Moderne—Sprache—Mathematik: Eine Geschichte des Streits um die Grundlagen der Disziplin und des Subjekts formaler Systeme.* Frankfurt am Main: Suhrkamp.

Mendenhall, Thomas C. 1902. "Henry A. Rowland, Commemorative Address." In *The Physical Papers of Henry Augustus Rowland.* Baltimore: Johns Hopkins Press.

Merchant, Carolyn. 1989. *The Death of Nature: Women, Ecology, and the Scientific Revolution.* San Fransisco: Harper and Row.

Mermin, N. David. 1996. "Reference Frame: The Golemization of Relativity." *Physics Today* 49, no. 4:11–13.

———. 1996. "Reference Frame: What's Wrong with This Sustaining Myth?" *Physics Today* 49, no. 3:11–13.

Michelson, Albert Abraham. 1878. "Experimental Determination of the Velocity of Light." *Proceedings of the American Association for the Advancement of Science* 27:71–77.

———. 1878. "On a Method of Measuring the Velocity of Light." *American Journal of Science* 15:394–95.

———. 1879. "Experimental Determination of the Velocity of Light." *U.S. Nautical Almanac Office Astronomical Papers* I, Part III:115–45.

———. 1881. "Letter to the Editor." *Nature* 24:460–61.

———. 1881. "The Relative Motion of the Earth and the Luminiferous Ether." *American Journal of Science* 22:120–29.

———. 1882. "Interference Phenomena in a New Form of Refractometer." *American Journal of Science* 23:395–400.

———. 1882. "Supplementary Measures of the Velocities of White and Colored Light in Air, Water, and Carbon Disulphide." In *Astronomical Papers Prepared for the Use of the American Ephemeris and Nautical Almanac*, 231–58.

———. 1882. "Sur le mouvement relatif de la terre et de l'ether." *Académie des sciences, Paris, Comptes rendus* 94:520–23.

———. 1884. "On the Velocity of Light in CS2 and the Difference in Velocity of Red and Blue Light in the Same." *Reports of the British Association for the Advancement of Science*:654.

———. 1889. "A Plea for Light Waves." *Proceedings of the American Association for the Advancement of Science* 37:67–78.

———. 1890. "Measurement by Light-Waves." *American Journal of Science* 39:115–21.

———. 1890. "On the Application of Interference Methods to Astronomical Measurements." *Philosophical Magazine* 30:1–21.

———. 1891. "On the Application of Interference Methods to Spectroscopic Measurements, I." *Philosophical Magazine* 31:338–45.

———. 1892. "On the Application of Interference Methods to Spectroscopic Measurements, II." *Philosophical Magazine* 34:280–98.

———. 1893. "Comparison of the International Meter with the Wave Length of the Light of Cadmium." *Astronomy and Astrophysics* 12:556–60.

———. 1894. "Détermination expérimentale de la valeur du mètre en longueurs d'onde lumineuses." *Travaux et Mémoires du Bureau International des poids et mesures* 11.
———. 1894. "Light Waves and Their Application to Metrology." *Astronomy and Astrophysics* 13:92–104.
———. 1895. "On the Broadening of Spectral Lines." *Astrophysical Journal* 2:251–63.
———. 1897. "On the Relative Motion of the Earth and the Ether." *American Journal of Science* 3:475–78.
———. 1897. "Radiation in a Magnetic Field." *Astrophysical Journal* 6:48–54.
———. 1898. "The Echelon Spectroscope." *Proceedings of the American Academy for the Arts and Sciences* 35:111–19.
———. 1898. "Radiation in a Magnetic Field." *Philosophical Magazine* 45:348–56.
———. 1899. "Radiation in a Magnetic Field." *Nature* 59:440–41.
———. 1902. *Light Waves and Their Uses*. Chicago: Univ. of Chicago Press.
———. 1902. "The Velocity of Light." *Philosophical Magazine* 3:330–37.
———. 1904. "Relative Motion of Earth and Ether." *Philosophical Magazine* 8:716–19.
———. 1905. "Report of Progress in Ruling Diffraction Gratings." *American Physical Society* 20:389–91.
———. 1906. "Form Analysis." *Proceedings of the American Philosophical Society* 45:110–16.
———. 1909. "Biographical Sketch." In *Les Prix Nobel en 1907*. Stockholm: P. A. Norstedt and Söner.
———. 1909. "Recent Advances in Spectroscopy Nobel Address, Royal Academy of Science, Stockholm." In *Les Prix Nobel en 1907*. Stockholm: P. A. Norstedt and Söner.
———. 1915. "The Ruling and Performance of a Ten-Inch Diffraction Grating." *Proceedings of the American Philosophical Society* 54:137–42.
———. 1927. "Measurement of the Velocity of Light between Mount Wilson and Mount San Antonio." *Astrophysical Journal* 65:1–22.
———. 1928. "Experimental Determination of the Velocity of Light." *Scientific Monthly* 27:562–65.
———. 1928. "Report, Conference on the Michelson-Morley Experiment Held at the Mount Wilson Observatory, Pasadena, California, February 4 and 5, 1927." *Astrophysical Journal* 68, no. 5:342–45.
———. 1967 [1909]. "Recent Advances in Spectroscopy: Nobel Lecture, December 12, 1907." In *Nobel Lectures: Including Presentation Speeches and Laureates' Biographies: Physics, 1901–1921*, 166–78. Amsterdam, London, New York: Elsevier Publishing Co. for the Nobel Foundation.
———. 1995 [1927]. *Studies in Optics*. New York: Dover.
Michelson, Albert Abraham, and Henry G. Gale. 1925. "The Effect of the Earth's Rotation on the Velocity of Light. Part II." *Astrophysical Journal* 61:140–45.

Michelson, Albert Abraham, and Edward Williams Morley. 1886. "Influence of Motion of the Medium on the Velocity of Light." *American Journal of Science* 31:377–86.

———. 1887. "On a Method of Making the Wave Length of Sodium Light the Actual and Practical Standard of Length." *American Journal of Science* 34:427–30.

———. 1887. "On the Relative Motion of the Earth and the Luminiferous Ether." *American Journal of Science* 34:333–45.

———. 1889. "On the Feasibility of Establishing a Light-Wave as the Ultimate Standard of Length." *American Journal of Science* 38:181–86.

Michelson, Albert Abraham, F. G. Pease, and F. Pearson. 1935. "Measurement of the Velocity of Light in a Partial Vacuum." *Astrophysical Journal* 82:26–61.

Michelson, Albert Abraham, and S. W. Stratton. 1898. "A New Harmonic Analyser." *American Journal of Science* 5:1–13.

Miller, Arthur I. 1981. *Albert Einstein's Special Theory of Relativity: Emergence (1905) and Early Interpretation (1905–1911)*. Reading, MA: Addison Wesley.

———. 1982. "On the Origins, Methods, and Legacy of Ludwig Boltzmann's Mechanics." In *Ludwig Boltzmann: Gesamtausgabe*, edited by Roman Sexl and John Blackmore, 231–61. Graz and Braunschweig: Akademische Druck- und Verlagsanstalt and Vieweg.

———. 1986. *Frontiers of Physics, 1900–1911: Selected Essays with an Original Prologue and Postscript*. Boston: Birkhäuser.

Miller, Dayton C. 1933. "The Ether-Drift Experiment and the Determination of the Absolute Motion of the Earth." *Reviews of Modern Physics* 5:203–42.

Milne, E. A. 1952. *Sir James Jeans: A Biography*. Cambridge: Cambridge Univ. Press.

Minkowski, Hermann. 1908. "Die Grundgleichungen für die elektromagnetischen Vorgänge in bewegten Körpern." *Königliche Gesellschaft der Wissenschaften zu Göttingen. Mathematisch-Physikalische Klasse. Nachrichten*:53–111.

———. 1909. "Raum und Zeit." *Physikalische Zeitschrift* 10:104–11.

———. 1974. "Raum und Zeit." In *Das Relativitätsprinzip: Eine Sammlung von Abhandlungen (1913)*, by H. A. Lorentz, Albert Einstein, Hermann Minkowski, and Hermann Weyl, 54–71. Leipzig and Berlin: Teubner.

———. n.d. "Space and Time." In *The Principle of Relativity*, edited by H. A. Lorentz, Albert Eins.tein, Hermann Minkowski, and Hermann Weyl, 75–91. New York: Dover.

Mirimanoff, D. 1908. "Bemerkung zur Notiz von A. Einstein: 'Bemerkung zu der Arbeit von D. Mirimanoff...'" *Annalen der Physik* 28:1088.

———. 1908. "Über die Grundgleichungen der Elektrodynamik bewegter Körper von Lorentz und das Prinzip der Relativität." *Annalen der Physik* 28:192–98.

Morand, Paul. 1931. *1900 A.D.* Translated by Romilly Fedden. New York: W. F. Payson.

Morley, Edward Williams. 1879. "On a System of Standards of Time for the Whole Country." *Journal of the American Electrical Society* 4:102–6.

———. 1902. "Biographical Memoir of William Augustus Rogers, 1832–1898." *Biographical Memoirs of the National Academy of Sciences* 4:187–99.

Morley, Edward Williams, and William A. Rogers. 1896. "On the Measurement of the Expansion of Metals by the Interferential Method." *Physical Review* 4:1–22, 106–27.

Morus, Iwan Rhys. 1998. *Frankenstein's Children: Electricity, Exhibition, and Experiment in Early-Nineteenth-Century London.* Princeton, NJ: Princeton Univ. Press.

Moyer, Albert E. 1992. *A Scientist's Voice in American Culture: Simon Newcomb and the Rhetoric of Scientific Method.* Berkeley: Univ. of California Press.

Muthesius, Hermann. 1900. "Der monumentale Eingang zum Weltausstellungsgelände in Paris." *Centralblatt der Bauverwaltung* 20:269–70.

Naumann, Friedrich. 1909. *Ausstellungsbriefe.* Berlin: Hilfe.

Nebel, Heinz. 1900. "Die Weltausstellung und ihre Kataloge." In *Die Pariser Weltausstellung in Wort und Bild*, edited by Georg Malkowsky and Paul Apostol, 339–40. Berlin: Kirchhoff.

Needell, Allan A. 1980. "Irreversibility and the Failure of Classical Dynamics: Max Planck's Work on the Quantum Theory, 1900–1915." PhD dissertation, Yale University.

———. 1988. "Introduction." In *The Theory of Heat Radiation*, by Max Planck, xi–xlv. Los Angeles: Tomash; New York: American Institute of Physics.

Nelson, Daniel. 1980. *Frederick W. Taylor and the Rise of Scientific Management.* Madison: Univ. of Wisconsin Press.

Nernst, Walther. 1906. "Antrittsrede." *Königlich preussische Akademie der Wissenschaften (Berlin). Sitzungsberichte* 1:549–52.

———. 1910. "Untersuchungen über die spezifische Wärme bei tiefen Temperaturen. II." *Königlich preussische Akademie der Wissenschaften (Berlin). Sitzungsberichte* 1:262–82.

———. 1911. "Über neuere Probleme der Wärmetheorie." *Königlich preussische Akademie der Wissenschaften (Berlin). Sitzungsberichte* 1:65–90.

———. 1911. "Untersuchungen über die spezifische Wärme bei tiefen Temperaturen. III." *Königlich preussische Akademie der Wissenschaften (Berlin). Sitzungsberichte* I:306–15.

———. 1911. "Zur Theorie der spezifischen Wärme und über die Anwendung der Lehre von den Energiequanten auf physikalisch-chemische Fragen überhaupt." *Zeitschrift für Elektrochemie und angewandte physikalische Chemie* 17:265–75.

Neuburger, Albert. 1900. "Chemie und Physik auf der Weltaustellung." In *Das XIX. Jahrhundert in Wort und Bild. Politische und Kultur-Geschichte*, edited by Hans Kraemer, 175–233. Berlin: Bong.

Newcomb, Simon. 1881. "Report of Progress in the Experiments on the Velocity of Light." *Reports of the National Academy of Science*:17–19.

———. 1882. "Measures of the Velocity of Light Made under Direction of the Secretary to the Navy during the Years 1880–'82." In *Astronomical Papers*

Prepared for the Use of the American Ephemeris and Nautical Almanac. Washington, DC: Navy Department.

Noether, Fritz. 1909. "Zur Kinematik des starren Körpers." *Annalen der Physik* 31:919–44.

Norberg, Arthur L. 1978. "Simon Newcomb's Early Astronomical Career." *Isis* 69:209–25.

Norton, John D. 2004. "Einstein's Investigations of Galilean Covariant Electrodynamics Prior to 1905." *Archive for History of Exact Sciences* 59:45–105.

Nye, David E. 1990. *Electrifying America: Social Meanings of a New Technology, 1880–1940.* Cambridge, MA: MIT Press.

Nye, Mary Jo. 1972. *Molecular Reality: A Perspective on the Scientific Work of Jean Perrin.* London: Macdonald; New York: American Elsevier.

Olesko, Kathryn M. 1988. "Michelson and the Reform of Physics Instruction at the Naval Academy in the 1870s." In *The Michelson Era in American Science: 1870–1930*, edited by Stanley Goldberg and Roger H. Stuewer, 111–32. New York: American Institute of Physics.

———. 1991. *Physics as a Calling: Discipline and Practice in the Königsberg Seminar for Physics.* Ithaca, NY: Cornell Univ. Press.

———. 1996. "Precision, Tolerance, and Consensus: Local Cultures in German and British Resistance Standards." In *Scientific Credibility and Technical Standards in 19th and Early 20th Century Germany and Britain*, edited by Jed Z. Buchwald, 117–56. Dordrecht: Kluwer.

Ostwald, Wilhelm. 1909. *Grosse Männer. Studien zur Biologie des Genies.* Leipzig: Akademische Verlagsgesellschaft.

———. 1927. *Lebenslinien: Eine selbstbiographie.* 3 vols. Vol. 2. Berlin: Klasing and Co., GMBH.

Pais, Abraham. 1982. *"Subtle Is the Lord . . . ": The Science and the Life of Albert Einstein.* New York: Oxford Univ. Press.

Parshall, Karen V. H., and David E. Rowe. 1993. "Embedded in the Culture: Mathematics at the World's Columbian Exhibition of 1893." *Mathematical Intelligencer* 15, no. 2:40–45.

Peirce, C. S. 1879. "Note on the Progress of Experiments for Comparing a Wave-Length with a Meter." *American Journal of Science* 28:51.

———. 1879. "'On Ghosts in Diffraction Spectra' and 'Comparison of Wave-Lengths with the Metre,' papers presented to the U.S. National Academy of Sciences." *Nature* 20:99.

Pellat, H. 1900. "Les laboratoires nationaux physico-techniques." In *Rapports présentés au Congrès international de physique réuni à Paris en 1900 sous les auspices de la Société française de physique.* Tome I. *Questions générales. Métrologie. Physique mécanique. Physique moléculaire*, edited by Lucien Poincaré and Charles-Éduoard Guillaume, 101–7. Paris: Gauthier-Villars.

Pershey, Edward Jay. 1984. "The Early Telescopes of Warner & Swasey." *Sky and Telescope* 67:309–11.

———. 1990. "Warner and Swasey at the Naval Observatory: A View of the Science-Technology Relationship." In *Beyond History of Science: Essays in*

Honor of Robert E. Schofield, edited by Elizabeth Garber, 220–30. Bethlehem, PA: Lehigh Univ. Press.

Picard, Alfred, ed. 1903. *Exposition universelle internationale de 1900 à Paris. Rapport général administratif et technique. Pièces annexes: Actes officiels. Tableaux statistiques et financiers*. 8 vols. Paris: Imprimerie nationale.

———. 1903. *Exposition universelle internationale de 1900 à Paris. Rapport général administratif et technique*. Vol. 6: *7. ptie. Congrès de l'Exposition universelle internationale de 1900. Concours d'exercices physiques et de sports. Cérémonies et fêtes. Auditions musicales. Matinées littéraires et dramatiques. 8. ptie. Visiteurs de l'Exposition universelle internationale de 1900*. 8 vols. Paris: Imprimerie nationale.

Pippard, Brian. 1999. "Dark on a Light Subject: Nineteenth-Century Physicists Couldn't Agree about the Speed of Light." *Nature* 402:123.

Planck, Max. 1900. "Zur Theorie des Gesetzes der Energieverteilung im Normalspektrum." *Deutsche Physikalische Gesellschaft. Verhandlungen* 2:237–45.

———. 1906. "Das Prinzip der Relativität und die Grundgleichungen der Mechanik." *Deutsche Physikalische Gesellschaft. Verhandlungen* 8:136–41.

———. 1906. "Die Kaufmannschen Messungen der Ablenkbarkeit der ß-Strahlen in ihren Bedeutung für die Dynamik der Elektronen." *Physikalische Zeitschrift* 7:753–61.

———. 1906. *Vorlesungen über die Theorie der Wärmestrahlung*. Leipzig: J. A. Barth.

———. 1907. "Nachtrag zu der Besprechung der Kaufmannschen Ablenkungsmessungen." *Deutsche Physikalische Gesellschaft. Verhandlungen* 9:301–5.

———. 1907. "Zur Dynamik bewegte Systeme." *Königliche Preussische Akademie der Wissenschaften (Berlin). Sitzungsberichte* 29:542–70.

———. 1908. "Bemerkungen zum Prinzip der Aktion und Reaktion der allgemeinen Dynamik." *Deutsche Physikalische Gesellschaft. Verhandlungen* 10:728–32.

———. 1908. "Zur Dynamik bewegter Systeme." *Annalen der Physik* 26:1–34.

———. 1909. "Die Einheit des physikalischen Weltbildes." *Physikalische Zeitschrift* 10:62–75.

———. 1910. *Acht Vorlesungen über Theoretische Physik gehalten an der Columbia University in the City of New York im Frühjahr 1909*. Leipzig: Hirzel.

———. 1910. "Die Stellung der neueren Physik zur mechanischen Naturanschaungen." *Physikalische Zeitschrift* 11:922–32.

———. 1910. "Gleichförmige Rotation und Lorentz-Kontraktion." *Physikalische Zeitschrift* 11:294.

———. 1910. "Zur Theorie der Wärmestrahlung." *Annalen der Physik* 31:758–68.

———. 1913. "Die Gesetze der Wärmestrahlung und die Hypothese der elementaren Wirkungsquanten." In *Die Theorie der Strahlung und der Quanten. Verhandlungen auf einer von E. Solvay einberufenen Zusammenkunft (30. Oktober bis 3. November 1911)*, edited by Arnold Eucken, 77–94. Berlin: Verlag Chemie GMBH.

———. 1958 [1920]. "Die Entstehung und bisherige Entwicklung der Quantentheorie (Noble Lecture)." In *Physikalische Abhandlungen und Vörtrage*, 121–36. Braunschweig.

———. 1960 [1909]. "The Unity of the Physical World Picture." In *A Survey of Physical Theory*, 1–26. New York: Dover.

———. 1960 [1910]. "The Place of Modern Physics in the Mechanical View of Nature." In *A Survey of Physical Theory*, 27–44. New York: Dover.

Plum, William B. 1954. "The Michelson Museum." *American Journal of Physics* 22:177–81.

Poincaré, Henri. 1900. "Relations entre la physique expérimentale et la physique mathématique." In *Rapports présentés au Congrès international de physique réuni à Paris en 1900 sous les auspices de la Société française de physique.* Tome I. *Questions générales. Métrologie. Physique mécanique. Physique moléculaire*, edited by Lucien Poincaré and Charles-Éduoard Guillaume, 1–29. Paris: Gauthier-Villars.

———. 1900. "Über die Beziehungen zwischen der experimentellen und der mathematischen Physik." *Physikalische Zeitschrift* 2:166–71, 182–86, 196–201.

———. 1902. *La science et l'hypothèse*. Paris: Flammarion.

———. 1904. *Wissenschaft und Hypothese*. Translated by F. and L. Lindemann. Leipzig: Teubner.

———. 1905. *Science and Hypothesis*. Translated by William John Greenstreet. London: Walter Scott.

———. 1910. *Die neue Mechanik. Vortrag gehalten von Henri Poincaré am 13. Oktober 1910 im "Wissenschaftlicher Verein" zu Berlin*. Berlin: G. Bernstein.

———. 1914. *Science and Method*. Translated by F. Maitland. London: Nelson.

———. 1954 [1912]. "L'hypothèse des quanta." In *Œuvres de Henri Poincaré*, 654–68. Paris: Gauthier-Villars.

———. 1954 [1912]. "Sur la théorie des quanta." In *Œuvres de Henri Poincaré*, 626–53. Paris: Gauthier-Villars.

———. 1986 [1905]. "The Principles of Mathematical Physics." In *Physics for a New Century: Papers Presented at the 1904 St. Louis Congress. A Compilation Selected and a Preface by Katherine R. Sopka. Introduction by Albert E. Moyer*, edited by Katherine R. Sopka and Albert E. Moyer, 281–99. Los Angeles: Tomash; New York: American Institute of Physics.

———. 2001 [1908]. "Science and Method." In *The Value of Science: Essential Writings of Henri Poincaré*. New York: Modern Library.

Poincaré, Lucien, and Charles-Édouard Guillaume, eds. 1900. *Rapports présentés au Congrès international de physique réuni à Paris en 1900 sous les auspices de la Société française de physique.* Tome 1: *Questions générales. Métrologie. Physique mécanique. Physique moléculaire*. 4 vols. Vol. 1. Paris: Gauthier-Villars.

———. 1900. *Rapports présentés au Congrès international de physique réuni à Paris en 1900 sous les auspices de la Société française de physique.* Tome II. *Optique. Électricité. Magnétisme*. 4 vols. Vol. 2. Paris: Gauthier-Villars.

———. 1900. *Rapports présentés au Congrès international de physique réuni à Paris en 1900 sous les auspices de la Société française de physique.* Tome III. *Électro-*

optique et ionisation. Applications. Physique cosmique. Physique biologique. 4 vols. Vol. 3. Paris: Gauthier-Villars.

———. 1901. *Travaux du Congrès international de physique réuni à Paris en 1900 sous les auspices de la Société française de physique.* Tome IV. *Procès-verbaux. Annexes. Liste des membres.* 4 vols. Vol. 4. Paris: Gauthier-Villars.

Polanyi, Michael. 1958. *Personal Knowledge: Towards a Post-Critical Philosophy.* Chicago: Univ. of Chicago Press.

Porter, Theodore M. 1994. "The Death of the Object: *Fin de Siècle* Philosophy of Physics." In *Modernist Impulses in the Human Sciences, 1870–1930*, edited by Dorothy Ross, 128–51. Baltimore: Johns Hopkins Univ. Press.

Pratt & Whitney. 1930. *Accuracy for Seventy Years, 1860–1930.* Hartford, CT: Pratt & Whitney.

Preston, Thomas. 1895. *The Theory of Light.* 2nd ed. London: Macmillan.

———. 1898. "Radiation Phenomena in the Magnetic Field." *Philosophical Magazine* 45:325–39.

———. 1899. "Radiation Phenomena in the Magnetic Field." *Nature* 53:224–29.

———. 1899. "Radiation Phenomena in the Magnetic Field (Letter to the Editor)." *Nature* 59:485.

———. 1899. "The Interferometer (Letter to the Editor)." *Nature* 59:605.

Pringsheim, Ernst. 1898. "Ueber ein Interferenzmikroskop nach Sirks." *Verhandlungen der Physikalischen Gesellschaft zu Berlin* 17:152–56.

———. 1900. "Sur les l'émissions des gaz." In *Rapports présentés au Congrès international de physique réuni à Paris en 1900 sous les auspices de la Société française de physique.* Tome II. *Optique. Électricité. Magnétisme*, edited by Lucien Poincaré and Charles-Édouard Guillaume, 100–132. Paris: Gauthier-Villars.

"Procès-verbaux." 1901. In *Travaux du Congrès international de physique réuni à Paris en 1900 sous les auspices de la Société française de physique.* Tome IV. *Procès-verbaux. Annexes. Liste des membres*, edited by Lucien Poincaré and Charles-Édouard Guillaume, 5–60. Paris: Gauthier-Villars.

"Procès-Verbaux Sommaries." 1900. In *Rapports présentés au Congrès international de physique réuni à Paris en 1900 sous les auspices de la Société française de physique.* Tome I. *Questions générales. Métrologie. Physique mécanique. Physique moléculaire*, edited by Lucien Poincaré and Charles-Édouard Guillaume. Paris: Gauthier-Villars.

Pyatt, Edward. 1983. *The National Physical Laboratory: A History.* Bristol: Adam Hilger.

Pyenson, Lewis. 1973. "The Göttingen Reception of Einstein's General Theory of Relativity." PhD dissertation, Johns Hopkins University.

———. 1983. *Neohumanism and the Persistence of Pure Mathematics in Wilhelmian Germany.* Philadelphia: American Philosophical Society.

———. 1985. "Einstein's Early Scientific Collaboration." In *The Young Einstein: The Advent of Relativity*, 215–46. Bristol and Boston: Adam Hilger.

———. 1985. "Physical Sense in Relativity: Max Planck Edits the *Annalen der Physik*, 1906–1918." In *The Young Einstein: The Advent of Relativity*, 194–214. Bristol and Boston: Adam Hilger.

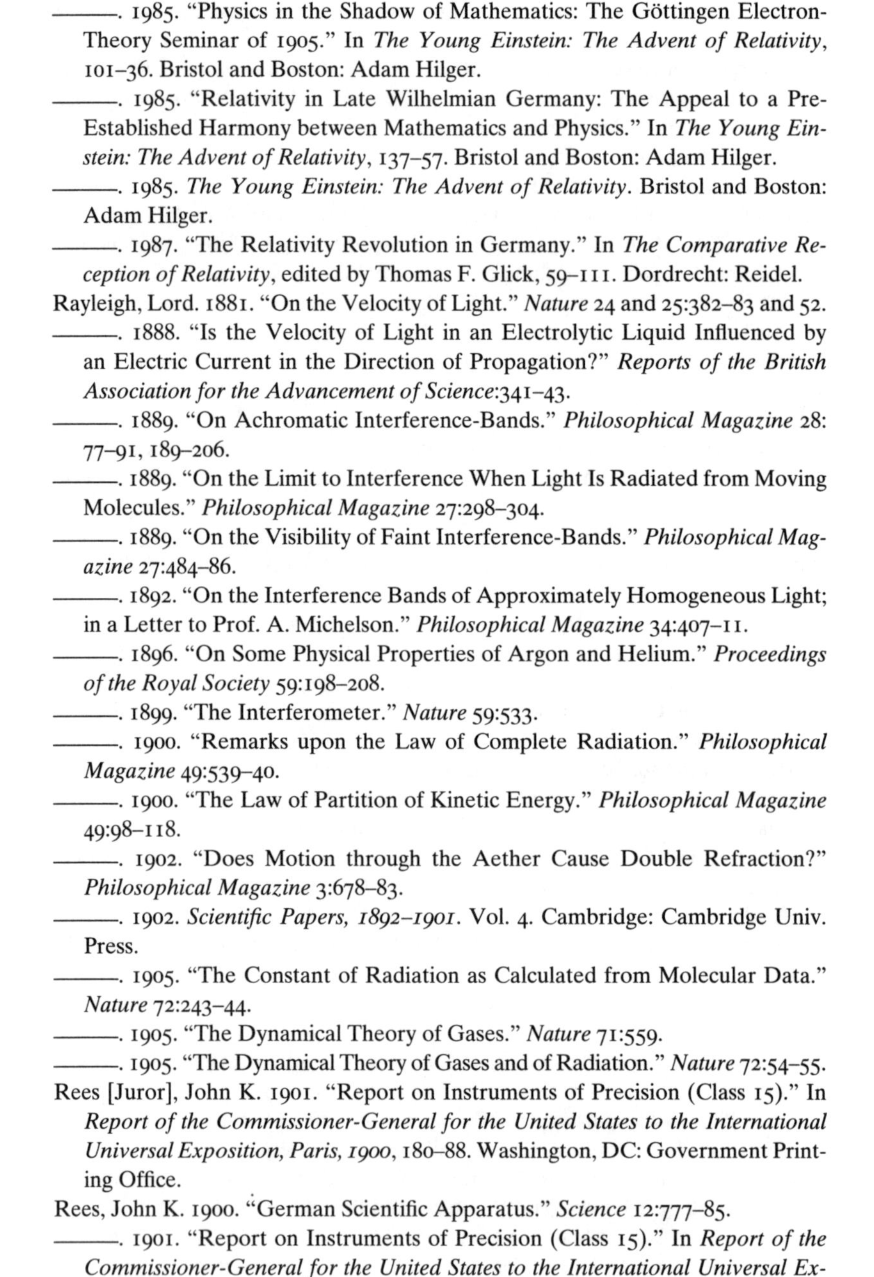

———. 1985. "Physics in the Shadow of Mathematics: The Göttingen Electron-Theory Seminar of 1905." In *The Young Einstein: The Advent of Relativity*, 101–36. Bristol and Boston: Adam Hilger.

———. 1985. "Relativity in Late Wilhelmian Germany: The Appeal to a Pre-Established Harmony between Mathematics and Physics." In *The Young Einstein: The Advent of Relativity*, 137–57. Bristol and Boston: Adam Hilger.

———. 1985. *The Young Einstein: The Advent of Relativity*. Bristol and Boston: Adam Hilger.

———. 1987. "The Relativity Revolution in Germany." In *The Comparative Reception of Relativity*, edited by Thomas F. Glick, 59–111. Dordrecht: Reidel.

Rayleigh, Lord. 1881. "On the Velocity of Light." *Nature* 24 and 25:382–83 and 52.

———. 1888. "Is the Velocity of Light in an Electrolytic Liquid Influenced by an Electric Current in the Direction of Propagation?" *Reports of the British Association for the Advancement of Science*:341–43.

———. 1889. "On Achromatic Interference-Bands." *Philosophical Magazine* 28: 77–91, 189–206.

———. 1889. "On the Limit to Interference When Light Is Radiated from Moving Molecules." *Philosophical Magazine* 27:298–304.

———. 1889. "On the Visibility of Faint Interference-Bands." *Philosophical Magazine* 27:484–86.

———. 1892. "On the Interference Bands of Approximately Homogeneous Light; in a Letter to Prof. A. Michelson." *Philosophical Magazine* 34:407–11.

———. 1896. "On Some Physical Properties of Argon and Helium." *Proceedings of the Royal Society* 59:198–208.

———. 1899. "The Interferometer." *Nature* 59:533.

———. 1900. "Remarks upon the Law of Complete Radiation." *Philosophical Magazine* 49:539–40.

———. 1900. "The Law of Partition of Kinetic Energy." *Philosophical Magazine* 49:98–118.

———. 1902. "Does Motion through the Aether Cause Double Refraction?" *Philosophical Magazine* 3:678–83.

———. 1902. *Scientific Papers, 1892–1901*. Vol. 4. Cambridge: Cambridge Univ. Press.

———. 1905. "The Constant of Radiation as Calculated from Molecular Data." *Nature* 72:243–44.

———. 1905. "The Dynamical Theory of Gases." *Nature* 71:559.

———. 1905. "The Dynamical Theory of Gases and of Radiation." *Nature* 72:54–55.

Rees [Juror], John K. 1901. "Report on Instruments of Precision (Class 15)." In *Report of the Commissioner-General for the United States to the International Universal Exposition, Paris, 1900*, 180–88. Washington, DC: Government Printing Office.

Rees, John K. 1900. "German Scientific Apparatus." *Science* 12:777–85.

———. 1901. "Report on Instruments of Precision (Class 15)." In *Report of the Commissioner-General for the United States to the International Universal Exposition, Paris, 1900*, 180–88. Washington, DC: Government Printing Office.

Reingold, Nathan, ed. 1964. *Science in Nineteenth-Century America: A Documentary History*. New York: Hill and Wang.

———. 1988. "Physics and Engineering in the United States, 1945–1965: A Study of Pride and Prejudice." In *The Michelson Era in American Science, 1870–1930*, edited by Stanley Goldberg and Roger H. Stuewer, 288–98. New York: American Institute of Physics, 1988.

Renn, Jürgen. 2000. "Einstein's Controversy with Drude and the Origin of Statistical Mechanics: A New Glimpse from the 'Love Letters.'" In *Einstein: The Formative Years, 1879–1909*, edited by Don Howard and John Stachel, 107–57. Boston, Basel, Berlin: Birkhäuser.

———. 2005. "Einstein's Invention of Brownian Motion." *Annalen der Physik* 14, Supplement:23–37.

"Report of Professor Wm. A. Rogers." 1887. In *Standards of Length and Their Practical Application*, edited by George M. Bond, 1–48. Hartford, CT: Pratt & Whitney.

"Report of the Committee of the Master Car-Builders' Association on Standard Screw Threads. Presented at the Annual Convention held in Philadelphia, June, 1882." 1887 [1882]. In *Standards of Length and Their Practical Application*, edited by George M. Bond, 59–82. Hartford, CT: Pratt & Whitney.

Riecke, Eduard. 1911. "Zur Theorie des Interferenzversuches von Michelson." *Königliche Gesellschaft der Wissenschaften zu Göttingen. Mathematisch-Physikalische Klasse. Nachrichten*:271–77.

Righi, Auguste. 1900. "Les ondes hertziennes." In *Rapports présentés au Congrès international de physique réuni à Paris en 1900 sous les auspices de la Société française de physique*. Tome II. *Optique. Électricité. Magnétisme*, edited by Lucien Poincaré and Charles-Édouard Guillaume, 301–24. Paris: Gauthier-Villars.

Robert John Strutt, Fourth Baron Rayleigh. 1924. *Life of John William Strutt, Third Baron Rayleigh, O.M., F.R.S.* New York and London: Longmans Green/Edward Arnold.

Rogers, William A. 1875–76. "On a Possible Explanation of the Method Employed by Norbert in Ruling His Test Plates." *Proceedings of the American Academy of Arts and Sciences* 11:237–55.

———. 1880. "On the First Results from a New Diffraction Ruling Engine." *American Journal of Science* 19:54–59.

———. 1884. "On a Practical Solution to the Perfect Screw Problem." *Transactions of the American Society of Mechanical Engineers* 5:216–35, discussion 236–59.

———. 1885. "Determination of the Absolute Length of Eight Rowland Gratings at 62° Fahr." *Proceedings of the American Society of Microscopists Eighth Annual Meeting* 7:151–98.

Rolt, F. H. 1929. *Gauges and Fine Measurements*. Vol. I, *Standards of Length, Measuring Machines, Comparators*. 2 vols. London: Macmillan.

Rosenberg, Nathan. 1963. "Technological Change in the Machine Tool Industry, 1840–1910." *Journal of Economic History*:414–46.

Rossiter, Margaret W. 1982. *Women Scientists in America: Struggles and Strategies to 1940*. Baltimore: Johns Hopkins Univ. Press.

Rowe, David E. 1985. "Essay Review." *Historia Mathematica* 12:278–91.

———. 1989. "Klein, Hilbert, and the Göttingen Mathematical Tradition." *Osiris: A Research Journal Devoted to the History of Science and Its Cultural Influences* 5:186–213.

Rowland, Henry Augustus. 1883. "A Plea for Pure Science." *Proceedings of the American Association for the Advancement of Science* 32:105–26.

Rubens, Heinrich. 1900. "Le spectre infra-rouge." In *Rapports présentés au Congrès international de physique réuni à Paris en 1900 sous les auspices de la Société française de physique*. Tome II. *Optique. Électricité. Magnétisme*, edited by Lucien Poincaré and Charles-Édouard Guillaume, 141–74. Paris: Gauthier-Villars.

Runge, Carl. 1903. "Ueber die elektromagnetische Masse der Elektronen." *Königliche Gesellschaft der Wissenschaften zu Göttingen. Mathematisch-Physikalische Klasse. Nachrichten*:326–30.

Rynasiewicz, Robert. 2000. "The Construction of the Special Theory: Some Queries and Considerations." In *Einstein: The Formative Years, 1879–1909*, edited by Don Howard and John Stachel, 159–201. Boston, Basel, Berlin: Birkhäuser.

Rynasiewicz, Robert, and Jürgen Renn. 2006. "The Turning Point for Einstein's *Annus mirabilis*." *Studies in History and Philosophy of Modern Physics* 37:5–35.

Schaffer, Simon. 1991. "Utopia Unlimited: On the End of Science." *Strategies: A Journal of Theory, Culture, and Politics* 4/5:151–81.

———. 1992. "Late Victorian Metrology and Its Instrumentation: A Manufactory of Ohms." In *Invisible Connections: Instruments, Institutions, and Science*, edited by Robert Bud and Susan E. Cozzens, 23–56. Bellingham, WA: SPIE Optical Engineering Press.

———. 1994. "Empires of Physics." In *The Physics of Empire: Public Lectures*, edited by Richard Staley, 97–109. Cambridge: Whipple Museum of the History of Science.

———. 1997. "Metrology, Metrication, and Victorian Values." In *Victorian Science in Context*, edited by Bernard Lightman, 438–74. Chicago: Univ. of Chicago Press.

———. 2006. "Time Machines." In *The Whipple Museum of the History of Science: Instruments and Interpretations, to Celebrate the Sixtieth Anniversary of R. S. Whipple's Gift to the University of Cambridge*, edited by Liba Taub and Frances Willmoth, 345–66. Cambridge: Whipple Museum of the History of Science.

Schilpp, Paul Arthur, ed. 1979. *Albert Einstein: Autobiographical Notes. A Centennial Edition*. La Salle, IL: Open Court.

Schivelbusch, Wolfgang. 1992. *Licht, Schein und Wahn: Auftritte der elektrischen Beleuchtung im 20. Jahrhundert*. Berlin: Ernst &t Sohn.

Schorske, Carl E. 1981. *Fin-de-Siècle Vienna: Politics and Culture*. 1st Vintage Book edition. New York: Vintage Books.

Schuster, Arthur. 1904. *An Introduction to the Theory of Optics*. London: E. Arnold.

———. 1909. *An Introduction to the Theory of Optics*. 2nd ed. London: E. Arnold.
Schwarzschild, Karl. 1896. "Über Messung von Doppelsternen durch Interferenzen." *Astronomische Nachrichten* 139:353–60.
"Scientific Notes and News." 1899. *Science* 10, no. 244:300–304.
Serway, Raymond A., and John W. Jewett Jr. 2004. *Physics for Scientists and Engineers, with Modern Physics*. 6th ed. Belmont, CA: Thomson-Brooks/Cole.
Seth, Suman. 2003. "Principles and Problems: Constructions of Theoretical Physics in Imperial Germany, 1890–1918." Ph.D. dissertation, Princeton University.
———. 2004. "Quantum Theory and the Electromagnetic World View." *Historical Studies in the Physical and Biological Sciences* 35:67–93.
———. 2005. "Response to Shaul Katzir: 'On the Electromagnetic World-View.'" *Historical Studies in the Physical and Biological Sciences* 36, no. 1:193–96.
———. 2007. "Crisis and the Construction of Modern Theoretical Physics." *British Journal for the History of Science* 40:25–51.
Sexl, Roman, and John Blackmore, eds. 1982. *Internationale Tagung, 5.-8. September 1981. Ausgewählte Abhandlungen*. Vol. 8, Ludwig Boltzmann: Gesamtausgabe. Graz and Braunschweig: Akademische Druck- und Verlagsanstalt and Vieweg.
Shankland, R. S. 1964. "Michelson-Morley Experiment." *American Journal of Physics* 19:16–35.
———. 1967. "Rayleigh and Michelson." *Isis* 58:86–88.
Shapin, Steven. 1984. "Talking History: Reflections on Discourse Analysis." *Isis* 75:125–28.
———. 1989. "The Invisible Technician." *American Scientist* 77:554–63.
———. 1992. "Discipline and Bounding: The History and Sociology of Science as Seen through the Externalism-Internalism Debate." *History of Science* 30:333–69.
———. 1994. *A Social History of Truth: Civility and Science in Seventeenth-Century England*. Science and Its Conceptual Foundations. Chicago: Univ. of Chicago Press.
Shapin, Steven, and Simon Schaffer. 1985. *Leviathan and the Air-Pump: Hobbes, Boyle, and the Experimental Life. Including a Translation of Thomas Hobbes, Dialogus physicus de natura aeris by Simon Schaffer*. Princeton, NJ: Princeton Univ. Press.
Sharpe, Henry Dexter. 1949. "Joseph R. Brown, Mechanic, and the Beginnings of Brown & Sharpe." *Newcomen Society of England in America: Addresses* 67, no. 13:5–28.
Sharpe, Lucian. 1892. "'Letter of December 10, 1887' printed as Development of the Micrometer Gauge." *American Machinist*, no. 15, December.
Shedd, John Cutler. 1899. "An Interferometer Study of Radiations in a Magnetic Field. I." *Physical Reveiw* 9:1–19.
———. 1899. "An Interferometer Study of Radiations in a Magnetic Field. II." *Physical Reveiw* 9:86–115.
———. 1900. "Untersuchung der Strahlung in einem magnetischen Felde mittelst des Interferometers." *Physikalische Zeitschrift* 1:270–72.

———. 1901. “Die Anwendung des Interferometers beim Studium des Zeemaneffektes.” *Physikalische Zeitschrift* 2:278–83.

———. 1902. “Über die Formen der von dem Michelsonschen Interferometer gelieferten Kurven.” *Physikalische Zeitschrift* 3:47–51.

Shinn, Terry. 2001. “The Research Technology Matrix: German Origins, 1860–1900.” In *Instrumentation: Between Science, State, and Industry*, edited by Bernward Joerges and Terry Shinn, 29–48. Dordrecht and Boston: Kluwer.

Sibum, H. Otto. 1995. “Reworking the Mechanical Value of Heat: Instruments of Precision and Gestures of Accuracy in Early Victorian England.” *Studies in History and Philosophy of Science* 26:73–106.

———. 2002. “Exploring the Margins of Precision.” In *Instruments, Travel, and Science: Itineraries of Precision from the Seventeenth to the Twentieth Century*, edited by Marie-Noelle Bourget, Christian Licoppe, and H. Otto Sibum, 216–42. London: Routledge.

Sigurdsson, Skúli. 1992. “Equivalence, Pragmatic Platonism, and Discovery of Calculus.” In *The Invention of Physical Science: Intersections of Mathematics, Theology, and Natural Philosophy since the Seventeenth Century; Essays in Honor of Erwin N. Hiebert*, edited by Mary Jo Nye, Joan L. Richards, and Roger Stuewer, 97–116. Dordrecht: Kluwer.

Silberstein, Ludwig. 1914. *The Theory of Relativity*. London: Macmillan.

Simon, Hermann Theodor. 1900. “Vom internationalen Physikerkongresse zu Paris [6.–12. August 1900.].” *Physikalische Zeitschrift* 1:571–72.

Smith, Crosbie. 1998. *The Science of Energy: A Cultural History of Energy Physics in Victorian Britain*. Chicago: Univ. of Chicago Press.

Smith, Crosbie, and M. Norton Wise. 1989. *Energy and Empire: A Biographical Study of Lord Kelvin*. Cambridge: Cambridge Univ. Press.

Smith, George E. 2001. “J. J. Thomson and the Electron, 1897–1899.” In *Histories of the Electron: The Birth of Microphysics*, edited by Jed Z. Buchwald and Andrew Warwick, 21–76. Cambridge, MA: MIT Press.

Sommerfeld, Arnold. 1910. “Zur Relativitätstheorie I: Vierdimensional Vektoralgebra.” *Annalen der Physik*:749–76.

———. 1910. “Zur Relativitätstheorie II: Vierdimensional Vektoralgebra.” *Annalen der Physik*:649–89.

———. 1911. “Das Plancksche Wirkungsquanten und seine allgemeine Bedeutung für die Molekularphysik.” *Physikalische Zeitschrift* 12:1057–68.

———. 2000. *Wissenschaftlicher Briefwechsel*. Vol. 1, *1892–1918*. Edited by Michael Eckert and Karl Märker. Berlin, Diepholz, Munich: Deutsches Museum and Verlag für Geschichte der Naturwissenschaften und der Technik.

Sopka, Katherine R., and Albert E. Moyer. 1986. *Physics for a New Century: Papers Presented at the 1904 St. Louis Congress. A Compilation Selected and a Preface by Katherine R. Sopka. Introduction by Albert E. Moyer*. Los Angeles: Tomash; New York: American Institute of Physics.

Sorrenson, Richard. 1999. “George Graham, Visible Technician.” *British Journal for the History of Science* 32:203–21.

Sponsel, Alistair. 2002. "Constructing a 'Revolution in Science': The Campaign to Promote a Favourable Reception for the 1919 Solar Eclipse Experiments." *British Journal for the History of Science* 35:439–67.

Stacey, D. N. 1994. "Rayleigh's Legacy to Modern Physics: High Resolution Spectroscopy." *European Journal of Physics* 15:235–42.

Stachel, John. 1995. "History of Relativity." In *Twentieth Century Physics*, edited by Laurie M. Brown, Abraham Pais, and Brian Pippard, 249–356. Bristol, Philadelphia, New York: Institute of Physics Publishing and American Institute of Physics Press.

———. 2002. "Einstein and Michelson: The Context of Discovery and the Context of Justification." In *Einstein from 'B' to 'Z,'* 177–90. Boston, Basel, Berlin: Birkhäuser.

———. 2002. *Einstein from "B" to "Z."* Vol. 9, Einstein Studies. Boston, Basel, Berlin: Birkhäuser.

———. 2002. "The Rigidly Rotating Disk as the 'Missing Link' in the History of General Relativity." In *Einstein from 'B' to 'Z,'* 245–60. Boston, Basel, Berlin: Birkhäuser.

Staley, Richard. 1992. "Max Born and the German Physics Community: The Education of a Physicist." PhD dissertation, University of Cambridge.

———. 1998. "On the Histories of Relativity: The Propagation and Elaboration of Relativity Theory in Participant Histories in Germany, 1905–1911." *Isis* 89:263–99.

———. 2002. "Interstitial Instruments (Review of Bernward Joerges and Terry Shinn [eds], *Instrumentation: Between Science, State, and Industry*)." *Social Studies of Science* 32:469–76.

———. 2002. "Travelling Light." In *Instruments, Travel, and Science: Itineraries of Precision from the Seventeenth to the Twentieth Century*, edited by Marie-Noelle Bourget, Christian Licoppe, and H. Otto Sibum, 243–72. London: Routledge.

———. 2003. "The Interferometer and the Spectroscope: Michelson's Standards and the Spectroscopic Community." *Nuncius* 18, no. 2:779–801.

———. 2003. "Interdisciplinary Atomism? Exploring Twentieth-Century Culture through Einstein (Essay Review)." *British Journal for the History of Science* 36, no. 2:221–30.

———. 2005. "On the Co-Creation of Classical and Modern Physics." *Isis* 96:530–58.

———. 2007. "Conspiracies of Proof and Diversity of Judgment in Astronomy and Physics: On Physicists' Attempts to Time Light's Wings and Solve Astronomy's Noblest Problem." *Cahiers François Viète* 11–12:83–97.

———. Forthcoming. "Michelson and the Observatory: Physics and the Astronomical Community in Nineteenth-Century America." In *The Heavens on Earth: Observatory Techniques in Nineteenth-Century Science*, edited by David Aubin, Charlotte Bigg, and H. Otto Sibum. Durham, NC: Duke Univ. Press.

Stanley, Matthew. 2003. "'An Expedition to Heal the Wounds of War': The 1919 Eclipse and Eddington as Quaker Adventurer." *Isis* 94:57–89.

Stapleton, Darwin H. 1988. "The Context of Science: The Community of Industry and Higher Education in Cleveland in the 1880s." In *The Michelson Era in American Science: 1870–1930*, edited by Stanley Goldberg and Roger H. Stuewer, 13–22. New York: American Institute of Physics.

Stark, Johannes. 1900. "J. Macé de Lépinay, Massbestimmungen mit den Interferenzmethoden. 24 Seiten." *Physikalische Zeitschrift* 1:489–90.

———. 1908. "Bemerkung zu Herrn Kaufmanns Antwort auf einen Einwand von Herrn Planck." *Deutsche Physikalische Gesellschaft. Verhandlungen* 10:14–16.

"Statistique de l'Exposition." 1900. *La Nature* 2:407–9.

Strassmann, W. Paul. 1959. *Risk and Technological Innovation: American Manufacturing Methods during the Nineteenth Century*. Ithaca, NY: Cornell Univ. Press.

Susalla, Peter. 2006. "The Old School in a Progressive Science: George Cary Comstock and the Middle Ground between 'Old' and 'New' Astronomy, 1879–1922." Master's dissertation, University of Wisconsin–Madison.

Sutherland, William. 1898. "Relative Motion of the Earth and Aether." *Philosphical Magazine* 45:23–31.

Sweetnam, George Kean. 2000. *The Command of Light: Rowland's School of Physics and the Spectrum*. Vol. 238, Memoirs of the American Philosophical Society Held at Philadelphia for Promoting Useful Knowledge. Philadelphia: American Philosophical Society.

Swenson, Loyd S., Jr. 1972. *The Ethereal Aether: A History of the Michelson-Morley-Miller Aether-Drift Experiments, 1880–1930*. Austin: Univ. of Texas Press.

———. 1988. "Michelson-Morley, Einstein, and Interferometry." In *The Michelson Era in American Science: 1870–1930*, edited by Stanley Goldberg and Roger H. Stuewer, 235–45. New York: American Institute of Physics.

"Tagesereignisse." 1900. *Physikalische Zeitschrift* 1:391–92.

"The International Congress of Mathematicians." 1900. *Nature* 62:418–20.

"The Third International Conference on a Catalogue of Scientific Literature." 1900. *Science* 12:77–78.

Thompson, Silvanus P. 1910. *The Life of William Thomson, Baron Kelvin of Largs*. Vol. 2. London: Macmillan.

Thomson, George Paget. 1966. *J. J. Thomson: Discoverer of the Electron*. Science Study Series. Garden City, NY: Anchor Books.

Thomson, J. J. 1899. "On the Masses of the Ions in Gases at Low Pressures." *Philosophical Magazine* 48:547–67.

———. 1900. "Indications relatives a la constitution de la matière fournies par les recherches récentes sur le passage de l'électricité a travers les gaz." In *Rapports présentés au Congrès international de physique réuni à Paris en 1900 sous les auspices de la Société française de physique*. Tome III. *Électro-optique et ionisation. Applications. Physique cosmique. Physique biologique*, edited by Lucien Poincaré and Charles-Édouard Guillaume, 138–51. Paris: Gauthier-Villars.

———. 1967. "Carriers of Negative Electricity: Nobel Lecture, December 11, 1906." In *Nobel Lectures: Including Presentation Speeches and Laureates'*

Biographies: Physics, 1901–1921, 145–53. Amsterdam, London, New York: Elsevier Publishing Co. for the Nobel Foundation.

Thomson, William (Lord Kelvin). 1889 [1883]. "Electrical Units of Measurement." In *Popular Lectures and Addresses*. Vol. 1: *Constitution of Matter*, 73–136. London and New York: Macmillan.

Todd, D. P. 1880. "Solar Parallax from the Velocity of Light." *American Journal of Science* 19:59–64.

Turner, Steven. 1991. "William Würdemann: First Mechanician of the U.S. Coast Survey." *Rittenhouse* 5, no. 20:97–110.

Uselding, Paul. 1981. "Measuring Techniques and Manufacturing Practice." In *Yankee Enterprise: The Rise of the American System of Manufactures*, edited by Otto Mayr and Robert C. Post, 103–26. Washington, DC: Smithsonian Institution Press.

Usselman, Steven W. 1984. "Air Brakes for Freight Trains: Technological Innovation on American Railroads, 1869–1900." *Business History Review* 58, Spring:30–50.

———. 1991. "Patents Purloined: Railroads, Inventors, and the Diffusion of Innovation in 19th-Century America." *Technology and Culture* 32:1047–75.

Vaupel, Elisabeth. 2000. "Chemie für die Massen: Weltausstellungen und die Chemieabteilung im Deutschen Museum." *Kultur & Technik* 24, no. 3:46–51.

Villard, P. 1900. "Les rayons cathodiques." In *Rapports présentés au Congrès international de physique réuni à Paris en 1900 sous les auspices de la Société française de physique*. Tome III. *Électro-optique et ionisation. Applications. Physique cosmique. Physique biologique*, edited by Lucien Poincaré and Charles-Édouard Guillaume, 115–37. Paris: Gauthier-Villars.

Volkmann, Paul. 1900. *Einfuhrung in das Studium der theoretischen Physik, insbesondere in das der analytischen Mechanik mit einer Einleitung in die Theorie der physikalischen Erkenntniss. Vorlesungen*. Leipzig: Teubner.

———. 1901. "Die gewöhnliche Darstellung der Mechanik und ihre Kritik durch Hertz." *Zeitschrift für den physikalischen und chemischen Unterricht* 14:266–83.

von Kossel, W. 1947. "Walter Kaufmann." *Die Naturwissenschaften* 34:33–34.

Voss, A. 1901. "Die Prinzipien der rationellen Mechanik." In *Encyclopädie der mathematischen Wissenschaft*. Vol. 4, *Mechanik*, 1. Teil, edited by Felix Klein and Conrad Müller, 3–121. Leipzig: B. G. Teubner.

Wadsworth, F. L. O. 1894. "On the Manufacture of Very Accurate Straight Edges." *Journal of the Franklin Institute*, July:1–20.

Walter, Scott. 1999. "Minkowski, Mathematicians, and the Mathematical Theory of Relativity." In *The Expanding Worlds of General Relativity*, edited by Hubert Goenner, Jürgen Renn, Jim Ritter, and Tilman Sauer, 45–86. Boston: Birkhäuser.

Warner, Deborah Jean. 1971. "Lewis M. Rutherfurd: Pioneer Astronomical Photographer and Spectroscopist." *Technology and Culture* 12, no. 2:190–216.

———. 1986. "Rowland's Gratings: Contemporary Technology." *Vistas in Astronomy* 29:125–30.

———. 1995. "Octave Leon Petitdidier: Precision Optician." *Rittenhouse* 9:54–58.

Warwick, Andrew. 1989. "The Electrodynamics of Moving Bodies and the Principle of Relativity in British Physics, 1894–1919." PhD dissertation, University of Cambridge.

———. 1992. "Cambridge Mathematics and Cavendish Physics: Cunningham, Campbell and Einstein's Relativity 1905–1911. Part I: The Uses of Theory." *Studies in History and Philosophy of Science* 23:625–56.

———. 1993. "Cambridge Mathematics and Cavendish Physics: Cunningham, Campbell, and Einstein's Relativity, 1905–1911. Part 2: Comparing Traditions in Cambridge Physics." *Studies in History and Philosophy of Science* 24:1–25.

———. 2003. *Masters of Theory: Cambridge and the Rise of Mathematical Physics.* Chicago: Univ. of Chicago Press.

Weaire, Denis, and Seamus O'Connor. 1987. "Unfulfilled Renown: Thomas Preston (1860–1900) and the Anomolous Zeeman Effect." *Annals of Science* 44:617–44.

Weart, Spencer R. 1976. "The Rise of 'Prostituted' Physics." *Nature* 262:13–17.

Weltausstellung in Paris 1900: Sonderkatalog der deutschen Kollektivausstellung für Mechanik und Optik. 1900. Berlin: Reichsdruckerei.

Whittaker, Edmund. 1960. *A History of the Theories of Aether and Electricity*, 2 vols. Vol. 2, *The Modern Theories.* New York: Harper.

Wien, Wilhelm. 1900. "Les lois théoriques du rayonnement." In *Rapports présentés au Congrès international de physique réuni à Paris en 1900 sous les auspices de la Société française de physique.* Tome II. *Optique. Électricité. Magnétisme*, edited by Lucien Poincaré and Charles-Édouard Guillaume, 23–41. Paris: Gauthier-Villars.

Williams, Howard R. 1957. *Edward Williams Morley: His Influence on Science in America.* Easton, PA: Chemical Education Publishing.

Wise, M. Norton, ed. 1995. *The Values of Precision.* Princeton, NJ: Princeton Univ. Press.

Witt, Otto N. 1902. *Die chemische Industrie auf der Internationalen Weltausstellung zu Paris 1900.* Berlin: R. Gärtners Verlagsbuchhandlung.

Wood, Robert W. 1905. *Physical Optics.* New York: Macmillan.

Woolgar, S. W. 1976. "Writing an Intellectual History of Scientific Development: The Use of Discovery Accounts." *Social Studies of Science* 6:359–422.

Young, J., and G. Forbes. 1881. "Experimental Determination of the Velocity of White and Coloured Light." *Nature* 24:303–4.

Zahar, Elie. 1996. "Poincaré's Structural Realism and His Logic of Discovery." In *Science et philosophie. Science and Philosophy. Wissenschaft und Philosophie*, edited by Jean-Louis Greffe, Gerhard Heinzmann, and Kuno Lorenz, 45–68. Berlin: Akademie Verlag; Paris: Albert Blanchard.

———. 2001. *Poincaré's Philosophy: From Conventionalism to Phenomenology.* Chicago and La Salle: Open Court.

Zahn, C. T., and A. A. Spees. 1938. "A Critical Analysis of the Classical Experiments on the Relativistic Variation of Electron Mass." *Physical Review* 53:511–21.

Zeeman, Pieter. 1897. "Doublets and Triplets in the Spectrum Produced by External Magnetic Forces (II.)." *Philosophical Magazine* 44:255–59.

———. 1902. "Some Observations on the Resolving Power of the Michelson Echelon Spectroscope." *Astrophysical Journal* 15:218–22.

———. 1913. *Researches in Magneto-Optics: With Special Reference to the Magnetic Resolution of Spectrum Lines*. Macmillan's Science Monographs. London: Macmillan.

Zehnder, L. 1891. "Ein neuer Interferenzrefraktor." *Zeitschrift für Instrumentenkunde* 11:275–85.

Index

The letter f *following a page number denotes a figure (or figures if following a page range).*